Disaster Risk Reduction

Bupinder Zutshi · Akbaruddin Ahmad
Ananda Babu Srungarapati
Editors

Disaster Risk Reduction

Community Resilience and Responses

palgrave
macmillan

Editors
Bupinder Zutshi
Centre for the Study of Regional Development
Jawaharlal Nehru University (JNU)
New Delhi, India

Ananda Babu Srungarapati
Disaster Management Initiatives and Convergence Society (DMICS)
Hyderabad, India

Akbaruddin Ahmad
Darul Ihsan University
Dhaka, Bangladesh

ISBN 978-981-10-8844-5 ISBN 978-981-10-8845-2 (eBook)
https://doi.org/10.1007/978-981-10-8845-2

Library of Congress Control Number: 2018943266

Printed on acid-free paper

This Palgrave Macmillan imprint is published by the registered company Springer Nature Singapore Pte Ltd.
The registered company address is: 152 Beach Road, #21-01/04 Gateway East, Singapore 189721, Singapore

Foreword

Disaster Research is "transdisciplinary" and contemporary in nature, bridging the gap between the natural and the social sciences that eventually contributes to better governance and national progress. Global experiences have indicated that poor people suffer the most from disasters, as they lack the capacity and resources to effectively cope with disasters, therefore, efforts should be made to build capacities of communities to adopt disaster risk-reduction strategies. Disaster risk-reduction management (DRM) strategies through community resilience, responses, and participation has been major policy shift recommended by the Sendai Framework of Action 2015–2030 (SFDRR).

Government of India launched a comprehensive National Disaster Management Plan, (NDMP) prepared by the National Disaster Management Authority of India (NDMA) in June 2016. The document was released by Prime Minister of India and it endorses disaster risk-reduction measures which are in sync with Sendai Framework of Action 2015. The new NDMP focusses to reduce disaster risk, disaster damage and losses, and to create a disaster resilient society focusing on the well-being of the people.

This book discusses the interconnected, complex, and emerging risks in today's societies and deliberates on the various aspects of disaster risk-reduction strategies especially through community resilience and responses. It consists of selected papers presented at the World Congress on Disaster Management, which focused on community resilience and responses toward disaster risk-reduction based on South Asian experiences, and closely examines the coordinated research activities involving all stakeholders, especially the communities at risk. Further, it narrates the experiences of disaster risk-reduction in different communities that have policy implications for mitigation of future disaster risks in the societies affected by these types of disasters. Written from the social science perspective to disasters rather than an engineering approach, the book helps development and governance institutions to prioritize disasters as a problem of development rather than being parallel to it.

(M. Venkaiah Naidu)

New Delhi, India

M. Venkaiah Naidu
Vice-President of India

Preface

Special Centre for Disaster Research Series

This volume is part of a series of books that the Special Centre for Disaster Research, Jawaharlal Nehru University, New Delhi, India, is planning to publish under Springer Nature's Palgrave imprint. The series editor for these works is Prof. Amita Singh, Chairperson of the Centre.

The transdisciplinary Centre has been working upon a social sciences approach to disaster research and has undertaken three concerns very genuinely—first, the lack of non-western, local and indigenous literature in India and in the Asia; second, the gulf existing due to the language of law which prevents appropriate implementation of Disaster Risk Reduction (DRR) laws; and third, an inherently patriarchal and dysfunctional structures of disaster governance.

The philosophical pretense in a culturing of human beings is much provoked and stimulated from the particularities of relationships in any society. Research which enables the discovery of an idea of responsibility towards nature, which naturalizes laws and streamlines the structures of governance in accordance with the geospatial warnings have greater possibility to reconnect decision makers to the ecological processes. Nature and natural processes are beyond human contrivance. The reconstructive postmodern approaches in social sciences also highlight the fact the world is but language and an inherent part of any structures of governance. Therefore, the most justiciable prevention of disasters is to bridge the fissures of social life in a lawmakers' understanding of disaster and the administrative ability to implement DRR policies. The undertaking of this Palgrave series is to create a temporal model of institutional framework and decision making which could at best be a guideline to those who wish to study disasters beyond existing engineering approaches. The pedagogy of disaster studies would most appropriately focus upon the deontology or the moral rightness of doing one's duty towards the social system and communitarian traditions which we live in. This series attempts to assemble experience, spawn community initiatives and highlight the gaps in the implementation processes which throw communities into recurrent disasters. Some of the forthcoming volumes in the series are: *Development and Disaster Management*, *Disaster Risk Reduction* and *Rural Communities, Agriculture and Disasters.*

New Delhi, India — Bupinder Zutshi
Dhaka, Bangladesh — Akbaruddin Ahmad
Hyderabad, India — Ananda Babu Srungarapati

About the Book

The book is a collection of the selected papers presented at the Second World Congress on Disaster Management held at Vishakhapatnam in November 2015. The papers explore the experiences of disaster risk-reduction measures adopted by different communities in several countries. Major focus of the papers is to study experiences of community participation, resilience, and responses for disaster risk-reduction management in Coastal Regions.

This book discusses the interconnected, complex, and emerging risks in today's societies and deliberates on the various aspects of disaster risk-reduction strategies especially through community resilience and responses. It consists of selected papers presented at the World Congress on Disaster Management, which focused on community resilience and responses toward disaster risk reduction based on South Asian experiences, and closely examines the coordinated research activities involving all stakeholders, especially the communities at risk. Further, it narrates the experiences of disaster risk-reduction in different communities that have policy implications for mitigation of future disaster risks in the societies affected by these types of disasters. Written from the social science perspective to disasters rather than an engineering approach, the book helps development and governance institutions to prioritize disasters as a problem of development rather than being parallel to it.

Global experiences have indicated that poor people suffer the most from disasters, as they lack the capacity and resources to effectively cope with disasters, therefore efforts should be made to build capacities of communities to adopt disaster risk-reduction strategies. Disaster risk-reduction management (DRM) strategies through community resilience, responses, and participation have been a major policy shift recommended by the Sendai Framework of Action 2015–2030 (SFDRR). The Sendai Framework states four priority areas for immediate implementation by the countries. These four areas are, understanding disaster risk, Strengthening disaster risk governance to effectively manage disaster risks, Investing in disaster risk reduction resilience measures, and enhancing disaster preparedness for effective response and to build capacities for immediate and effective recovery, rehabilitation, and reconstruction. The Sendai Framework of Action is now accepted as a global policy framework to represent the focused and coherent directions, undertaking the complexities of the present world vulnerabilities to disasters. DRM is the key to minimize the human fatalities/casualties, loss of building and other construction and infrastructure structures, agricultural crops, loss of animal and livestock. Mainstreaming disaster risk-reduction management within the policies and programs of different sectors ensures, that the effects of disasters are minimized. At the same time, it enables governments to ensure that these policies and programs do not put people at risk.

Government of India launched a comprehensive National Disaster Management Plan, (NDMP) prepared by the National Disaster Management Authority of India (NDMA) in June 2016. The document was released by Prime Minister of India and it endorses disaster risk reduction measures which are in sync with Sendai Framework of Action 2015. The new NDMP focuses to reduce disaster risk, disaster damage and losses and to create a disaster resilient society focusing on the well-being of the people.

The book is divided into four parts. Part-I examines **Disaster Risk Reduction Laws and their Implementation issues.** Four papers are presented in this part. Akbaruddin Ahmad and Md Zahir Ahmed paper entitled Chapter "Law Can Set a Stage for Disaster Risk Reduction: A Study on Nepal" aims at identifying the apertures and areas of specious practice in addressing legislative issues to operate post-disaster management operation with implementing the regional relevant to disasters in Nepal. Marzina Begum in her paper entitled Chapter "Coordination Does Matter for Effective Disaster Management in the Coastal Areas in Bangladesh" examines the concept of coordination and investigates its components in terms of disaster management in the coastal areas in

Bangladesh. She highlights the importance of coordination in disaster management. The focal point of this paper is on coordination challenges and how the complexity of the existing institutional framework affects the capacity for disaster management activities in the coastal areas in Bangladesh. Rajendra Baikady, Heng Shengli, and Channaveer R. M. in their paper entitled Chapter "Integrating Disaster Management Content in to Social Work Curriculum-Voices of Social Work Students" aims to *understand the perception and beliefs of Master of Social Work students toward disaster management. The study explores, different perspectives of social work students toward disaster management and its Implication for teaching disaster management concepts in future studies of social work. The paper also integrates disaster management concepts in the social work curriculum and emphasizes the importance of student's interest to take up disaster management in social work Practice. Mahfuzul Haque in his paper entitled* Chapter "Indigenous Knowledge and Practices in Disaster Management: Experiences of the Coastal People of Bangladesh" emphasized proper documentation of success stories of disaster management experiences in Bangladesh especially local wisdom of communities in disaster risk-reduction strategies.

Part-II of the book is devoted to **Disaster: Identification and Use of Technology for Risk-Reduction.** This part consists of seven papers. Bupinder Zutshi, Homolata Borah, and Priya Bhakat in their paper entitled Chapter "Mapping Vulnerable Earthquake Disaster Class-I cities for Disaster Risk-Reduction and Community Resilience in India" attempts to *capture the vulnerability of Class-I cities especially their built environment to an elevated risk of earthquakes and also posing threats to a large number of population dwelling in these cities. Rajesh Kumar in his paper entitled* Chapter "Mainstreaming Nuclear Risk-Reduction in India" *analyses India's preparedness in case of a nuclear disaster and suggests to enhance awareness among people through training and preparedness in general, and among people of Amritsar city in particular. It focuses the role of education and knowledge, specially, among youth of India in building a culture of resilience and enhancing the participatory role of communities along with the administration in mainstreaming nuclear disaster risks. Nivedita Haran in her paper entitled* Chapter "Bridging the Gap Between Academicians and Bureaucrats: The NYSAF Way" *emphasizes the role of coordination and co-operation between research scholars who conduct research in the field and prepare recommendations based on deeper analysis of data and administrators who implement the policies.*

Such coordination is need of the hour for effective management of disaster risk-reduction strategies. Neena Joseph in her paper entitled Chapter "Sustainability of Community Managed Disaster Risk Reduction Projects" *analyses the sustainability of CMDRR plan and project handed over in 2013 to Elamakunnapuzha, Grama panchayat by the NGO, Ernakulam Social Service Society situated in Kerala, India. The paper concludes that stand-alone modality makes an NGO-based CMDRR unsustainable. What is required is coordinated venture of multiple stakeholders with mutually agreed roles, rules, rights and authority, maintaining structural and functional convergence with government and local bodies. Sanghamitra Nath in her paper entitled* Chapter "A Sociological Perspective on "Institutional Coordination" for Disaster Risk Reduction in India" *examines institutional coordination sociologically, especially from the functionalist perspective. It explores the multifarious functions served by institutional coordination (manifest, moral, and latent) which contribute to pre- and post-disaster management, deference to ethical principles of duty and willingness, and preservation of social order and values. The paper also enquires about elements which may strengthen institutional coordination. Namrita Agrawal in her paper entitled* Chapter "Innovative Use of Technology (GIS) For Disaster Risk Reduction: A Case Study" *suggests for adopting a multidimensional endeavour involving Government and Corporate Sectors to use the ICT facility specially GIS and Remote Sensing effectively and efficiently to tackle one of the most alarming problem in highly disaster prone-States. The study finds that if there would have been an effective GIS and Remote Sensing System in J&K in 2008, many more victims/lives would have been saved/timely received the relief and resources.* Marlon de Luna Era in his paper entitled Chapter "From Response to Responsibility: An Academe-Industry Partnership on Solid Waste Management in the Philippines" *attempts to take an in-depth look on the roles, participation, and accountability of non-state actors in the implementation of the solid waste management program. Further, the study also looks into the trajectory of environmental governance, the roles and the specific tasks particularly in the performance of private sector in its pursuit of Corporate Social Responsibility.*

Part-III of the book is devoted to **Disaster Risk-Reduction: Structural Inequality, Women and Child Resilience.** The part has six papers. Shishir Kumar Yadav in his paper attempts *to bring out the limitations and strengths for women during recurrent disasters. The author also tries to look at the experiences of the women in different structures,* *social and economic, and locations.* Binod Kumar in his paper entitled Chapter "Structural Inequality and Disaster Risk Reduction: An Indian Experience" emphasizes the importance of knowledge of structural inequalities before undertaking region specific disaster management policies, as sometimes *disaster is caused by insufficient development of means to avoid crisis and sometimes aspect of development itself becomes the reason for crisis. The paper underscores to develop a holistic view on disaster and structural inequality. Rahila Sikandar and Asif Khan in their paper entitled* Chapter "Linking Gender Vulnerability and Disasters" *focuses on some of the critical issues and concerns relating to the vulnerability of women at the times of disasters. The paper examines the impact of disasters on women. It also analysis the measures taken by the state and the required public policy for the women affected during the said period. Priyanka Jha in her paper entitled* Chapter "Women and the Gaps in Framework of Disaster Risk Reduction" *attempts to locate the discourse of Disaster Risk reduction from the lens of the women perspective. it attempts to not only bring out the problems, but also suggests outcomes that could be undertaken to resolve the gender blindness that policies both at national and international level faces with.* Isaias S. Sealza in her paper entitled Chapter "Post-Disaster Economic Vulnerability of Female-Headed Households" *pins down a point of entry for possible interventions from the outside to help the FHHs. It examines the relationship of this point of entry with a factor (income) that can serve as a measure of an intervention's performance or impact. The hypothesis that the "FHHs are likelier than MHHs to belong to the lower end of the income distribution" is supported by the data. The findings support the feminization of poverty thesis and look at opportunities for diversified sources of income as a potential entry point for interventions. Ray Kancharla in his paper entitled* Chapter "Child Centered Disaster Risk Reduction (CC DRR)/Resilience" *C-R2R suggests that disasters have the biggest impact on the poorest communities and the most vulnerable people which children. He examines how children face problems of food insecurity, malnutrition, and trafficking.*

Part-IV of the book is devoted to **Socioeconomic Impact and Post-Disaster Recovery.** The part presents four papers. Mondira Dutta in her paper entitled Chapter "Disaster Risk Reduction and Resilience Building Among the Coastal Zone Communities in Lakshadweep and Kerala" attempts to explore the disaster preparedness among the lives of coastal zone communities when disasters hit the regions of Lakshadweep and

Kerala coasts. It attempts to study how far community resilience in the region has been successful in mitigating disaster risk reduction among the fishing community where the main sources of livelihood are limited to coconut plantation and fishing. It is based on a primary field visit undertaken during the month of October 2017 in the islands of Agatti, Karavatti in Lakshadweep and parts of disaster-affected areas in Kerala for an understanding of the measures adopted by the coastal community in handling disasters. *Md Zahir Ahmed in his paper* Chapter "Posttraumatic Stress Disorder and Depression Among the Survivors of Nepal Earthquake: A Need Assessment Study" examines the post-traumatic stress disorders among the disaster victims and survivors. The *Study shows that people who have lost their amorous one and wealth are possibly to suffer from significant severe psychological disorders compared to those who have not. These sorts of disorders not only affect the victims but it may also develop a kind of mental disorder in the rescue workers, medical, and support teams. Nasim Banu in her paper entitled* Chapter "Cyclone Aila: Immediate Challenges and Impact on Socio-economy" identifies the *challenges faced in rescue and relief operations on the event of cyclone Aila and impact analysis on socio-economic conditions of the Alia effected people.* R. Lalitha S. Fernando and M. S. Dimuthu Kumari ion their paper entitled Chapter "Recovery after Disasters—Problems and Prospects: The Case of Koslanda-Meeriyabedda Landslide in Sri Lanka" examines *how the government intervened to the recovery processes in Sri Lanka and also explores issues and problems during the recovery process from the perspectives of both the victims and the administration and in the end the paper propose suitable strategies to improve their living standards.*

Contents

Part I Disaster Risk Reduction: Laws and Management

Law Can Set a Stage for Disaster Risk Reduction: A Study on Nepal 3
Akbaruddin Ahmad and Md Zahir Ahmed

Coordination Does Matter for Disaster Management in Bangladesh 19
Mst. Marzina Begum and Md. Nurul Momen

Integrating Disaster Management Content into Social Work Curriculum-Voices of Social Work Students 37
Rajendra Baikady, Cheng Shengli and R. M. Channaveer

Indigenous Knowledge and Practices in Disaster Management: Experiences of the Coastal People of Bangladesh 59
Mahfuzul Haque

Part II Disaster: Identification and Use of Technology for Risk-Reduction

Mapping Vulnerable Earthquake Disaster Class-I Cities for Disaster Risk-Reduction and Community Resilience in India 75
Bupinder Zutshi, Homolata Borah and Priya Bhakat

Mainstreaming Nuclear Disaster Risk Reduction in India 95
Rajesh Kumar

Bridging the Gap Between Academicians and Bureaucrats: The NYSAF Way 113
Nivedita Haran

Sustainability of Community Managed Disaster Risk Reduction Projects 119
Neena Joseph

Innovative Use of Technology (GIS) for Disaster Risk Reduction: A Case Study 137
Namrata Agrawal

A Sociological Perspective on "Institutional Coordination" for Disaster Risk Reduction in India 157
Sanghamitra Nath

From Response to Responsibility: An Academe–Industry Partnership on Solid Waste Management in the Philippines 173
Marlon de Luna Era

Part III Disaster Risk-Reduction: Structural Inequality, Women and Child Resilience

The Struggles of Women: An Experience from the Flood-Prone Area 191
Shishir Kumar Yadav

Structural Inequality and Disaster Risk Reduction: An Indian Experience 205
Binod Kumar

Linking Gender Vulnerability and Disasters 215
Rahila Sikandar and Asif Khan

Women and the Gaps in Framework of Disaster Risk Reduction 231
Priyanka Jha

Post-disaster Economic Vulnerability of Female-Headed Households 241
Isaias S. Sealza

Child Centered Disaster Risk Reduction (CC DRR)/ Resilience 261
Ray Kancharla

Part IV Disaster: Socio-economic Impact and Post Disaster Recovery

Disaster Risk Reduction and Resilience Building Among the Coastal Zone Communities in Lakshwadeep and Kerala 279
Mondira Dutta

Posttraumatic Stress Disorder and Depression Among the Survivors of Nepal Earthquake: A Need Assessment Study 299
Md Zahir Ahmed

Cyclone Aila: Immediate Challenges and Impact on Socio-economy 319
Nasim Banu

Recovery After Disasters—Problems and Prospects: The Case of Koslanda-Meeriyabedda Landslide in Sri Lanka 335
R. Lalitha S. Fernando and M. S. Dimuthu Kumari

Index 357

Notes on Contributors

Dr. Namrata Agrawal is a Professor at National Institute of Financial Management, an Institute of Ministry of Finance, Government of India. She is former faculty member of Motilal Nehru National Institute of Technology, Allahabad (presently NIT of UP, India) and has more than 25 years of teaching, research and consultancy experience.

Prof. Dr. Akbaruddin Ahmad an outstanding scholar has authored papers and researched on disaster management, laws, mitigation, and climate change. He is a specialist with exposure in management of educational institutions of higher learning, financial institutions, banks, and civil society think tank. He is Chairman of the Policy Research Centre (PRC. bd), a civil society think tank, Director and Professor, Bircham International University, Spain, and former Vice Chancellor, Darul Ihsan University, Bangladesh. He is also involved as the Chairman (Admn), NAPSIPAG. He is a reputed educationist and a prolific writer and regular contributor to national dailies and peer reviewed journals. He has attended national, regional and international conferences, seminars and presented papers and authored several books and technical papers that include.

Md Zahir Ahmed is currently pursuing his Doctoral Degree at Northwest Normal University, China. Prior to that he was working with Policy Research Centre (PRC.bd) as Junior Research Coordinator. He had been Former Research Assistant, Government of People's Republic of Bangladesh. His research areas are Sexual Harassment, Disaster Management, Domestic Violence, E-Governance, Social development, and Youth Empowerment.

Rajendra Baikady is presently a Ph.D. student in social work at Central University of Karnataka, Gulbarga, working under the guidance of Dr. R. M. Channaveer (India) and Cheng Shengli of the Shandong University (China). Rajendra, through his doctoral work, is seeking an understanding of the impetus for social work education in a parliamentary democracy and communist regimes, such as India and China. He has recently spent 6 months collecting data at the Department of Social Work, Shandong University China and was awarded the prestigious INLAKS research travel grant 2015, the ICSSR fellowship grant 2015, and the Shandong University Tuition fee exemption Scholarship 2016, for his research work in China. Rajendra is undertaking some additional research work in international social work, peace, poverty, and issues concerning youth within South Asian contexts.

Prof. Nasim Banu is a researcher of Social Sciences with Ph.D. who did Masters in Public Administration; M Phil in Regional Planning. She is a Professor of Public Administration in Islamic University, Kushtia, Bangladesh.

Mst. Marzina Begum is an Associate Professor in the department of Public Administration of the university of Rajshahi in Bangladesh. She obtained her Doctor of Philosophy (Ph.D.) from the University of Pavia in Italy. Her main area of interest covers disaster management, environmental policy and governance, and climate change.

Ms. Priya Bhakat is a full-time Ph.D. scholar at the CSRD in Jawaharlal Nehru University. She was awarded Junior Research Fellowship by University Grants Commission in 2012. Presently she is working as a National Research Coordinator in National Coalition for Education, India. Her area of research includes Marriage, Fertility, Education, Gender, and Child Rights.

Homolata Borah is a Research Associate at the Special Center for Disaster Research and at the Center for the Study of Regional Development, Jawaharlal Nehru University. She has worked on a range of socio-economic and cultural issues including disaster risk management, gender aspects of disasters, domestic violence and public health. Her passion is to apply rigorous social science research to address development challenges and help create a better world.

Dr. R. M. Channaveer is an Associate Professor and Head in the Department of Social Work of Central University of Karnataka. He has been in post-graduate teaching for 17 years. He has carried research projects funded from various regional and national level agencies. He has published 30 research papers in national and international journals. He is known for his innovative community-based interventions, initiatives, and activities. His area of work is social development, international social work, community-based rehabilitation, and social entrepreneurship.

Prof. Mondira Dutta, M.A., M.Phil., Ph.D. is currently the Senior Fellow of ICSSR at the Special Centre for Disaster Research and former Professor, Jawaharlal Nehru University. Her fields of specialisation include Gender Studies, Disaster Mitigation and Human Security in South Asia, Afghanistan and Central Asia. With four decades of teaching and research, she has been a Consultant to UN and several other International and National organisations.

Dr. Marlon de Luna Era an Associate Professor at De La Salle University, Manila, Philippines has completed his Ph.D. on Development Studies at De La Salle University. He has enormous professional experience on solid waste management, gender and development, and disaster risk as engaged by national, local, and international organizations.

Prof. Mrs. R. Lalitha S. Fernando, Ph.D. serves as a Professor and Head of the Department of Public Administration, Faculty of Management Studies and Commerce of the University of Sri Jayewardenepura in Sri Lanka. Currently, she serves as the Secretary General of the NAPSIPAG (Network of Asia-Pacific Schools and Institutes of Public Administration and Governance).

Dr. Mahfuzul Haque a former Secretary of the Government of Bangladesh has been teaching at the Department of Development Studies, University of Dhaka and other private universities for a decade. His areas of interests are sustainable development; natural resource management; climate change; biodiversity conservation; disaster risk reduction etc. He joined civil service in 1981. During his decade-long stint at the Ministry of Environment and Forests, he led official Bangladesh delegation at the Conference of Parties of various conventions, including UNFCCC, CBD, and Montreal Protocol. He served as the President of Montreal Protocol Implementation Committee and a Bureau member for Convention on Biodiversity secretariat for two years.

Nivedita Haran joined the Indian Administrative Service in 1980. She worked in different developmental and regulatory departments including disaster management, revenue, home, and agriculture. She was part of the UN Peace-keeping Mission in Kosovo. She holds a Ph.D. in Sociology from IIT Delhi. Post-retirement she is engaged in activities linked to land management, good governance, and skill development among others.

Dr. Priyanka Jha teaches Political Science in Ambedkar University. Trained in Centre for Political Sciences in JNU and Lady Shri Ram, she works extensively on Nation, its thinkers, philosophical frames, its values and its communities.

Dr. Neena Joseph has recently retired as Professor in Institute of Management in Government, Regional Centre, Kochi after putting in 30 years of service. She had been conceiving, designing and implementing training programmes for government in the areas of Governance including Decentralized Governance, Gender, Training, Research Methodology, Disaster Management and General Management, and undertaking research programmes for Government.

Ray Kancharla is the National Manager of DRR CCA, Save the Children.

Co-author

Dr. Asif Khan, did his Ph.D. in Social Work from Aligarh Muslim University, Aligarh. Presently, working as Assistant Professor (Guest Faculty) at Delhi School of Social Work, University of Delhi. Earlier he served as an Academic Associate in Communication Area at Indian Institute of Management Kashipur, Uttarakhand. He has more than a year experience working as a Research Assistant in an ICSSR funded research project on Disaster mitigation at Indian Institute of Management Kashipur, Uttarakhand.

Mr. Binod Kumar is an Assistant Professor, Dr. Bhim Rao Ambedkar College, University of Delhi, New Delhi, India. His Ph.D. work locates disaster prevention and mitigation as a subject matter of law apart from developmental planning. Just as knowledge of weather, acquaintance of existing legal framework can also save the life and property of people from disasters. In this backdrop, his research navigates the intersection of disaster, law and community resilience in India.

Rajesh Kumar is an Assistant Professor of Political Science at School of Social Sciences, Guru Nanak Dev University, Amritsar, Punjab, India. He has Doctoral degree from Jawaharlal Nehru University, New Delhi. Dr. Kumar's teaching and research interests are in the areas of Indian government and politics, foreign and security affairs. His publications include the book Indo-USA Political and Strategic Relations besides several articles contributed in Journals like, Pakistan Horizon, World Focus, South Asian Affairs, and Journal of Punjab Politics. He is Alumni Summer Workshop, Regional Centre for Strategic Studies, Colombo, Sri Lanka. He is also Member of Network of Asia Pacific Schools and Institutes of Public Administration and Governance (NAPSIPAG), J.N.U., New Delhi.

Co-author

Ms. M. S. Dimuthu Kumari currently serves as a lecturer in the Department of Public Administration of University of Sri Jayewardenepura in Sri Lanka.

Md. Nurul Momen is an Associate Professor in the department of Public Administration of the university of Rajshahi in Bangladesh. He has published many articles in the international journals including chapters in different books, dealing in a range of debates in public policy, governance, and public sector reform in south Asia.

Ms. Sanghamitra Nath is an Assistant Professor and Head of the Department of Sociology at Bajkul Milani Mahavidyalaya, Vidyasagar University, West Bengal. She was selected by the West Bengal College Service Commission. She completed M. Phil on Disaster Risk Reduction from Centre for the Study of Law and Governance, Jawaharlal Nehru University, New Delhi.

Isaias S. Sealza is a full Professor, Dean of Graduate School, and Senior Research Associate at Xavier University-Ateneo de Cagayan. He has done projects for local and international organizations. He has served as resource person in seminars and workshops; and, published and refereed in publications of domestic and international circulation.

Cheng Shengli is a Professor in the Department of Social Work School of Philosophy and Social Development, Shandong University, Peoples Republic of China. He has been in post-graduate teaching for more than 25 years. Cheng has carried out research projects funded by international agencies such as UNESCO, Washington University in St. Louis in United States, the University of British Columbia in Canada and Taiwan Dongwu University. He was a visiting professor in many international university departments. Cheng has six books, 37 research articles (both in English and Chinese) and five international projects to his credit. His areas of work are urban poverty and social assistance, social policy and social welfare, social psychology, family, child, and youth.

Dr. Rahila Sikandar is currently working at Special Centre for Disaster Research, Jawaharlal Nehru University. taught Political Science at University of Delhi and also associated with Indian Institute of Management, Kashipur. Awarded Ph.D. in 2015 in Law and Governance from Centre for the Study of Law and Governance, Jawaharlal Nehru University, New Delhi. Interest area includes Gender and Development, Sustainable Development, Higher and Professional Education, Minority Studies, Disaster Studies, etc.

Dr. Ananda Babu Srungarapati President, Disaster Management Initiatives and Convergence Society (DMICS) is a societal awareness specialist and a scholar.

Shishir Kumar Yadav is a full-time Ph.D. research scholar at the Centre of Social Medicine and Community Health (CSMCH) in Jawaharlal Nehru University, New Delhi. By training he is an anthropologist.

Bupinder Zutshi, M.A, M.Phil, Ph.D. Professor, Centre for the Study of Regional Development, Jawaharlal Nehru University, New Delhi. He has more than 38 years of teaching and research experience at post-graduate and research level. He has published seven books, completed more than dozen research projects and written several research articles in research journals of repute published from India and abroad. He has organized several national and international seminars and conferences. His major research interests are issues related to human settlements and their changing habitat, disaster risk and vulnerability of human settlements. Other research interests are issues related to demographic changes and their consequences on population composition and population characteristics, education and child labor issues, studies on tribal communities. He has completed several research projects sponsored by national and international organizations, which include UN Women (UNIFEM), UNESCO (New Delhi), International Bureau of Education (IBE, Geneva), United Nations High Commissioner for Human Rights (Geneva), Ford Foundation (New Delhi), International Labour Organization (New Delhi), National Human Rights Commission of India, Election Commission of India, and Indian Council for Social Science and Research.

Glossary

CBDM	Community-Based Disaster management
CBDRM	Community-Based Disaster Risk management
CBO	Community-Based Organization
CC DRR	Child Centred Disaster Risk Reduction
CCA	Climate Change adaptation
CFS	Child Friendly Space
CG	Children Group
CPIE	Child Protection in Emergencies
CPWG	Child Protection Working Group
CRPC	Children Resilience and Protection Committee
CSO	Civil society organization
DFID	Department for International Development
DP	Disaster Preparedness
DRM	Disaster Risk management
DRR	Disaster Risk Reduction
EIE	Education in Emergencies
EW	Early Warning
EWS	Early Warning System
HFA	Hyogo Framework for Action
M&E	Monitoring and Evaluation
NGO	Non-governmental Organization
PTSD	Post-traumatic Stress disorder
PVCA	Participatory Vulnerability Capacity Analysis
SFA	Sendai Framework for Action (2015–2030)
UN ISDR	UN International Strategy for Disaster Reduction
UN OCHA	UN Office for Coordination of Humanitarian Affairs

UN	United Nations
UNCRC	United Nations Convention on the Rights of the Child
VCA	Vulnerability and Capacity Assessment/Analysis
VDMP	Village Disaster Management Plan
WASH	Water, Sanitation and Hygiene

List of Figures

Law Can Set a Stage for Disaster Risk Reduction: A Study on Nepal

Fig. 1 Tally of deaths after the earthquake of Nepal (*Sources* Nepal National Information Technology Center, U.S.G.S) 5

Fig. 2 Administrative map with the damage and death (*Source* CATDAT) 6

Integrating Disaster Management Content into Social Work Curriculum-Voices of Social Work Students

Fig. 1 Students understanding 46

Fig. 2 Disaster Management Content in Social Work Curriculum 47

Fig. 3 Knowledge on social workers role 49

Mapping Vulnerable Earthquake Disaster Class-I Cities for Disaster Risk-Reduction and Community Resilience in India

Map 1 Natural hazards in India (*Source* Poorest Areas Civil Society (PACS) Programme 2001–2008) 78

Map 2 Earthquake risk zones of India (*Source* Disaster Resistant Architecture www.slideshare.net) 84

Map 3 Class-I cities falling in very high damage earthquake zone (*Source* Research Work on Census of India, General Population Tables 2011 and NDMA Map) 85

Map 4 Class-I cities falling in high damage earthquake zone (*Source* Research Work on Census of India, General Population Tables 2011 and NDMA Map) 87

Map 5 Class-I cities falling in moderate damage earthquake zone (*Source* Research Work on Census of India, General Population Tables 2011 and NDMA Map) 88

Innovative Use of Technology (GIS) for Disaster Risk Reduction: A Case Study
Fig. 1 District Kupwara (*Source* Wikimedia) 142
Fig. 2 District Baramulla (*Source* Wikimedia) 143
Fig. 3 Data integration is the linking of information in different forms through GIS (*Source* US geological survey) 144
Fig. 4 Epicenter of the 2005 earthquake (*Source* Centre for Excellence in Disaster Management and Humanitarian Assistance) 146
Fig. 5 Total lives lost in villages 149
Fig. 6 Lives lost among sample data 150
Fig. 7 Total lives lost in villages due to delayed aid 151
Fig. 8 Reduction in reaction time during disaster (*Source* gisdevelopment.net) 152
Fig. 9 Gaps in claims and relief distributed 154

From Response to Responsibility: An Academe–Industry Partnership on Solid Waste Management in the Philippines
Fig. 1 Map of Laguna Province highlighting the City of Santa Rosa 175
Fig. 2 Accountability of the non-state actors in Academe–Industry Partnerships 186

Structural Inequality and Disaster Risk Reduction: An Indian Experience
Fig. 1 Reproducing caste inequality in different phases of disaster cycle 211

Linking Gender Vulnerability and Disasters
Fig. 1 Represents the understanding *about the gendered vulnerabilities and the areas of concern* which might be addressed accordingly (*Source* Ashraf and Azad 2015) 226
Chart 1 Represents limitations (*Source* Ariyanbandu and Wickramasinghe 2003) 220
Chart 2 Represents low access and response (*Source* Ariyanbandu and Wickramasinghe 2003) 221

Post-disaster Economic Vulnerability of Female-Headed Households
Fig. 1 Hypothesized relationships among the variables 243
Fig. 2 Tercile distribution of the total sample 251
Fig. 3 Tercile distribution of FHHs 252
Fig. 4 Tercile distribution of MHHs 253

Disaster Risk Reduction and Resilience Building Among the Coastal Zone Communities in Lakshwadeep and Kerala
Fig. 1 Kerala, Crime Against Women, (2007–2017). *Data Source* http://www.keralapolice.gov.in/public-information/crime-statistics/crime-against-women 293

List of Tables

Integrating Disaster Management Content into Social Work Curriculum-Voices of Social Work Students

Table 1 Students selection of courses 50

Mapping Vulnerable Earthquake Disaster Class-I Cities for Disaster Risk-Reduction and Community Resilience in India

Table 1 Distribution of earthquakes in India (1991–2015) 80
Table 2 Earthquake zone wise total number of class-I towns in India 86

Innovative Use of Technology (GIS) for Disaster Risk Reduction: A Case Study

Table 1 Villages visited during survey 146
Table 2 O/o the divisional commissioner of Kashmir (all amounts in Rs. lakhs) 148
Table 3 Figures of casualties in surveyed villages 150

Linking Gender Vulnerability and Disasters

Table 1 Indicates the disasters occurred in the last 25 years (1990–2015) 219
Table 2 Gender relations in society: Reflected in 223
Table 3 Causes for gender difference in disaster 225

Post-disaster Economic Vulnerability of Female-Headed Households

Table 1 Distribution of type of household by income group 253
Table 2 Association between type of household and the test factors 254
Table 3 Association between the test factors and income 254
Table 4 Type of household and income level under separate categories of work status 255

Disaster Risk Reduction and Resilience Building Among the Coastal Zone Communities in Lakshwadeep and Kerala
Table 1 Recent major natural disaster in Kerala 284

Posttraumatic Stress Disorder and Depression Among the Survivors of Nepal Earthquake: A Need Assessment Study
Table 1 Sample items from Nepali translated PTSD check list—civilian version (PCL-S) 306
Table 2 Sample items from Nepali translated national Stressful events survey PTSD Short Scale (NSESSS) 307
Table 3 Sample items from Nepali translated ZUNG self-rating depression scale 307
Table 4 Frequency and percentages of respondents diagnosed as having PTSD or not ($n = 100$) 309
Table 5 Percentages of respondents at severity level of PTSD ($n = 18$) 309
Table 6 Frequency and percentages of respondents diagnosed as having different levels of depression ($n = 100$) 310
Table 7 Distribution of respondents diagnosed as having PTSD or not by districts ($n = 100$) 310
Table 8 Gender differences in PTSD and depression 310

Recovery After Disasters—Problems and Prospects: The Case of Koslanda-Meeriyabedda Landslide in Sri Lanka
Table 1 Relief services provided by different parties 343
Table 2 Income status of the affected families 346

PART I

Disaster Risk Reduction: Laws and Management

Law Can Set a Stage for Disaster Risk Reduction: A Study on Nepal

Akbaruddin Ahmad and Md Zahir Ahmed

Introduction

Nepal is a landlocked nation, situated between China's Tibetan Autonomous Region in the north and India in the west, south, and east, with a land territory of 147,181 km^2. From east to west the nation extends for just about 900 km. Height in Nepal ascends from almost ocean. It is a precipitous nation situated in South Asia, covering a territory of 147,181 km^2. The nation fringes India and China, and spreads 145–241 km from north to south and 885 km from east to west. Nepal is partitioned into three geological districts: the Himalayan, slope, and plain locales. The Himalayan district covers 15 for every penny, the slopes 68 for each penny and the plain (Terai) covers 17 for every penny of the aggregate region of Nepal. The Terai is the nourishment crate of Nepal and is home to 48.4% of the populace. The aggregate populace of Nepal is around 27 million. For advancement and administration purposes, Nepal is isolated into 5 locales, 14 zones, and 75

A. Ahmad · M. Zahir Ahmed (✉)
Policy Research Centre.bd, Dhaka, Bangladesh

A. Ahmad
Darul Ihsan University, Dhaka, Bangladesh

B. Zutshi et al. (eds.), *Disaster Risk Reduction*,
https://doi.org/10.1007/978-981-10-8845-2_1

regions. There are 3913 Village Development Committees and 58 districts. Horticulture gives around 33 for each penny of the total national output, enterprises 10 for every penny, and administration parts like exchanging, development, and tourism share 57 for every penny (Central Bureau of Statistics 2009). Through and through, 14.2 for each penny of the aggregate populace live in urban zones. Movement from towns to urban regions has been expanding each year. The Human Development Report brings up that Nepal is among the poorest nations on the planet, with 33% of the populace living beneath the neediness line.

Nepal faces diverse sorts of intermittent debacles, for example, ecological debasement, deforestation, soil disintegration, avalanches, and surges. Surges are very regular in the fields, while avalanches happen in slopes and mountains, and torrential slides in the Himalayan district. Flames have caused deforestation and wellbeing plagues can make an incredible danger groups. The quantities of passings, wounds, and relocations are expanding each year. What's more, there are accounted for instances of seismic tremors in various parts of the nation. The most incessant calamities of Nepal are:

- Floods and landslides
- Earthquakes
- Fires
- Epidemics and
- Drought and avalanches

Among the several severe disasters in both forms of natural and human-induced; this present paper is focused on the recent earthquake.

On April 25 at 11:56 Nepalese time, a tremor of extent 7.8 struck Nepal, with an epicenter 77 km northwest of Nepal's capital Kathmandu in the Gorkha locale. As of May 15, 8316 individuals had been accounted for murdered and 17,866 individuals harmed. The generally shallow profundity of the quake, at 2 km, was related with especially solid powers scattered over a wide geographic range.

On May 12 at 12:50 Nepalese time, another seismic tremor (extent 7.3) hit with an epicenter 76 km upper east of Kathmandu in the Dolakha/Sindhupalchowk locale. This tremor influenced areas as of now unequivocally affected by the April 25 seismic tremor and new locale. As of May 15, 117 passings, and 1961 harmed have been accounted for because of this quake (Fig. 1 *Nepal Earthquakes 2015: Desk Review of Existing Information with Relevance to Mental Health & Psychosocial Support*).

If we sum up the Geographical and climatic aspects on Nepal then we have seen the followings:

- Officially, Nepal is sorted out into 5 formative areas (far west, mid-west, west, focal, and east), 14 zones, 75 locale, and 3915 town improvement councils, or VDCs (rustic), and 58 districts (urban), with VDCs comprising of between 9 and 35 wards.
- Nepal has high land assorted variety, covering three biological zones from practically ocean level to the most noteworthy point on earth.
- Nepal is inclined to characteristic risks, including quakes, streak surges in the rainstorm time frame, avalanches, blasting of frosty lakes, changes because of an Earth-wide temperature boost, and ecological debasement.
- Topographical and climatic highlights will hamper seismic tremor reaction, with Nepal having extremely poor street scope (Fig. 2 *Deadly Earthquake Nepal 25/04/2015—Archived Part Nr. 5 (April 29-00:00 until April 30-24:00). Earthquake-Report.com. N.p., 06 May 2015. Web*).

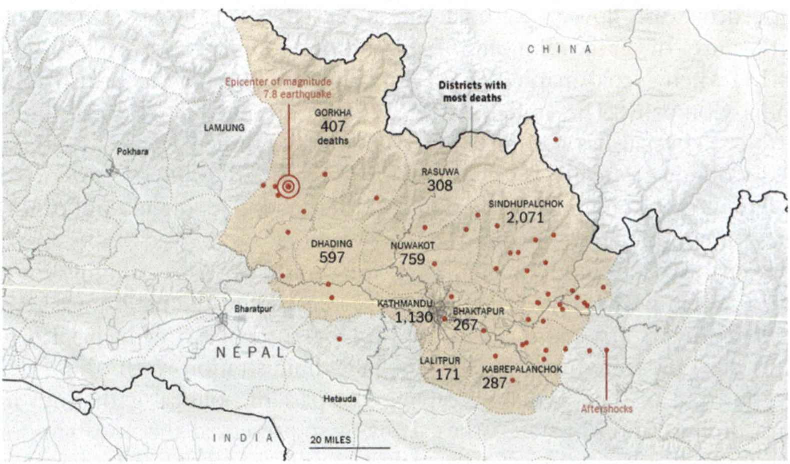

Fig. 1 Tally of deaths after the earthquake of Nepal (*Sources* Nepal National Information Technology Center, U.S.G.S)

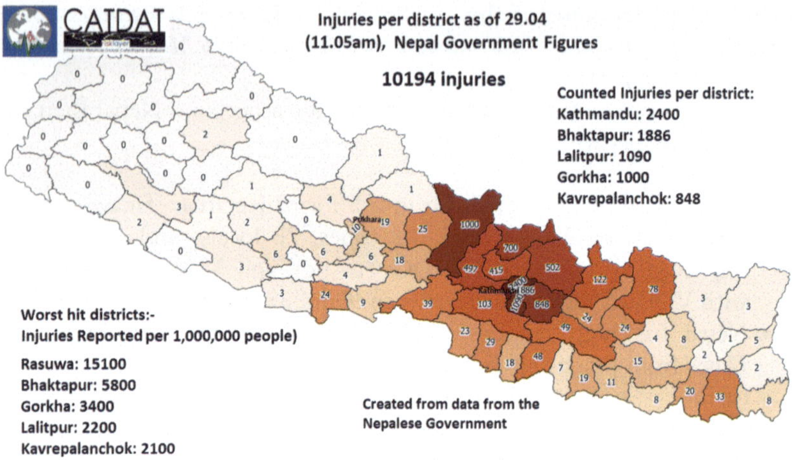

Fig. 2 Administrative map with the damage and death (*Source* CATDAT)

Administrative and Legislative Structure

Present day Nepal goes back to 1768, when King Prithvi Narayan Shah, leader of a little realm of Gorkha, added Kathmandu Valley as a major aspect of his unification crusade. He named the new nation Nepal, moved his cash flow to Kathmandu, and established the Shah line. The procedure of unification and regional development proceeded until 1815, when Nepal marked the Treaty of Sugauli with the British East India Company. The settlement settled the present day regional limits of Nepal. Nepal has had different types of government, state structures and working strategies from that point forward. In 2006, Nepal started debating the rebuilding of the state, and rethinking its organs and capacities for consolidation into another constitution.

Nepal started modernizing after the finish of Rana administer in 1951, when popular government was presented and the nation was represented by its first statute, the Interim Government Act of Nepal 1951. This law presented the parliamentary administration with the ruler as the head of state, while perceiving the part of political gatherings in the administration of state issues. It likewise presented the possibility of partition of forces between a lawmaking body, an official and a legal, in spite of the fact that by and by the ruler delighted in the most elevated official, authoritative, and legal specialist. Alongside this adjustment in the political framework, the part of government started to grow from ensuring national security, and keeping up peace to incorporate giving essential administrations and advancing improvement.

There is a Supreme Court as the peak body of the legal, with three levels of courts, in particular, locale courts, request courts, and the Supreme Court. The Interim Constitution of 2007 has redesigned the National Human Rights Commission from a statutory guard dog to a sacred body. The Authority Abuse Investigation Commission and an Auditor General are established elements that watch the administration's exercises and make suggestions if standards are damaged. Major rights are ensured by the Constitution and secured by the legal. Article 33(m) of the Interim Constitution of Nepal of 2007 has an arrangement for executing universal bargains and understandings, to which the state is a gathering. Article 9 of the Nepal Treaty Act of 1991 states that, "if any arrangement of the settlement to which Nepal is a gathering is conflicting with any law in compel in Nepal, the law to the degree of that irregularity should be void and the arrangements of the Treaty might win as law of Nepal." All bargains to which Nepal is a gathering and has confirmed are, in principle, straightforwardly enforceable. It ought to be noted here that standard worldwide law is not some portion of Nepalese law unless it is changed into local law.

Legislative Framework for Disaster Management

Prior to the appearance of Natural Disaster Relief Act (NDRA), 1982 A.D. there was no all-around organized calamity strategy in Nepal. Before 1982 A.D., alleviation and safeguard works were done as the social works as it were. Along these lines, understanding the need of a demonstration, NDRA, 1982 A.D. was planned which has just been corrected twice in 1989 A.D. what's more, 1992 A.D.

Be that as it may, Natural Disaster Relief Regulations (NDRR) couldn't yet be detailed which is extremely basic. Without NDRR, NDRA couldn't be completely powerful. NDRA, 1982 does not portray the obligations and duties of all the fiasco administration-related organizations other than the Ministry of Home Affairs. Obligations and duties of different other debacle administration offices must be reflected in NDRR as each and everything couldn't be stipulated in the Act. As indicated by NDRA, 1982 there has been the arrangement of Central Natural Disaster Relief Committee (CNDRC). Besides, there is additionally the arrangement of a Relief and Treatment Sub-Committee (RTSC), Supply, Shelter, and Rehabilitation Sub-Committee (SSRSC), Regional Natural Disaster Relief Committees (RNDRC), District Natural Disaster Relief Committees (DNDRC), and Local Natural Disaster Relief Committees (LNDRC) in the NDRA, 1982. RTSC and SSRSC should be activised in the season of intense cataclysmic events while CNDRC and DNDRC are dynamic constantly. This is so to speed up the quick protect and alleviation works. Nonetheless, RNDRC and LNDRC should be enacted as a when important, right now, these two offices are as lethargic organizations. Also, as the NDRA, 1982 does not depict the capacities and obligations of all region calamity administration-related organizations, the issue of collaboration, coordination, and common comprehension between different region fiasco administration-related offices are seen. Thusly, at times, debacle casualties don't get prompt, proficient and viable safeguard, and alleviation administrations. Postponed alleviation works frequently brings intense and obnoxious outcomes. What's more, here and there, duplication of alleviation works have additionally been experienced, for the most part, because of the nonattendance of exchange and common comprehension among catastrophe administration-related offices. Likewise, a portion of the locale calamity administration-related offices endeavor to move their duties to alternate as there is no obvious expected set of responsibilities in the Act. Along these lines, prompt plan of NDRR is exceptionally fundamental where obvious set of working responsibilities of all the catastrophe administration-related organizations ought to be made.

Administration of Nepal has declared different laws and approaches, including Natural Calamity (Relief) Act, 1982 and National Strategy on Disaster Risk Management (NSDRM), 2009. The demonstration has provisioned a Committee being led by the Minister of Home Affairs at the focal level, Regional Natural Disaster Relief Committee at local level, and District Natural Disaster Relief Committee at region level. Service of Home Affairs is functioning as a nodal organization of fiasco hazard administration, both at National and International level for Nepal. Alongside the Act, Government embraced National Strategy for Disaster Risk Management (NSDRM), 2009 in view of Hyogo Framework for Action and this new technique envelops aversion, relief, readiness, reaction, and recuperation. This technique has dispensed the unmistakable part of various Ministries for various periods of catastrophe. Fiasco Risk Reduction and Climate Change Adaptation have turned out to be a national need and being regulated to help manageable advancement in Nepal through the harmonization and mainstreaming process. The present 13th Five Year Plan (2013/14–2015/16) has underlined the fiasco hazard administration issues as an acquired character of manageable improvement and has concurred need to pre-catastrophe readiness to recuperation process.

While their new Disaster Management Act is as yet pending, there have been a few improvements in Nepal with respect to catastrophe chance administration as of late. These incorporate the creation in 2009 of the Nepal Risk Reduction Consortium, which unites philanthropic and improvement organizations with money-related foundations and the legislature of Nepal. Additionally, in 2009, the administration received a National Strategy for Disaster Risk Management, which diagrams the parts of various services amid a crisis and built up a National Emergency Operations Center in 2010. These instruments and structures are being operationalized in the present help and recuperation operations. The genuine trial of their prosperity will, however, be in the months and years ahead.

This present paper aims to find out the obstacles for entering in Nepal with the aid due to the legal constrains.

Barrier to the International Aid

Being a stimulant nation, Nepal needs adequate assets to give sufficient help to the cataclysmic event casualties. Basically, remaking and recovery programs have needed to endure because of such asset limitations. Notwithstanding extremely restricted assets and assets the Ministry of Home Affairs has been attempting to adapt to the cataclysmic events to its most extreme limit. A few Countries around the world have propelled broad guide operations to help casualties of the Nepal seismic tremor, yet conveyance of help faces challenges as Nepal is as yet attempting to grapple with the size of the catastrophe. With more than 4400 dead, 8000 harmed and 8 million individuals crosswise over Nepal influenced, various guide gatherings and no less than 16 countries have raced to send supplies and specialists to the stricken nation. However, the most vital thing is, Nepalese government and armed force who are driving the fiasco reaction confront another issue: How to adequately arrange and sort out the gigantic flood of philanthropic guide. We have discovered three noteworthy snags that are hampering endeavors of appropriation.

i. **Bottleneck of the Airport**

Nepal depends on just a single universal airplane terminal to get and convey help. Help associations say the landing area at Tribhuvan International Airport remains stick pressed with an extensive number of payload planes. A few airplane conveying basic supplies have been dismissed, or occupied to India and somewhere else. The air terminal is completely congested. Indeed, even the planes which have landed, products haven't been offloaded. A live air activity screen, demonstrates that approaching flights have needed to hover noticeable all around a few times previously having the capacity to arrive (Flightradar24). The United Nations Disaster Assessment and Coordination (UNDAC) group organizing the ground to oversee universal guide tended to these issues. Despite the fact that hunt and save group came to Nepal subsequent to flying a huge number of miles yet confronted troubles in landing. Moreover, the airplane terminal administration was confused with staff and military not having the capacity to clear runways and offload supplies viably.

ii. **Damaged Infrastructure**

The primary streets in Kathmandu are open, yet remote territories are to a great extent difficult to reach. Because of the rugged geology, framework harm, crumbled connects and harmed streets, access to a large number of the influenced ranges is accounted for to be greatly restricted. Additionally, the Nepalese government has not possessed the capacity to give alleviation to every single influenced territory because of challenges in transporting merchandise. The circumstance has crumbled to the point where alleviation specialists are on a par with stranded. Close by, more than 40,000 houses have been completely obliterated.

iii. **Ongoing Damage Assessment**

Following 48 hours of seismic tremor struck, Nepalese specialists driving the reaction are as yet attempting to completely comprehend which zones have been influenced, how they are influenced and what the needs are and that exacerbated things. In the event that it takes some of few days to institutionalize the supply lines, appropriation lines to settle then it'll be the genuine test. The operations of circulating the guide were to "impromptu" so it took more circumstances which prompts the tremendous harm toll.

Major Problems in Post Disaster Management

- Employees of help bunches have been influenced themselves as they sort out reactions to the calamity. The volunteers and the individuals from fiasco administration have lost their home so it was somewhat hard to arrange and start the operation with the national quality. So the entire operation relied upon the global guide.
- Engaging the group is the best alternative since individuals have been versatile. Be that as it may, it's been hard to get individuals and supplies into the nation, particularly to provincial regions a long way from Katmandu since they stay occupied to moderate their own troubles.
- More than 6 million individuals live in the territories of Nepal that have been influenced. Numerous people have either lost their homes or have been compelled to live in the city due to the risk of further delayed repercussions. The worldwide reaction is being composed by the Nepalese government through its National Emergency Operation Center.
- The most critical to recollect the calamity is the thickness. Kathmandu is a town that developed into a city. There is no adequate open space to suit the general population who got the uprooting.
- Since the rainstorm season starts in June so the dropped rain put the operation in much decline circumstance. Which was viewed as the "race against time".
- Still, on the grounds that a serious quake has for quite some time been anticipated by geologists, a few offices had made arrangements and fiasco designs. So there must have a preparation sense dependably.
- Experts stated, notwithstanding, that remote regions, some of which are reachable just by walking, may need to hold up longer to get offer assistance. Nepal's poor street organize, a set number of helicopters and planes to carry supplies to removed towns, and discontinuous correspondences all through the nation would likely exacerbate the present circumstance.

Recommendations

The debacle administration gets ready for underdeveloped nations like Nepal needs to fortify the catastrophe administration capacity by receiving different conceivable means through the appropriate authoritative structure. The arrangement and law visualize that an endeavor ought to be made for anticipation, moderation, and decrease of cataclysmic event through further developed geographical, hydrological, and meteorological innovation. Danger mapping, weakness evaluation, chance investigation, and early cautioning framework must be produced for which very much prepared and proficient labor ought to be made accessible. The arrangement and law likewise focus on the need to reinforce the capacity of flame detachment. The arrangement and law must be suggestive that the approaches and directions concerning fiasco administration ought to be revised by the need. The arrangement stresses the significance and requirement for national as well as worldwide help. Since there is the new Disaster Management Act is pending so its improvement, adjustment, and execution are essential. It has set compulsory commitments and duties on services and boards of trustees, and guarantees straightforwardness and responsibility in the general debacle administration framework. The prime goals of the respectable Act ought to be:

- Measures related to national policy and planning for making institutional arrangement, providing legal framework, adopting national policy, and plan on disaster management.
- Measures related to geological, hydrological, and meteorological hazard assessment and environmental engineering studies.
- Infrastructure specific and hazard-specific preparedness measures.
- Measures related to strengthening firefighting capabilities in fire-prone areas.
- Measures related to awareness raising, training, rehearsal, simulation activities.
- Measures related to establishment of disaster management information system and stock piling of emergency supply materials.
- Risk assessment for development planning.
- Policies on the role of NGOs, local community, private sector and people's participation especially women and socially disadvantaged groups.
- Incorporation of environment impact assessment for disaster reduction development planning.
- Promotion of regional and subregional cooperation between countries expecting same types of hazards.
- Establishment of documentation center on disaster activities.

- Substantial reduction of the overall risks of disasters to an acceptable level with appropriate risk reduction interventions.
- Effective to the post-disaster emergency response; rehabilitation and recovery measures.
- Insurance of the provision of emergency humanitarian assistance to the most vulnerable community people.
- Ensuring the strongest institutional capacity for effective coordination of disaster management involving government and nongovernment organizations and other international Aid organizations.
- Establishing a multidimensional disaster management system capable of dealing with all hazards for the country.
- The disaster management Act will help in promoting a comprehensive disaster management programme upholding the all-hazards, both natural and Man-Made.
- All-hazards and all-systems approach where risk reduction as a core element of disaster management has equal emphasis with emergency response management with greater focus on equitable and sustainable development.

Notwithstanding the difficult accomplishments in the enactment and the institutional structure, a few prompt activities should be useful in national and neighborhood level of debacle administration since we trust the exercises are no issues up to this point so a further upgrade is expected to make the Act genuine working. To pick up a definitive objective the catastrophe administration, the new Act needs to do following:

- To make the Act effective dissemination of the information is the most important through proper training and orientation to all both national and local level.
- Orientation for the different level disaster management committee to make them more informative and ensuring the strength.
- Since the National Disaster Management Plan was affirmed sometime before the endorsement of Disaster Management Act, along these lines the arrangement requests correction and overhauled in view of the Disaster Management Act to guarantee the general rule for the significant frameworks and debacle administration boards of trustees at all levels to get ready and actualize their parts as indicated by particular designs.

- Severally, appropriate rules for government, NGOs, catastrophe administration boards, and the common social orders need to place to help the Ministry of Disaster Management and Relief.
- Any research institute at the national level with other significant branches to the disaster-prone areas needs to functional to cope with the different nature's disasters along with strengthening the national disaster management system.
- Exchanging information and for resource sharing of the space-based technology; government can sign the MOU with the expert foreign country or agencies to make the research effective.
- Establishment of national volunteer unit and enhancing the capacities of the existing volunteer community to make the preparation phase smooth and the search and Rescue operation effective.
- Merging the urban volunteer units with the Civil Defense in order to make rapid response in all level of disaster management.
- Besides the conventional, a charitable disaster management fund needs to formulate with the proper legislation to raise the fund legally from both home and abroad.
- The charitable fund rising, specially from the abroad needs to be more free while releasing.
- The funds need to be disseminated to not only at the central level but also to the microlevels.
- Since the predetermined rule for the crisis administration is available in the Act, now it should be scattered to the nearby level.
- Local fiasco administration boards of trustees have been given the arrangement of asking for the representative chief to pronounce crisis for a specific catastrophe influenced region.
- In the Disaster Management Act, there is the arrangement of compensating the Individuals of associations in crisis administration. Other than this reward may spread to the media while and pre debacle administration through dispersing the data and cautioning.
- A solid control should be figured to characterize the way toward asking for the guide, administrations, vehicles, building, and some other crisis requests.
- Rules and checking must be put for the predefined horrible practices, for example,
  - Creation of issues in conveying administrations in calamity administration framework.

- Violation of government requests and law applicable to media and coordination gatherings.
- Making wrong and weak articulations and cases.
- Misappropriation of assets.
- Intentional value climb of vital wares in the calamity influenced zones.
- Obstructing water stream for individual advantage, for example, making saltiness and surge, making unlawful gaps in the dikes, etc.
- Skullduggery in crisis orders to the broad communications and others.
- Failure of government staff to release their obligation; abuse of position and power and so forth.

Conclusion

Nepal has started a few law and guidelines to catastrophe administration. Prior to the appearance of NDRA, 1982 A.D. there was no all around organized fiasco strategy in Nepal. Before 1982 A.D. alleviation and save works were done as the social works as it were. Hence, understanding the need of a demonstration, NDRA, 1982 A.D. was defined which has just been revised twice in 1989 A.D. also, 1992 A.D. Be that as it may, NDRR couldn't yet be detailed which is exceptionally basic. Without NDRR, NDRA couldn't be completely compelling. NDRA, 1982 does not depict the obligations and duties of all the catastrophe administration-related organizations other than the Ministry of Home Affairs. Obligations and duties of different other catastrophe administration organizations must be reflected in NDRR as each and everything couldn't be stipulated in the Act. As per NDRA, 1982 there has been the arrangement of CNDRC. Moreover, there is likewise the arrangement of a RTSC, SSRSC, RNDRC, DNDRC and LNDRC in the NDRA, 1982. RTSC and SSRSC might be activised in the season of intense cataclysmic events while CNDRC and DNDRC are dynamic constantly. This is so to speed up the quick save and help works. Be that as it may, RNDRC and LNDRC should be actuated as a when important, right now, these two offices are as torpid offices. Moreover, as the NDRA, 1982 does not depict the capacities and obligations of all region catastrophe administration-related offices, the issue of collaboration, coordination and common comprehension between different locale fiasco administration-related offices are seen. Thus, in some cases, catastrophe casualties don't get quick, proficient and successful safeguard and help administrations. Deferred alleviation works regularly brings intense and obnoxious outcomes. What's more, infrequently, duplication of help works have additionally been experienced, for the most part because of the non-appearance of discourse and shared comprehension among catastrophe administration-related offices. What's more, a portion of the locale calamity administration-related offices endeavor to move their obligations to alternate as there is no obvious set of working responsibilities in the Act. Subsequently, quick detailing of NDRR is extremely important where obvious set of working responsibilities of all the debacle administration-related organizations ought to be made.

So for smooth keep running on the calamity administration, the pending Disaster Management Act should be adjust and executed with all the advanced measures.

Bibliography

Abas, Azura. (2015, April 26). "Malaysia Ready to Help Nepal, Says Anifah". *New Straits Times*. Retrieved April 25, 2015.

"Estas son lasformas en quepuedesayudar a Nepal desde México" (in Spanish). CNN México. *Turner Broadcasting System*. 27 April 2015. Archived from the Original on 28 April 2015. Retrieved April 28, 2015.

"Everestotragedijąišgyvenęsvyras: maniau, kadesumiręs". *DELFI* (in Lithuanian). 26 April 2015. Retrieved April 26, 2015.

"Latvia Conveys Condolences to Nepal Over a Devastating Earthquake". Ministry of Foreign Affairs of the Republic of Latvia. 25 April 2015. Retreived April 27, 2015.

"LietuvosvadovaireiškiaužuojautądėlžemėsdrebėjimoNepale". *DELFI*. 25 April 2015. Retrieved April 26, 2015.

"Lithuanian President Offers Condolences Over Nepal Earthquake". *DELFI*. 25 April 2015. Retrieved April 26, 2015.

"Malaysia to Send Doctors, Rescuers to NEPAL: PM Najib". *Channel News Asia*. 26 April 2015. Retrieved April 25, 2015.

"Nepal Earthquake: Eerie Reminder of 1934 Tragedy".

Nepal Red Cross Society. (2015). "Regulatory Barriers to Providing Emergency and Transitional Shelter After Disasters—Country Case Study: Nepal".

New York Times. (2015). "Nepal Earthquake Poses Challenge to International Aid Agencies".

Poudyal Chhetri, Meen B. "Nepal Country Report 1999".

Rajghatta, Chidanand. (2015, April 26). "Is This the 'Big Himalayan Quake' We Feared?". *The Times of India*. Retrieved April 26, 2015.

Red Cross and Red Crescent Society. (2015). "Nepal—Looking Forward on Disaster Law".

Reliefweb. (2015). "Disaster Management Institution and System in Nepal".

Sugden, Joanna. (2015, April 26). "What 1934 Told Nepal to Expect About the Next Big Quake". *The Wall Street Journal.*

The Asia Foundation. (2012). "A Guide to Government in Nepal: Structures, Functions, and Practices".

"Timeline: Nepal 2015 to 1934, the Worst Quake Disasters in the Last 80 Years".

Tran, Cindy, and Quinn, Liam. (2015, April 25). "Desperate Families of DOZENS of Australians Unaccounted in the Deadly Himalayan Earthquake Disaster Take to Social Media with Pleas for Help Finding Their Loved Ones". *Daily Mail* (London). Retrieved April 26, 2015.

"URM: žinomaapie 60 Nepalebuvusiųpiliečių, nesusisiektasuketuriais". *DELFI* (in Lithuanian). 27 April 2015. Retrieved April 27, 2015.

Coordination Does Matter for Disaster Management in Bangladesh

Mst. Marzina Begum and Md. Nurul Momen

Introduction

Bangladesh has been administratively divided into 64 districts, covering 147,570 km^2. Among them, 19 districts are geographically located in coastal areas which covers nearly 20% of the total land area of the country. However, the length of the coastline is about 720 km, whereas 36.8 million population lives in the area, and it is estimated from a World Bank study that near about 12 million coastal people live in poverty, and people live in poor household and hazardous area, the simple answer is because they are poor that confronts challenges coming from disaster. However, it is important to note that the total coastal areas are divided into three regions of the country, such as eastern, western, and central (Islam 2008, cited by Parvin et al. 2008).

Coordination means integration of activities (Akrani, no date), and it is considered as the core function of management. Considering its importance, Mary Parker Follett argued that the principle of coordination start at an early stage in the management process (ibid.).

Mst. Marzina Begum (✉) · Md. Nurul Momen
Department of Public Administration, Rajshahi University, Rajshahi, Bangladesh

B. Zutshi et al. (eds.), *Disaster Risk Reduction*,
https://doi.org/10.1007/978-981-10-8845-2_2

So, coordination in disaster management is a set of systematic activities that guide to protect the vulnerable people. In order to achieve the objectives of public policy, the implementation functions performed by different organizations and agencies must be coordinated, otherwise, it may negatively affect its successful implementation. Pressman and Wildavsky rightly observed

> Much more has been said about coordination than actually has been done about it, and coordination remains a principal "philosophers' stone" in the analysis of good public administration? (Peters 2006)

Recent evidence shows that many governments in the developing world have expressed their serious concern over the problem of coordination in the implementation stage of any public policy or projects or programs. Poor coordination among the organizations also hampers the proper utilization of allocated funds for desired results. However, over the years, there have been plenty of empirical research on coordination that has been done in the field of social science, and many of the findings have been incorporated in the policy document of the government. Mintzberg (1988), analyzed the concept from the intra-organizational coordination, while Christensen and Lægreid (2008) analyzed coordination from both intra- and interorganizational perspectives. Again, coordination can be viewed from two angles such as horizontally or vertically. Horizontal or interorganizational coordination can be placed between organizations on the same line, while vertical or intra-organizational coordination takes place within the organization. Put it differently, coordination is concerned with managing the activities of different organizations and individuals or subunits of an organization are considered as inter and intra-organizational coordination respectively. To make things clear, Verhoest and Bouckaert (2005, cited by Christensen and Lægreid 2008) explain the interorganization coordination as network-based, while intra-organization coordination is considered as hierarchy-based (Christensen and Lægreid 2008).

Empirical research on coordination not only concentrates on successful public policy implementation, but also seeks answers to some unresolved questions with regard to coordination challenges in organization settings. The question is about how a proper coordination can be maintained in disaster management, in this regard, it requires knowledge and understanding coming out from the experiential learning through which it can be upheld. Robinson et al. (2000) expanded the notion of successful coordination mechanism may either be chosen based on the capacities, legitimacy, and expertize of those whose functions are subject to coordinate. The Canadian academic, Mintzberg describes six mechanisms for an organizational framework, such as mutual adjustment, and direct supervision as well as four kinds of standardization of work for coordination, such as processes, outputs, norms, and skills (Mintzberg 1988). Drawing on Gulick, McCurdy (1983) identifies some common mechanisms for coordination, such as structure of organization, written rules, systems of authority, defining duties and responsibilities, and purposes. Hence, this can be concluded by saying that lack of proper coordination and relationship between the organizations involved may adversely affect effective implementation of disaster management. These issues will be addressed in the following discussion by analyzing coordination and the process of response in the light of organizational settings involved in disaster management in Bangladesh.

The Scenario of Coastal Areas in Bangladesh

A natural disaster reveals the vulnerability of the people living in the disaster-prone areas, and it hampers the process of socioeconomic and environmental development. On a positive note, the coastal areas play an important role in the economy of Bangladesh, as the coastal areas are an important source of agriculture production, fisheries, forestation, marine transportation, and the future possibility of oil and gas exploration. Two seaports are located in the coastal area, namely Chittagong and Mongla, as a consequence, few numbers of industrial organizations are built there. The Sundarban mangrove forest located in the coastal area could contribute further to the national economy, as it is the largest mangrove forest in the world, and cover the areas of India and Bangladesh declared a World Heritage Site by UNESCO. The only newsprint mill is established in the Sundarban, which is based upon its basic raw materials drawn from the area (Ali, no date).

The devastating impact of the Asian tsunami in 2004, super cyclone SIDR in 2007 and Cyclone Alia in 2009 have shown to us the vulnerability of the coastal communities' due to natural disasters in Bangladesh. It is important to make a note that cyclone as a major natural disaster occurs most frequently over the northwest Pacific, over the southern end of the Bay of Bengal, coast of Indian and south of Bangladesh (UNESCAP and ADB 2000; Ali 1999; Huang 1999; Relly and Adger 2000).

Disaster Management Initiatives in Bangladesh

Analyzing the history of Bangladesh, a natural disaster has catastrophic different categorical impacts on livelihoods in the coastal areas. Damages to infrastructure and resources severely impede economic activity. However, Crowards (2000) points out that social impact of a natural disaster includes loss of life, injury, homelessness, and disruption of the normal life of communities. As environmental impacts, it may include environmental damage, and finally, all sorts of mental illness occurred due to damage of psychological conditions. Traditionally the ever-increasing effects on human and economic loses have received attention in disaster management approach, but in most cases, it ignores the impacts of environment or psychological damage. Sendai framework for Disaster Risk Reduction (2015–2030) further explain that disaster-prone countries are faced with increased levels of hidden costs of disaster recovery, and other challenges of natural disaster in order to meet financial obligations.

In order to reduce the negative effects on natural disaster, there were hardly any governmental efforts until the 1990s to address the people's vulnerability in managing natural disasters in Bangladesh. After the devastating cyclone in 1991, the country developed some regulatory framework, and organizational mechanisms to deal with different categories of natural disaster (Haque and Uddin 2013), such as Disaster Management Act (2012), National Disaster Management Policy (2008), Standing Orders on Disaster (1997).

After two years of the devastating cyclone, the GOB introduced a new project called "*Support to Comprehensive Disaster Management*" in 1993 with a goal of reducing the human suffering, and reducing environmental costs of natural disaster. The objective of the project was stated to enhance the capacity building of every coastal household, and helps to people in the disaster-affected areas by establishing Local Disaster Action Plans (LDAPs). However, as a mission, the LDAP work with any disaster situations at the local level, and train and create awareness in the disaster-prone areas. The LDAP was ended on June 30, 2001, and felt to continue and initiate for the formulation of a new program that is called as "*Comprehensive Disaster Management Program* (*CDMP*)". As a strategy of disaster management, the CDMP took a holistic approach to disaster risk management, and during the period of implementation, it received financial support from development partners and international agencies. Since then, the government handles all types of natural disaster based on the notion that has been evolved in the country commonly known as Community-Based Approach (CBA) which focuses on the active participation of all people facing natural disaster.

On January 1997, the government has approved the Standing Orders on Disaster (SOD) for the purpose of guiding and monitoring disaster management activities and programs in Bangladesh, although later, on April 2010, the GOB amended the SOD. On the other hand, the Disaster Management and Relief Division (DM&RD) of the GOB has the responsibility for coordinating the national disaster management efforts within all across the disaster management-related organizations. Furthermore, a number of public organizations are involved in disaster management, there are 12 committees at national level, while at the subnational levels, there are six committees that work for effective planning process, and coordination of disaster risk reduction and emergency response management (National Plan for Disaster Management, 2005–2015) in Bangladesh.

It is important to note that two bodies at the national level, *National Disaster Management Council* (*NDMC*) and *Inter-Ministerial Disaster Management Coordination Committee* (*IMDMCC*) maintain coordination and the process of response during the disaster, and its recovery stages. Coordination at the subnational level, for example, District, Thana and Union unit, will be managed and coordinated by the respective Disaster Management Committees. However, Department of Disaster Management (DDM) provides all assistance to national and subnational bodies by facilitating the disaster-related data and information (ibid.). It is also worth to mention that the GOB has approved the Disaster Management Act (DMA) on September 12, 2012 which presents a major breakthrough in the disaster management strategies in Bangladesh. Furthermore, the country has also National Plan for Disaster Management which runs from 2010 to 2015, which have emphasized broad-based strategies in minimizing disaster losses. Apart from that, the GOB has some other regulatory frameworks that specifically clarifying about the disaster management objectives, strategies, resources, and institutional framework for disaster management.

Coordination Challenges in Disaster Management

From the existing literature, coordination among public and private organizations at the local, national, and even international levels have become an increasingly dominant issue today in the disaster management efforts. It is worth to mention that for a better preparedness in order to face disaster requires an understanding of the stakeholders for different categories of hazards mapping, disaster planning, coordination among organizations in their activities, arrangement of relevant training for an effective national disaster management strategy, and finally development of leadership in each organizations involved in disaster management, as a consequence, it helps to promote coordination and collaboration at all levels of the existing framework of decentralized governance (Bentley and Waugh 2005; Cigler 2006, cited by Ross 2008: p. 76). Proper coordination in disaster management is an investment in minimizing future losses of lives and property, and this may contribute to achieve the sustainable development goals as declared by the UN. However, it is worth to mention that disaster management is not the function of any particular body, rather it requires well-coordinated efforts from all concerned local and national bodies for a strategic decision, mutual exchange of learning, and partnership development. In this regard, the Hyogo Framework for Action (2015–2030) has focused on the creating awareness building in the community among the people and institutions, justifying political roles and responsibilities in disaster, catalyzing a plan of action, and strengthening capacities to the other stakeholders involved at all levels of disaster management.

Coordination within and across all relevant stakeholders are required for the design and implementation of disaster policies and plans, as it requires the full engagement of all institutions in order to empower local government and communities through providing required financial support, and decision-making authority. Sendai Framework for Disaster Risk Reduction (2015–2030) also points out that in order to complement disaster management, there is a need to enhance a set of cooperation mechanisms among the disaster-prone countries, and cooperation between states and international organizations are also highlighted in the framework. Apart from that, Sendai framework also highlights to enhance good practices and strategies in disaster management ranges from the national to global level, although the framework focuses on national coordination with regard to disaster-related activities, such as disaster response, rehabilitation and reconstruction, and post-disaster recovery, and reconstruction supported by strengthened cooperation that could be managed by bi-lateral agreement (ibid.).

Coordination is needed to all involved organizations must work with cooperation, and to create opportunities for collaboration in an objective to make sure their roles and responsibilities for an integrated disaster management practices in order to have a desired outcome (ibid.). Keeping the importance in mind, there is national policy in Bangladesh to address these issues, although a lack of integration and overlapping of responsibilities prevails among the organizations working in the disaster-prone areas. Not only are the challenges of these issues, but also there is very little progress of implementation of activities with specific reference to disaster management monitoring program. All of these issues have been examined in the following section.

Disaster Management Plans

After the disaster, it is important to note that disaster relief and rehabilitation must be coordinated in order to avoid overlapping of activities, thus helps to avoid wastage of resources across the board of different multi-stakeholder, such as donors, the government, and NGOs. It has also been stressed that it is essentially required to ensure an effective system within the government and to develop connectivity with the relevant national, regional, and international organizations about the planning process in disaster management (SAARC Framework for Action (SFA) 2006–2015, National Plan of Government of Bangladesh). However, in reality, all the disaster management plans and programs in Bangladesh have been prepared through techno-bureaucratic approaches. The approach does not expect participation from the many appropriate stakeholders in the policy formulation process of disaster management. This finding was substantiated by another study conducted by Kurshid and Masroor (2007) that there are critical gaps in the disaster management activities, and found no clear direction about coordinating with other stakeholders; hence, it is rare to follow the coordination line among the stakeholders in a common national goal in disaster management.

The International Strategy for Disaster Reduction (ISDR) has underscored the absence of coordination as the main cause of the failure of disaster management efforts at the global level. When we come to the Hyogo Protocol, which was developed in the 2005 World Conference on Disaster Reduction, was held in Japan. The protocol asked the independent states to adopt disaster risk reduction approaches at national level (UNISDR, no date). However, the Hyogo protocol also points out the disaster reduction framework with regard to community awareness, education and training, planning, and management, also shows us the required initiatives, and processes in order to achieve a successful disaster management program, but the ISDR and the Hyogo protocol depend on the binding states (who agreed to comply with it), and support at the period of implementation.

Disaster Management Information Centre (DMIC) is a virtual forum where various stakeholders participate and exchange information, but in practice, there is a limited maintenance of formal communication platform on disaster preparedness. Furthermore, as a coordinating body of the government directed by the NDMC, which is headed by the Prime Minister and IMDMCC, the Ministry of Disaster Management and Relief (MODMR) plays different kinds of role in dealing with disaster. Apart from that, DDM is the central point for disaster preparedness at the national level, whereas various disaster management committees are responsible for disaster preparedness at district, subdistrict, and local level. Due to many organizations involved in disaster management, it is observed that there is very often overlapping of activities between the central level and local level organizations.

However, DM&RD is the lead coordinating agency for disaster preparedness. DM&RD receives support from warning issuance center such as Bangladesh Meteorological Department (BMD), Flood Forecasting Warning Centre (FFWC); warning dissemination hubs, such as the Disaster Management and Information Centre, Emergency Operation Centre (EOC), National Electronic and Print Media, Cyclone Preparedness Programme (CPP) and DDM, Food Division, Armed Forces' Division, Directorate of Relief and rehabilitation, Directorate of Food, Department of Public Health Engineering, and others. It appears that many supporting agencies works in the same agenda, however, there is an accusation against the supporting agencies that the officials from DM&RD don't get support and message from then on time during and after a cyclone.

Another challenge for the effective implementation of disaster management strategies are cooperation amongst international organizations and/or bilateral countries. During the cyclone, Sidr in 2007, British Broadcasting Corporation (BBC) presented a news report, "*Plenty of agencies, but not enough aid- too little, too late*," and the BBC further quoted a comment made by a professional working in an coastal area, "*The reason why these people are not receiving enough help is because there is no coordination between the government and aid agencies*" (Haque and Uddin 2013). Hence, it is not beyond the question about the effectiveness of national disaster management strategy to deal with the huge emergency condition.

Some Examples of Poor Governance

There are some striking examples of poor management, and coordination in disaster managment efforts in Bangladesh. For instance, Southkhali village in Shoronkhola upazila of Bagerhat district was one of the deadlist affected areas in cyclone Sidr 2007. In an official visit after the cyclone, the Indian foreign minister had a commitment to build all the houses in these affected village areas. Since his commitment, no Indian government efforts were taken to give shelters to the affected coastal people in the said area. It is very pity that, due to commitment from the end of Indian government, an official ban was in effect on other interested aid agencies however, the committed Indian assistance did not come in due time, as a consequence, coastal people in these areas were compelled to live in the open ground, this cruel incident happened due to consequences from the lack of bilateral coordination between Bangladesh and India. More specifically, it could be blamed for the failure of bureaucracy on both parties of these two countries, for not reaching an agreement for building houses to the affected people (ibid.).

When it comes to an early warning before disaster, CPP under the Ministry of Disaster Management and Relief publicize early warning to coastal people with the help of community-based voluntary networks working in cyclone-prone areas. It is observed that in many cases, coordination gap arises due to lack of community network in the coastal areas. However, there is a theoretical underpinning about recognizing community as the first responder to any humanitarian crises that from disaster preparedness to relief and rehabilitation; until the process of reconstruction of locals are achieved. In the process of decentralized governance, the call for coordination and response at the lowest level of governance, namely, village level need to be activated in a proper way.

Furthermore, "Early Warning Centre" and Centre for Environmental and Geographic Information Services works under the Ministry of Water Resources. The center disseminates weather forecasts to different concerned agencies in advance through using telecommunication services containing information on the height of water during the cyclone.

On the other hand, Centre for Environmental and Geographic Information Services on a regular basis produce a yearly report on natural calamities. Interestingly, it is very common to find both the report is somewhat conflicting about the intensity of cyclone.

Damage Assessment Among Agencies

Assessments of damage due to disaster measured by various public and private agencies tend to be uncoordinated and inconsistent in Bangladesh. For example, the current assessment practices, and measurement tools what has been developed in the country still face constraints due to lack of baseline survey data. Other factors are also responsible for early assessment of damage, such as absence of capacity of accessing to developed satellite images, and no connectivity between Geographical Information System (GIS) and disaster-related data and information (UN-SPIDER 2010). Inconsistency is also found, when it comes to a major disaster in the context of Bangladesh, most of the relief and rehabilitation support during recovery stages come from donors outside. After the damage assessment of a major disaster, funding to emergency relief programs starts soon, and many NGOs are observed as "*over-inundated by funds*" and "*beyond their management capacity and remained unutilized*" (Cutler et al. 1989, cited by Matin 2002), although financing for disaster risk reduction and response has not been a major area of operation for the NGOs. In the case of disaster funds received from external sources, NGOs lack motivation in disaster management efforts, while the government has a problem of accountability in distributing relief. For example, Bangladesh chapter of Transparency International (TIB) published a report on the fund for climate changes that found government pressure, favoritism, and corruption in case of selection of NGOs to carry out project for adaptation program in the coastal areas (Islam 2013).

Matin et al. (2002) found that disaster management is slowly incorporating in the operation manual of the government, and this also happens to the case of some NGOs, although disaster management still has not become as a development program to many NGOs. Funding from external sources for disaster management activities in Bangladesh lacks focus on the disaster preparedness and early damage assessment; therefore, it requires broad-based disaster management approach, and long-term plan of action incorporating with other stakeholders involved in the process. However, the problem is also observed within the government about coordination, and the process of response to disaster preparedness, it always remains as reactive in approach and short-term in nature.

Comprehension of Warning Message

Early warning is one of the fundamental components of disaster risk management. It is especially important to those stakeholders who are engaged in disaster risk management. If stakeholders somehow fail to understand the warning announcement, it could create a negative consequence on the understanding of the concerned organizations and the affected people, and as a consequence, its impacts will be worse to their role while in disaster preparedness. From an empirical finding, for example, continuous technological development of warning message made a significant impact on agencies in coordination in the disaster management efforts, be it man-made or natural disasters, in the State of Hawaii, USA (HGICC 2006, cited by Ross 2008: p. 76). Islam et al. (2004) identifies that existing cyclone warning message in Bangladesh has been highly criticized by common citizens, as it is through the electronic media are being disseminated using a very formal language, and with some meteorological technical terms; however, there are many dialects in the country. Many people, particularly, the elderly and women, living in the coastal areas are not familiar with the language used in weather bulletin. As an attempt to understand the causes of low understanding among the affected people about the early warning with regard to the level of education. In a survey, it is found that about 35% illiterate people fail to follow the warning message clearly, while about 30% people having primary and secondary education could only partially understand it. Furthermore, more than 85% people having higher secondary and above can clearly understand these bulletins. The findings clearly indicate the strong positive relationship between the level of education and the understanding of the warning signals in the coastal areas in Bangladesh (Islam et al. 2004). So, it could be concluded by pointing out that the illiterate and less educated people face difficulties in understanding warning message properly and the electronic media transmitted cyclone warning signals for a number of reasons.

However, reducing the suffering of livelihood caused by the natural disaster is now receiving a worldwide attention. In order to take measures therefore, current cyclone warning message needs to be disseminated according to the needs and understanding of the local people, and attention should be given to transmit the warning bulletins in local dialects. In this process, electronic and print media can play an effective role in creating awareness to the affected people. But the very pity is that news disseminated by radio and television lost the authenticity, although they are used to receive the warning message from the government meteorological office. However, media and meteorological office are equally responsible for their humorous announcement that lost the touch with reality. Given the context, it is strongly argued that warning message can be transmitted in a way that will create awareness among the affected people and year-round programs on disaster preparedness, and mitigation need to be published, and regular broadcast by print and electronic media in order to reduce understanding gap of the people living in disaster-prone areas. Although there is a great challenges of poverty-stricken people who used to live in the remote areas have limited access to media in Bangladesh, as access to the media depends upon their purchasing ability, on a particular note, nearly 45 million people live below the poverty line in Bangladesh, which is estimated as one-third of the total population.

There is also communication gap due to delayed delivery of warnings that hampers decision-making process of the involved organizations and affected people in Bangladesh. For example, flood in 2017 forecasting and warning fail to predict, as a result, 5 million people in the north, northeast and south-western parts of the country suffered, and make their lives miserable during a disaster. As the citizens of the country are used to face multiple hazards in every year, for example, their process of responses to cyclone warnings depends on the intensity, and power of wind speed. Furthermore, if the earlier event of natural disasters, and local customary belief correspond with the cyclone warning of BMD of about six to seven on an average, in that case, they only start to evacuate or decide to leave their homes for a designated cyclone shelter. Beforehand, they adopt a "*wait-and-see*" approach, observing whether the cyclone intensity is rising (ibid.). Furthermore, it is observed that people live in the disaster-prone areas have very little level of confidence over the weather forecasts, and warning signal disseminated by the BMD, There is a perception among the common people that weather news, in many cases, treated as unreliable, since many times weather forecast declared by the BMD did not happen in reality. As a consequence, still many coastal people who usually depend upon kindness from God.

It is worth to mention that cyclone intensity, devastation and its route are collected from satellite images from space. The warning signal uses a scale ranges from 1 to 11 for the seaports, while 1–4 for river ports. However, there is general comprehension among the cyclone-prone areas that the higher the number means that the more urgent the need to response to cyclone (ibid.). There has been an exception during the Cyclone Sidr on November 1, 2007, the affected people were deliberately nonresponsive to cyclone warnings, even warning signal to cyclone raised to 10 for them. Why does the response low among them? In replying to this question, Alam (2010) found that affected people did not have a similar major cyclone since 1970 having warning signal 10. During Cyclone Sidr, it was very pity that most of the households only left their homes, when they saw water coming very fast, and near to them (Alam 2010). According to the GOB report (2008) as an effect of cyclone Sidr 2007, 2.3 million households affected to some extent, among them, about 1 million were badly affected by the cyclone in the coastal region of the country, while it is estimated that more than 3000 number of deaths caused by Sidr.

For an effective coordination, disaster response, and recovery efforts require interaction and data and information sharing policies among the organizations involved in disaster management in order to save lives and property. Lack of coordination among the organizations delays effective disaster management with an increased possibility of duplication of efforts. Therefore, it could be argued that coordination is necessary between organizations of Ministry of Defense (technology providers) and MODMR (technology users), thus it may enhance a set of cooperation among the agencies, such as DDM, DRR, SPARRSO, and BMD with a view of creating a common agenda for disaster management (UN-SPIDER 2010). However, capacity building of these agencies involved in disaster management is required to develop an interface between the technology provider organizations, and disaster management organizations. For example, as a disaster management organization, DDM will require services offered by SPARRSO, BMD, and other organizations like CEGIS (ibid.).

Sustainable Plan for Disaster Management

For a comprehensive addressing of disaster risks, the National Plan for Disaster Management (NPDM) 2010–2015 was developed in 2010 by the MODMR, as a response to the national and international demand.

The plan focused on the organizational accountability in implementing disaster management plans at different levels of the country. For the first time in the history of disaster management efforts, the NPDM incorporates to both disaster risk reduction and hazard-specific multisectoral plans in order to create effective mechanisms for reducing disaster risk, thus it might help to achieve sustainable development. It is worth to mention that the plan also covers some other disaster-related issues, such as threat analysis of disaster, measuring the magnitude of vulnerability, reducing measures of disaster, assignment of responsibilities, recovery provision during relief, cost management, and the process for coordination in disaster management.

Currently, for rapid assessment of damage and recovery, there is very limited use of space-based information for disaster risk management, and emergency response, although there is the presence of some centers of excellence like SPARRSO and CEGIS (ibid.). However, from the history of Bangladesh suffered stronger earthquake, but the unfortunate reality is that measuring its magnitude, epicenter and time, the country still have to depend on the report received from the US geological survey. Given the context, it is essentially required to develop a capacity building of the organizations involved with the assessment of disaster risk and needs trained staff affiliated with the DDM, and DRR in order to full utilization of the services received from the organizations such as SPARRSO and CEGIS (ibid.). At present, SPARRSO mainly receives data from the extremely low-resolution satellite images, and another challenge compounds about the existing capacity of the staff related to image processing that is not adequate to explore the full potential of the space-based resources. Therefore, the training on image processing, and the process of collected data, is essentially required for an effective risk assessment. On the other hand, as jurisdiction, BTRC is to make affordable delivery of telecommunication services all across the country with an acceptable quality. It is also mentioned in their list of function that it needs to resume emergency telecommunication services immediately, as the past experiences show that during the cyclone all existing telecommunication services are used to become nonfunctional (ibid.).

Sendai framework for disaster risk reduction (2015–2030) recognizes that disaster-prone countries require a risk-sensitive approach to ensure sustainability in disaster management. Keeping the approach in mind, Bangladesh also kept their commitment, and progress to address disaster risk reduction as a strategy into policies, plans, and programs at all levels. These initiatives have included both the risks and consequences of disaster risk, and its management. In the said approach that involvement of local government bodies, and local communities have been considered as an important agenda of disaster management strategy. Despite of the policy and programs mentioned above, successful implementation of disaster management still face challenges in Bangladesh.

However, from the discussion, the GOB has introduced wide-ranging programs, and integrated approach in disaster management in their course of action. But in the absence of proper coordination, an integrated and comprehensive disaster management approach remains largely on paper although, to some extent, it is found from the analysis that the GOB and the various involved organizations are working to reduce the sufferings of the affected people struck by natural disaster.

Bibliography

Akrani, Gaurav. (no date). "What is Co-ordination? Meanings and Articles on Co-ordination". Available at http://kalyan-city.blogspot.com/2011/05/what-is-coordination-meaning-and.html (Last accessed on 11 January 2015).

Alam, Edris. (2010). "Cyclone Disaster Vulnerability and Response Experiences in Coastal Bangladesh". Available at http://www.ncbi.nlm.nih.gov/pubmed/20561338 (Last accessed on 10 May 2015).

Alam, Khurshid, and Siddiqi, Masroor-ul Haq. (2007, June). *Disaster Preparedness for Natural Hazards: Current Status in Bangladesh*. International Centre for Integrated Mountain Development (ICIMOD), Kathmandu, Hill Side Press (P) Ltd. Nepal.

Ali, A. (1999). *Ghurnijhar (Cyclone)*. Dhaka: Bangla Academy.

Ali, Anwar. (no date). *Vulnerability of Bangladesh Coastal Region to Climate Change with Adaptation Options*. Dhaka: Bangladesh Space Research and Remote Sensing Organization (SPARRSO).

Christensen, T., and Lægreid, P. (2008). "The Challenge of Coordination in Central Government Organizations: The Norwegian Case". *Public Organization Review* 8(2): 97–116.

Crowards, Tom. (2000, May). "Comparative Vulnerability to Natural Disasters in the Caribbean". Caribbean Development Bank, Staff Working Paper No. 1/00. Social and Economic Research Unit, Economics and Programming Department. Paper presented at the OAS/USDE-NOAA/CSC Workshop on Vulnerability Assessment Techniques, Charleston, SC, March 20–22, 2000.

GOB. (2008). "Cyclone Sidr in Bangladesh Damage, Loss, and Needs Assessment for Disaster Recovery and Reconstruction". A Report Prepared by the Government of Bangladesh Assisted by the International Development Community with Financial Support from the European Commission, April 2008.

Haque, Emdad, and Uddin, Salim M. (2013). "Disaster Management Discourse in Bangladesh: A Shift from Post-event Response to the Preparedness and Mitigation Approach through Institutional Partnerships". *Approaches to Disaster Management-Examining the Implications of Hazards, Emergencies and Disasters.*

Huang, Z. G. (1999). *Sea Level Change in Guangdong and Its Impacts.* Guangzhou: Guangdong Science and Technology Press (original in Chinese).

Islam, Syful. (2013, October 15). *Transparency International Research Reveals Political Influence, Nepotism and Corruption in the Selection of NGOs to Carry Out Climate Change Adaptation Projects.* Thomson Reuters Foundation. Available at http://news.trust.org//item/20131015100349-6sbou (Last accessed on 23 September 2016).

Islam, M. Shahidul, Ullah, Mohammad Sahid, and Paul, Alak. (2004). "Community Response to Broadcast Media for Cyclone Warning and Disaster Mitigation: A Perception Study of Coastal People with Special Reference to Meghna Estuary in Bangladesh". *Bangladesh Asian Journal of Water, Environment and Pollution* 1(1 & 2): 55–64.

Matin, Nilufar. (2002, July). "Corporate Social Responsibility and Natural Disaster Reduction: Insights from Bangladesh". DFID. Available at https://assets.publishing.service.gov.uk/media/57a08d3e40f0b6497400173e/R7893CSRBangladesh.pdf (Last accessed on 10 June 2018).

Matin, Nilufar, Muhammad, Taher, and Matin, Nilufar. (2002, July). "Corporate Social Responsibility and Natural Disaster Reduction: Insights from Bangladesh". Available at https://assets.publishing.service.gov.uk/media/57a08d3e40f0b6497400173e/R7893CSRBangladesh.pdf (Last accessed on 12 December 2015).

McCurdy, H. E. (1983). "Coordination". In T. D. Lynch (Ed.), *Organization Theory and Management.* New York: Marcel Dekker, pp. 114–121.

Mintzberg, H. (1988). "A Typology of Organizational Structure". In D. Miller and P. Friesen (Eds.), *Organizations: A Quantum View.* London: Prentice-Hall International, pp. 278–304.

Parvin, Gulsan Ara, Takahashi, Fumito, and Shaw, Rajib. (November 2008). "Coastal Hazards and Community-Coping Methods in Bangladesh". *Journal of Coastal Conservation* 12(4): 181–193.

Peters, B. G. (2006). "Concepts and Theories of Horizontal Policy Management". In B. G. Peters and J. Pierre (Eds.), *Handbook of Public Policy.* London: Sage, pp. 115–138, Chapter 6.

Relly, P. M., and Adger, W. N. (2000). "Theory and Practice in Assessing Vulnerabilities to Climate Change". *Facilitating Management* 20(4): 527–39.

Robinson, D., Hewitt, T., and Harriss, J. (2000). "Why Inter-organizational Relationships Matter". In D. Robinson, T. Hewitt, and J. Harriss (Eds.), *Managing Development: Understanding Inter-organizational Relationships.* London: Sage, pp. 1–16.

Ross, Prizzia. (2008). "The Role of Coordination in Disaster Management", Chapter 5, pp. 76–96, in edited book by Pinkowski, Jack (2008). "Disaster Management Handbook", Public Administration and Public Policy/138 (A Comprehensive Publication Program), CRC Press, Taylor & Francis Group.

UN-SPIDER. (2010). *Report of the Technical Advisory Mission to Bangladesh 19–23 June, 2010.* United Nations Platform for Space-based Information for Disaster Management and Emergency Response (UN-SPIDER).

UNESCAP, and ADB. (2000). *State of the Environment in Asia and Pacific 2000.* Economic and Social Commission for Asia and the Pacific and Asian Development Bank, New York, United Nations, Available at http://www.un.org/earthwatch/asia/ (Last accessed on 12 December 2015).

Integrating Disaster Management Content into Social Work Curriculum-Voices of Social Work Students

Rajendra Baikady, Cheng Shengli and R. M. Channaveer

Introduction

There is a long-standing connection between disaster and social work profession. Social workers were actively participating in the disaster recovery and rehabilitation process across the world. As Desai (2007) noted social work profession has a long history in helping the disaster victims but Dominelli (2015) noted that the contribution of social workers in the disaster recovery and rehabilitation was rarely recognized by the larger public (Dominelli 2015). The term disaster is defined as "A serious disruption of the functioning of a community or a society involving widespread human, material, economic or environmental losses and impacts, which exceeds the ability of the affected community or society to cope using its own resources (Unisdrorg 2015)". The outcome of disaster is varied according to the nature of its occurrence.

R. Baikady (✉) · R. M. Channaveer
Department of Social Work, Central University of Karnataka, Gulbarga, India

C. Shengli
Department of Social Work, Shandong University China, Jinan, China

B. Zutshi et al. (eds.), *Disaster Risk Reduction*,
https://doi.org/10.1007/978-981-10-8845-2_3

The disaster at the household level may bring a major illness, death, a substantial economic, or social misfortune but the outcome of the disaster at the community level will have much more hazardous impact. The disaster at the community level could be a flood, a fire, a collapse of buildings in an earthquake, the destruction of livelihoods, an epidemic, or displacement through conflict. Disasters happening in many levels, such as district, national, or international levels differ in their affect and number of people affected and loss of wealth, infrastructure, and human life will always more when the disaster is at the national level or geographically large. Further, most of the disasters result in many kind of disability and makes people unable to perform daily activities and victims become dependents and leads to burden and dependency toward outside assistance.

Disasters are of two types of disasters, i.e., manmade disaster and natural disaster. Every country in the world faces one or the other disaster at some point of time. Further, the effects and the implications of these disasters depend on the geographical location of the country and type of disaster and the severity of the disaster. Joseph Kwok (2008) notes some major disasters that occurred in Asian region, i.e., Kobe earthquake in Japan in 1995, the Taiwan earthquake 2000, the frequent massive flooding in China and Bangladesh, the 2004 Tsunami, and 2008 devastating snowy weather in China are some among these which are known for massive distraction. Further, Asian region is also well known for the manmade disasters. Some major man-made disasters that caused large human and infrastructural loss are 1998 Indonesia's massive racial confrontations which targeted at ethnic Chinese, frequent racial, and religious conflicts and armed confrontations happening in Malaysia, Thailand, East Timor, terrorist insurgence in the Philippines, and border confrontations. In addition to this Asia is also known for the political upheavals that brought many changes in the Asian society and people's life. The major political events in the history of Asia are people's movements in the Philippines which toppled two presidents, the 2006s Nepal's people's movement that resulted in a new constitution and a new government to the country, 2006s massive demonstration against the governments corruption in Thailand and Taiwan, and most interestingly a military coup that occurred in Thailand in September 2006, which resulted in the relocation of the military rule in the country (after 19 years of constitutional democracy) until the end of 2007 (Kwok 2008).

This article is an attempt to explore the possibility of developing a new area "disaster social work" in social work education, more particularly in social work curriculum. Disasters and Development are the paradigms that social work must address as two sides of the same coin. Where development is happening, disasters occur; in addition, there are development-induced disasters. Masters of Social Work education empowers youth in becoming a confident community worker and shape the development thinking and practice. Disaster management needs to help protect the development investment. In the light of this, how social work as a human service profession illumines the path right from educating the social work students in disaster social work in the formative stage of their education is important. A specialized curriculum for disaster social work is the need of the day in Indian context to train social workers for the effective- and skill-based disaster management.

Review of Literature

According to National Governors' Association (1978), there are four phases in disaster management: mitigation, preparedness, response, and recovery. However, the recovery after disaster involves short- and long-term efforts in order to rebuild disaster-affected communities. The first step in the disaster recovery is meeting the basic needs of the survivors. The reconstruction of the lost infrastructure begins once the basic needs of survivors' are met (Berke and Karetz; Gardoni and Murphy 2008). Whereas, Tierney (1993) argued that the recovery process includes reconstructing and restoring the disaster-stricken area and providing basic facilities to the victims and the affected community. It also includes the process of mitigating the future hazards such as medical and communicable diseases (Tierney 1993). However, there are many debates regarding the definition of disaster recovery and management. In fact, the scholars have not yet come to a conclusive stand to define disaster recovery in a more precise form. As Phillips (2009) argues there are several theories for explaining the disaster recovery, such as systems theory, vulnerability theory, sociopolitical-ecology theory, feminist theory, and emergent-norm theory to explain disaster recovery. Different approaches to disaster recovery have also been proposed, such as a model of sustainable disaster recovery (Mileti 1999) and a capabilities-based approach to recovery (Gardoni and Murphy 2008). The social work profession has long been involved in disaster recovery and contributes to psychosocial support for survivors (Gillespie 2008).

The global Definition of the Social Work Profession approved by the IFSW general meeting and the IASSW General Assembly in July 2014 defined social work:

> as a practice-based profession and an academic discipline that promotes social change and development, social cohesion, and the empowerment and liberation of people. Principles of social justice, human rights, collective responsibility and respect for diversities are central to social work. Underpinned by theories of social work, social sciences, humanities and indigenous knowledge, social work engages people and structures to address life challenges and enhance wellbeing. (Ifsworg 2015)

This definition of social work addresses enhancing the wellbeing of the people as one of its core area of intervention. Supporting to this view Dominelli (2015) stated that social work profession had contributed much for the disaster management and disaster mitigation. Social workers were involved in the services whether it's about reducing risks, mitigating disaster, providing relief, or long-term reconstruction. Further, she recommends that the United Nations should include social workers more centrally within its humanitarian remit because social workers are professionally responsible for enhancing human well-being holistically. However, a group of scholars (Padgett 2002; Yanay and Benjamin 2005) noted the urgent need to create a cadre of professional social workers who can respond to the disasters in a much greater manner as social workers are known to play an important role in all phases of disaster management. Supporting to this Javadian (2007), Zakour (2006) argued that the social workers can play a contributing role in disaster recovery via micro-, meso-, and macrolevel interventions (Javadian 2007; Zakour 2006). Whereas, Mathbor (2007) noted that different roles played by social workers have helped over the years to address the special needs of vulnerable populations that are affected by disasters and in the response phase of recovery (Dominelli 2009; Javadian 2007) and struggling to build social capital in disaster risk reduction (Mathbor 2007). However, Javadian (2007) noted despite of all these roles by social work profession in disaster management, the profession is yet unable to create a specialization on disaster management, Dominelli (2012) noted that social workers' voices have rarely been heard in the media and rarely been engaged in discussing issues of practice and policy.

Social work profession had greatly contributing for the disaster management and disaster risk reduction in the recent past (Gillespie 2008; Ku et al. 2009). But, there is a little doubt about the integration of disaster management in the social work (Gillespie 2008; Ku et al. 2009). Hence, there is an urgent need for social work academia to think about incorporating the disaster management contents in its curriculum and training social workers extensively on disaster recovery work. Dominelli (2012) gave an urgent call to social work fraternity for setting up "Green" social work, thought this she advocated for the rights of peoples and developing new concepts and models for disaster management which would also involve a range of stakeholders to develop locality-specific and culturally relevant responses to the major environmental and social problems. Further Dominelli (2012) suggested social work academia to look into three core areas of disaster management that is challenging the existence of disaster social work practice and education. The three challenges that Dominelli (2012) noted are: inadequate disaster social work training, unclear roles and mandates in disaster management, and understanding and optimizing local cultures and resilience after a disaster.

Despite social works decades of involvement in disaster recovery, social workers are inadequately trained. Scholars have noted their concern about inadequate evaluation and consolidation in disaster management practice and also expressed the concern regarding continued involvement of social work in disaster management (Drumm et al. 2003; Javadian 2007; Rogge 2003). Further Rock and Corbin (2007) and Mathbor (2007) argued there is a lack of systematic training and educational programmes in social work with concentration to disaster management. Further supporting to this Becker (2007a) states social work academia is too far in developing a systematic curriculum in disaster social work and only a few can be seen internationally. The most disturbing practice of disaster management is the "quick and dirty" short-term training in disaster social work in the face of recurring disasters. Supporting to this view Becker (2007b) noted that observations in the past years have proven that some social workers received last-minute three-day training during the 2004 Indian Ocean earthquake and tsunami.

Ang (2007) recommended social work academia to incorporate the basic concepts of social work and social justice in the disaster management-related trainings. Along with traditional social work modules in casework, group work, and community work social work in disaster management need to incorporate the concepts related to human rights, public education, and social rehabilitation targeted at the marginal. Becker (2007a), Newhill and Sites (2000), Mathbor (2007) supported the recommendations by Ang (2007) and noted that the disaster management in the present situation is a very urgent and deliberate need of the developed society. The trainings related to disaster management and disaster mitigation are gaining importance with more frequent disasters occurring across the world. Further Becker (2007a), Newhill and Sites (2000), Mathbor (2007) suggested for training social workers with cross-disciplinary training in disaster social work. However, disaster social work in this article refers to an educational programme to train social work students with intensive social work and disaster management knowledge and skills at the macro, mezzo, and microlevels.

There is both need for developing specialized training in social work and incorporating one or two modules of disaster management into the general social work course curriculum. Iravani and Ghojavand (2005) suggested social work academia to think of including at least one to two sessions into general social work teaching and conduct some discussion on the issues, such as working with disaster survivors and communities, and providing basic recovery services and practicing social work, such as providing salient information, dealing with victim family members sensitively. Further Iravani and Ghojavand (2005) noted that developing an entire course or module in disaster social work may not be feasible given the lack of resources and experienced educators and trainers to implement the educational teaching and training. However, contradicting to this view several other social work educators and researchers have suggested the development of comprehensive disaster social work curricula and training formats (Chou 2003; Dominelli 2012; Inter-Agency Standing Committee (IASC) 2007; Jordan 2006; Padgett 2002; Rock and Corbin 2007). One important phenomenon need to be noted in this context is the unclear role of social work in the disaster management and undefined responsibilities for social workers in disaster recovery or rehabilitation. Further, the mandate for the social work profession in disaster management is completely missing and that's what creating a dilemma for social work practioners. MacCormack (2007), Mathbor (2007) argued that Social work profession is known for its expert knowledge and skills in mobilizing, linking and coordinating resources, services, and people. This particular strength of social workers can be effectively used for rehabilitation end speedy recovery of the disaster situations across the world.

However, the social workers role in disaster recovery is not free of hassle, social work professionals involved in the disaster recovery and rehabilitation tasks face many challenges like dealing with Dynamic local cultures, need to contend with the challenges of understanding and utilizing the local capacities and cultures adequately in post-disaster situations. Cahill (2007) has observed when the rescue and recovery programmes in a community are not planned according to the needs of the community it creates more problems in addition to that of created by disaster. Another challenge for social work practice in disaster recovery as noted by whereas Bormann and colleagues (2007) is being observant about specific social, cultural, and historical factors when carrying out an assessment in disaster situations, and look out for economic, social, and political injustices that may complicate rescue and responses in post-disaster situations. Further, they also noted that at presently many social work professionals involved in disaster-related services may not be trained in these responsibilities, and this need to be taken as a serious concern for developing specific and focused training and educational courses in disaster management for social workers.

A search of the literature reveals that the social work professionals are not adequately trained in disaster social work and the absence of disaster social work curriculum in social work academia is effecting in an adverse way. In response to these questions, the present study was planned to understand the perception and beliefs of the master of social work students toward disaster management. The study had the following objectives- (i) To study the perspectives of social work students toward disaster management (ii) To study the implications for teaching disaster management course in future studies of social work, (iii) integration of disaster management concepts into social work curriculum, and (iv) students interest to take up disaster management in social work practice.

Research Methods

The aim of the study was to understand the knowledge, perception, and beliefs of Master of social work students toward the concepts disaster management and its integration into social work curriculum. In the view of limited knowledge available on this topic and its exploratory nature, the qualitative research method supplemented with quantitative method was chosen because it enables the researcher to understand the perception of student social workers (Creswell 2008).

A grounded theory approach was used to enable the researcher to generate concepts to understand the perception of social work students (Charmaz 2006).

Participants

The participants of the study were from four schools of social work in the Sothern parts of India. In two schools of social work focus groups were conducted to the Master of social work students. A semi-structured focus group discussion guide was developed around the themes (i) various perspectives on how disaster and disaster management are perceived. (ii) Implications for teaching disaster management in the future study of social work, and (iii) students interests in taking up disaster management as academic course and profession. A semi-structured questionnaire was used to collect the responses as supplementary to the qualitative data.

Data Collection

Data were collected through focus group discussion in 2015 October by the First author of the paper. This research did not go through ethical review. The heads of four schools of social work were contacted and their oral consent to conduct research was obtained before the data collection. Participants were told the purpose of the study before the data collection. In the case of focus groups students studying in their second-year master of social work were included. The number of the students in the focus group was restricted to 10 in each school. The responses were recorded after obtaining the oral consent of the respondent and each focus group lasted about 45 minutes. An unstructured focus group schedule was prepared and the questions were updated after each focus group.

A semi-structured questionnaire was used to collect the quantitative responses from the social work students in the two schools of social work. The heads of the departments were contacted and obtained the permission to collect the data. The student social workers in their second year of course were explained the aim and the purpose of the study and oral consent was obtained to fill the questionnaire. The questionnaire had 30 closed-ended questions with two possible responses in most of the questions and four possible responses in some questions. In an average, each student took 15 minutes to fill the questionnaire.

Data Analysis

The transcripts of the focus groups were read and analyzed in Kannada the first language of the researchers to avoid losing any meaning of the dialog. Line-by-line review of the transcript was conducted and was coded by highlighting the keywords. The themes were determined based on the highlighted keywords and were established through phrases, sentences, and segments of text. Based on the principles of saturation (Corbin and Strauss 2008) data were analyzed respondent by respondent until no new theme emerged. Quotes representing the themes were selected from the transcripts and translated from Kannada to English.

Results

The themes identified in the transcripts of the focus groups are presented in the following. The responses of students from each school of social work are presented under separate themes, because such a presentation provides a clear picture of the perception of the students toward the concept of disaster and disaster management and their exposure to the concept in teaching and learning methods including the contents of the course curriculum.

Reflection on Meaning of Disaster and Disaster Management

In one school the master of social work students were asked to define disaster and disaster management:

> One student defined disaster as "a situation where distraction is caused by heavy wind and heavy rain fall further he added earthquake is the most dangerous and the powerful disaster that I believe".

Most of the students participated in the focus group discussion were not very clear about the concept of disaster in the first instance. So when they were asked to define disaster many of them had nothing to say, some students had some basic knowledge but lacked scientific understanding. Figure 1 shows the different levels of students understanding about disaster and disaster management. 32% of the respondents had some basic conceptual understanding about the disaster. Whereas 85% of the

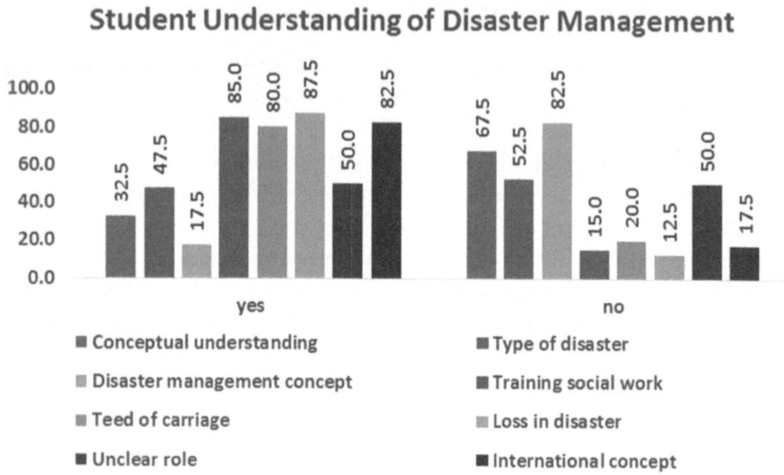

Fig. 1 Students understanding

respondents opinioned that the disaster management training is necessary for the social work students. For 52.5% of respondents disaster is always not natural, whereas 20% of the respondents believe that to undergo disaster management course one needs courage rest 80% supports the view that anybody interested in disaster management can undergo disaster management training. 82.5% of the respondents agreed that disaster management is an International concept and training and practice in disaster management will offer them a good employment opportunity.

However, the students were having some level of understanding on types of disaster. They were not very clear about which are the classification of disaster. But, they had basic knowledge of manmade disaster and natural disaster. As one student explained:

> Disaster in today's context is manmade and human interaction in the natural cycle creates disturbances. For example the heavy rain fall some times and drought some time is an example of natural cycle imbalance because of human interference.

Majority of the students had misunderstanding between pollution and the disaster, for example, one student while explaining what disaster is for him:

> The carbon dioxide coming out of the factories and the motor vehicles in our environment is creating much disaster.

Reflection on the Disaster Management Content in the Curriculum

Regarding disaster management students understanding is very unscientific. Majority of the students are from different streams and they did not have any prior experience of learning the concepts related to disaster management or disaster mitigation during their under graduation.

> I am from commerce background and during my undergraduate degree I did not have any concepts related to disaster management in my college curriculum all I studied is related to commerce and management which is nothing to do with disaster management.

> I come from science streams and I really did not have any prior experience of studying disaster management concepts in my under graduation. Same thing continued here in my master of social work in all the four semesters we don't have any special concentration on disaster management or disaster mitigation.

Figure 2 shows the students understanding with regards to disaster content in social work curriculum. 80% of the students agree that studying

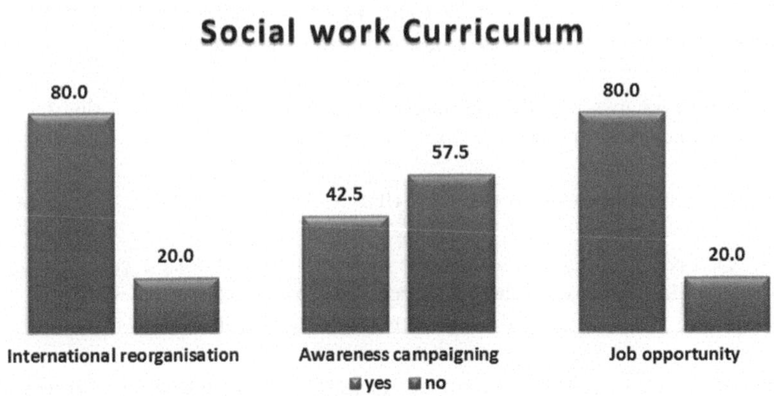

Fig. 2 Disaster Management Content in Social Work Curriculum

disaster management in their social work curriculum may give them international employment opportunity. Whereas 47.5% disaster management content in the curriculum may help to create awareness about the disaster and disaster management. 80% of the students are confident that by including the disaster management content in the curriculum and training students for disaster management will increase the employability of the students.

Reflection on Student's Interest to Take up Disaster Management Course

When students were asked whether they are interested to study disaster management and its relevance to social work majority of the participants in the focus groups had a positive response.

> I really would like to get trained in disaster management. I heard that some institutions/schools of social work in India as well as abroad provides disaster management as concentration/specialization/majors in master of social work courses. In addition some other institutions offer a full MA/MSc in disaster management. I personally believe that training in disaster management would be a good opportunity to practice true social work principles.

Some students have a basic understanding about the international organizations working for the disaster management. Majority of them are confident that to get an employment in these organizations they need training and theoretical knowledge of disaster management.

> I want to see myself in the international organizations working for disaster management and mitigation services at global level.

Another student expressed the view that

> In the present developing world, the disaster is an unavoidable yet unpredictable. In this situation there is a need as well as space for social work profession to practice with in the bounders of social work principles and values.

The students are quite confident about getting an employment opportunity after getting trained in disaster social work.

We can only try to prevent the man-made disaster and try to reduce the loss and damage caused by the natural disaster. Hence the human resource trained in managing the disaster situation is always in demand. Further the rehabilitation, recovery and the resettlement are the area where social work needs to intervene and for that one need to be trained during their academic course.

Figure 3 shows the understanding of the social work students with relation to the skills required for social work students to deal with disaster management task. 92% of the respondents agreed that social work skills are needed for disaster management. 95% of respondents agreed crisis intervention skill requirement whereas 97.5% agreed that social workers with enriching public awareness, 95% agreed excessive training is required for social workers on disaster management. 60 opinioned there is no need of social workers in the psychological intervention skills, 72.5% agreed one must need overall social work skills, 80% agrees that social work has its role in disaster management whereas 85% says academic training for social workers is very crucial for serving as a disaster management team member. 40% respondents are of the opinion that today's social workers are getting poor quality or inadequate training in social work related to disaster management.

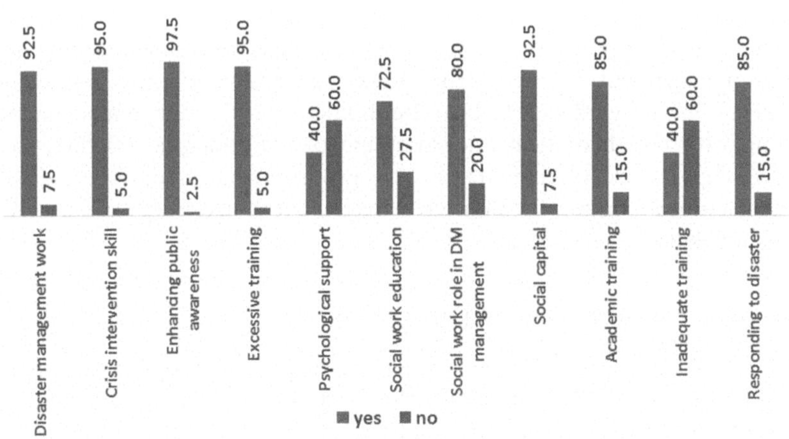

Fig. 3 Knowledge on social workers role

Table 1 Students selection of courses

Year	*Courses*
I year MSW	1. Contents on training disaster management
	2. Contents on disaster prevention
	3. Disaster and the vulnerable population
	4. Government and nongovernmental programmes for disaster management
	5. Organizations working for disaster management in India and abroad
	6. Disaster and economic injustice
	7. Disaster and political injustice
	8. Disaster and cultural differences
II year MSW	1. Human rights
	2. Social rehabilitation
	3. Public education
	4. Cross-disciplinary training
	5. Problems of disaster survivor's and communities
	6. Developing rescue plan
	7. Dealing with vulnerable family members
	8. International and transnational disaster management

Source Field data (FDG)

The students were asked to select the courses which they want to study as part of disaster management specialization during their Master of Social Work course. The contents suggested by the students were given in the Table 1.

An overall response of the students was: social work must adopt disaster management as one of the course subjects within the two years course. Social workers need to be trained for disaster management along with the theoretical understanding of the concepts. Further, the internships/block placements in the governmental/nongovernmental organizations working for disaster management is necessary to learn the practice skills.

Discussion

This study indicates that social work student's knowledge and understanding of the concepts related to disaster and disaster management is unscientific and irrational. But students are interested in learning disaster management and many of them wish to pursue their future career in disaster management field. As students expressed their views that the contents related to the disaster in their master of social work curriculum is very minimal in some cases and absolutely nil in some other cases. The students expressed their view on disaster as the conflicting meaning of pollution, it clearly shows the students understanding and conceptual knowledge is premature. Scott Harding (2007) explains disaster as follows:

> The idea of disaster is usually associated with human suffering caused by natural events: tsunamis, hurricanes, earthquakes, floods. However, human-created disasters stemming from war, the conduct of repressive regimes, the use of sanctions, and economic and social policies represent an equally important dimension of disaster. These practices have political goals and produce a range of negative social and economic conditions beyond their stated aims. They cause human suffering, especially among vulnerable groups, disproportionately women, children, older people and poor people.

The need to train social workers in disaster management and integrating disaster management concepts in social work curriculum is something related to the basic values and the principles of the profession. Social work profession is predicated on the values of social justice and elimination of all forms of oppression, discrimination, and inequality. To advance this perspective and confront human-made disaster, the profession should promote social development strategies and human rights principles through political practice and within social work education (Harding 2007).

The study proves that the students who have participated in the study are knowledgeable about the natural disasters to some extent but their knowledge on man-made disaster is very restricted and the result of man-made disaster is unknown to the students. The disasters like War and the armed forces are in the peak in the present society and because of the advancement of the science and technology the production of war weapons and the distracting instruments may lead to the disasters which are not less than any historic natural disasters which took away a number of lives in many incidents. As Krause and Mutimer (2005) argue that the man-made disasters like war divert needed resources to military means, displace a large amount of populations, disrupt economic networks, and thereby affects the economic stability and contribute to environmental degradation. The effects of the disasters can also be seen on food production, water quality, and living conditions. Further, the Violent conflict that created by war and war-related activities results in disrupting public health and other basic infrastructure services, in addition to these incidents results in outbreak of famine and disease, often killing more people indirectly than those who die from combat (Krause and Mutimer 2005). Hence, the knowledge on these incidents and the clearer understanding on these disasters need to be taught to the social work students in line with the professional goals and values. War and political violence create both direct and indirect negative effects (Oakes and Lucas 2001) and these effects offer results in diverse problems in the society.

The concept of disaster and disaster management has closer connections as the core principle of social work speaks about the human development and human wellbeing. Teaching and training disaster management to social work students and making them effective human service professionals is more relevant in line with the opinion given by Zakour. Zakour (1996) says that the social work interventions in disasters have focused on the events like working with effected individuals, families, organizations, and communities. Further, the areas of concern are traumatic stress, resources for disadvantaged and vulnerable populations, and coordination of various intervention systems. In his opinion, the disaster practice is arguably a reflection of the mission of social work itself (Zakour 1996).

Human-made disasters provoke a serious disruption of the economy, agriculture, and healthcare sectors of a society, typically producing long-lasting effects that perpetuate underdevelopment. Dealing with the ill effects of the human-made disaster needs the clear understanding of what human made disaster is and how it effects the development of the country. A lack of basic understanding and rational knowledge on disaster, human-made disaster, and its effects were found among the Participants of the present study. However, there are some other areas need to be investigated in line with the present study. What should be the contents of curriculum on disaster management? What needs to be included in the social work disaster management curriculum? What should be the course structure? Is it an independent course or a concentration in social work? These questions demand further investigations in this context.

Limitations and Direction of Future Research

The limitation of this study need to be addressed first, only four schools of social work were included in the study and the schools of social work were selected purposively. Out of four schools of social work one school has a specialized social work training ranging from Family and Child welfare, Human Resource Management, Social Development and Medic and Psychiatric social work, and other three schools had a generic social work curriculum. Hence, the understanding of the students may differ because of the course content.

Second, this study was conducted only to the social work student's community which only brings the students perception about the issue and not the educators. Future studies with educators should be conducted as the perception of educators shapes the perception of students.

Third, the present study was focused only on master of social work students. Hence, their finding reflects only on the master of social work course curriculum and its contents. The future studies in the same should be conducted to other levels of course.

Fourth, the present study was conducted in the one particular geographical area with socio political and cultural similarity. The conceptual understanding of disaster and the disaster management has different implications in different socio, political, and economic situations. Hence, the future studies should be conducted in other societies with cultural, political, and economic variations to understand the differences and the influences of these factors.

Conclusion

This study explored the perception of master of social work students on disaster and disaster management. The result shows that the students are not adequately taught about the concepts of disaster management and many of the schools of social work do not have any content or lessons related to disaster in their course curriculum. Social work has a major responsibility of working for the wellbeing of the society, and hence, the social workers need to be equipped with the services and the management of disaster which is actual hurdles to the development of the nations. These concepts should not be overlapped in the curriculum of the helping profession and adequate teaching method for making students to understand the concept. By taking these findings seriously considerable headways can be made in supporting the views of incorporating disaster management contents into social work curriculum.

Acknowledgements The authors of the paper are thankful to Head Department of Social Work Central University of Karnataka, Head Department of Social Work Gulbarga University, Principal Sri Saiprasad School of Social Work, Principal Mahagawankar College of social work, and all the student participants for their support in the study.

Bibliography

Ang, J. (August 2007). "Disaster Preparedness and Social Work". In *International Federation of Social Workers (IFSW) Asia Pacific Regional Symposium*, Kuala Lumpur, West Malaysia, 27–29.

Banerjee, M. M., and Gillespie D. F. (1994). "Linking Preparedness and Organizational Disaster Response Effectiveness". *Journal of Community Practice* 1(3): 129–142.

Becker, S. M. (2007a). "Psychosocial Care for Adult and Child Survivors of the Tsunami Disaster in India". *Journal of Child and Adolescent Psychiatric Nursing* 20(3): 148–155.

Becker, S. M. (2007b). "Communal Rebuilding After Destruction: The World Trade Centre Children's Mural Project". *Psychoanalytic Social Work* 10(1): 27–41.

Berke, P. R., Karetz, J., and Wenger, D. (1993). "Recovery After Disaster: Achieving Sustainable Development, Mitigation, and Equity". *Disasters* 17(2): 93–109.

Bormann, S., Klassen, M., and Spatscheck, C. (2007). *International Social Work: Social Problems, Cultural Issues and Social Work Education*. Opladen: Barbara Budrich Publishers.

Cahill, K. M. (2007). *The Pulse of Humanitarian Assistance*. New York: Fordham University Press and The Center for International Humanitarian Cooperation.

Charmaz, K. (2006). *Constructing Grounded Theory: A Practical Guide Through Qualitative Analysis*. Thousand Oak, CA: Sage.

Chou, Y. C. (2003). "Social Workers Involvement in Taiwan's 1999 Earthquake Disaster Aid: Implications for Social Work Education". *Social Work & Society* 1(1): 14–36.

Corbin, J., and Strauss, A. (2008). *Basics of Qualitative Research* (3rd ed.). Thousand Oaks, CA: Sage.

Creswell, J. W. (2008). *Research Design: Qualitative, Quantitative, and Mixed Methods Approaches* (3rd ed.). Thousand Oaks, CA: Sage.

Desai, A. (2007). "Disaster and Social Work Responses". In L. Dominelli (Ed.), *Revitalising Communities in a Globalising World* (pp. 297–314).

Dodds, S., and Nuehring, E. (1996). A Primer for Social Work Research on Disaster. *Journal of Social Service Research* 22(1): 27–56.

Dominelli, L. (2009). *Introducing Social Work*. Cambridge: Polity Press.

Dominelli, L. (2012). *Green Social Work: From Environmental Crises to Environmental Justice*. Cambridge: Polity Press.

Dominelli, L. (2015). "The Opportunities and Challenges of Social Work Interventions in Disaster Situations". *International Social Work* 58(5): 659–672. https://doi.org/10.1177/0020872815598353.

Drumm, R. D., Pittman, S. W., and Perry, S. (2003). "Social Work Interventions in Refugee Camps: An Ecosystems Approach". *Journal of Social Service Research* 30(2): 67–92.

Gardoni, P., and Murphy, C. (2008). "Recovery from Natural and Man-Made Disasters as Capabilities Restoration and Enhancement". *International Journal of Sustainable Development and Planning* 3(4): 317–333.

Gillespie, D. F, and Murty, S. A. (1994). "Cracks in a Post Disaster Service Delivery Network". *American Journal of Community Psychology* 22(5): 639–660.

Gillespie, D. F. (2008). "Disaster". In T. Mizrahi and L. E. David (Eds.), *Encyclopedia of Social Work* (Vol. 2, 20th ed., pp. 60–65). Washington, DC: NASW Press.

Gray, D. E. (2009). *Doing Research in the Real World* (2nd ed.). Thousand Oaks, CA: Sage.

Harding, S. (2007). "Man-Made Disaster and Development: The Case of Iraq". *International Social Work* 50: 295.

Ifsworg. (2015). Ifsworg. Retrieved November 14, 2015 from http://ifsw.org/policies/definition-of-social-work/.

Inter-Agency Standing Committee (IASC). (2007). *IASC Guidelines on Mental Health and Psychosocial Support in Emergency Settings*. Geneva: IASC.

Iravani, M. R., and Ghojavand, K. (2005). "Social Work Skills in Working with Survivors of Earthquake: A Social Work Intervention—IRAN". *Social Work & Society* 3(2): 265–272.

Javadian, R. (2007). "Social Work Responses to Earthquake Disasters: A Social Work Intervention in Bam, Iran". *International Social Work* 50(3): 334–346.

Jordan. (2006). "A Case Study: How a Disaster Mental Health Volunteer Provided Spiritually, Culturally, and Historically Sensitive Trauma Training to Teacher-Counselors and Other Mental Health Professionals in Sri Lanka, 4 Weeks After the Tsunami". *Brief Treatment and Crisis Intervention* 6(4): 316–325.

Kwok, Joseph. (2008). "Regional Perspectives... From Asia: Social Work and Social Development in Asia". *International Social Work* 51: 699. https://doi.org/10.1177/0020872808093349.

Krause, K., and Mutimer, D. (2005). "Introduction". In *Small Arms Survey 2005*. Geneva: Graduate Institute of International Studies.

Ku, H. B., Ip, D., and Xiong, Y. G. (2009). "Social Work in Disaster Intervention: Accounts from the Grounds of Sichuan". *China Journal of Social Work* 2(3): 145–149.

MacCormack, C. F. (2007). "Coordination and Collaboration: An NGO View". In K. M. Cahill (Ed.), *The Pulse of Humanitarian Assistance*. New York: Fordham University Press, The Center for International Humanitarian Cooperation, pp. 243–261.

Mathbor, G. M. (2007). "Enhancement of Community Preparedness for Natural Disasters: The Role of Social Work in Building Social Capital for Sustainable Disaster Relief and Management". *International Social Work* 50(3): 357–369.

Mileti, D. S. (1999). *Disasters by Design: A Reassessment of Natural Hazards in the United States*. Washington, DC: John Henry.

National Governors' Association. (1978). *Emergency Preparedness Project: Final Report*. Washington, DC: NGA.

Newhill, C. E., and Sites, E. W. (2000). "Identifying Human Remains Following an Air Disaster: The Role of Social Work". *Social Work in Health Care* 31(4): 85–10.

Oakes, M. G., and Lucas, F. (2001). "How War Affects Daily Life: Adjustments in Salvadoran Social Networks". *Journal of Social Work Research* 2(2): 143–155.

Padgett, D. K. (2002). "Social Work Research on Disasters in the Aftermath of the September 11 Tragedy: Reflections from New York City". *Social Work Research* 26(3): 185–192.

Patton, M. Q. (1990). *Qualitative Evaluation and Research Methods*. Newbury Park, CA: Sage.

Phillips, B. D. (2009). *Disaster Recovery*. Baca Raton, FL: Auerbach Publications.

Rock, L. F., and Corbin, C. A. (2007). "Social Work Students and Practitioners' Views on the Need for Training Caribbean Social Workers in Disaster Management". *International Social Work* 50(3): 383–394.

Rogge, M. E. (2003). "Children, Poverty, and Environmental Degradation: Protecting Current and Future Generations". *Social Development Issues* 22(2/3): 46–53.

Tierney, K. J. (1993). Disaster Preparedness and Response: Research Findings and Guidance from the Social Science Literature. Retrieved from http://dspace.udel.edu:8080/dspace/bitstream/handle/19716/579/PP193.pdf?sequence=3.

Tumelty, D. (1990). *Social Work in the Wake of Disaster*. London: Jessica Kingsley Publishers.

Twigg, J. (2004). *Disaster Risk Reduction: Mitigation and Preparedness in Development and Emergency Programming*. London: Humanitarian Practice Network (HPN), Overseas Development Institute.

Unisdrorg. (2015). Unisdrorg. Retrieved November 14, 2015 from http://www.unisdr.org/we/inform/terminology.

Webster, S. A. (1995). "Disaster and Disaster Aid". In R. L. Edwards (Ed.), *Encyclopedia of Social Work* (Vol. 1, 19th ed., pp. 761–771). Washington, DC: NASW Press.

Yanay, U., and Benjamin, S. (2005). "The Role of Social Workers in Disasters: The Jerusalem Experience". *International Social Work* 48(3): 263–276.

Zakour, M. J. (1996). "Disaster Research in Social Work". *Journal of Social Service Research, 22*(1/2): 7–25.

Zakour, M. J. (2006). "Social Work and Disasters". In D. A. McEntire (Ed.), *Disciplines, Disasters and Emergency Management* (pp. 124–140). Washington, DC: Federal Emergency Management Agency, Emergency Management Institute.

Indigenous Knowledge and Practices in Disaster Management: Experiences of the Coastal People of Bangladesh

Mahfuzul Haque

Introduction

Indigenous knowledge (IK) means the functional knowledge of local people inhabiting in a particular ethno-cultural and agro-ecological condition. IK develops through experience sharing and normally passes through generations by oral expressions and it operates in all aspects of community life and persists therein as unwritten forms (Haque 2013a). Although, it is often said that unwritten knowledge is vulnerable to being lost, local wisdom has a habit of persisting in the villages, old towns, markets, and other places, where people continue to put it into

---

Former Secretary, Government of Bangladesh and currently involved in teaching at the Department of Development Studies, University of Dhaka, Bangladesh. His areas of interests are environment and development; sustainable development; environmental governance; natural resource management; biodiversity conservation; climate change; desertification; disaster risk reduction, etc.

M. Haque (✉)
Department of Development Studies, University of Dhaka, Dhaka, Bangladesh

B. Zutshi et al. (eds.), *Disaster Risk Reduction*,
https://doi.org/10.1007/978-981-10-8845-2_4

practice. IK usually connotes a holistic system of knowledge, comprising of values, concepts, beliefs, and perceptions, which is naturally located among people living in a local (often rural) environment (Warner 1991; Sillitoe et al. 1998). Walker et al. (1991) define it as the knowledge held collectively by a defined community.

IK is not generally codified or written down in formal language or forum. It remains mostly in verbal form and passes through generation to generation. It is increasingly recognized as a resource that should be mobilized to complement scientific knowledge in order to promote appropriate plans and interventions for rural development. IK refers to knowledge including values, concepts, perceptions, and beliefs of a particular local community. It is inherently diverse and multifaceted knowledge system with varied expressions. Sillitoe (2000) stated that instead of "top-down" imposition of interventions, IK has been developed through a process of "grassroots" participatory way. Mostly found in rural areas in unwritten form and in informal conversations, IK is differentiated by scientific knowledge, which is more formal, written, and codified. It is socially and culturally specific, constructed, and bound. It is typically developed in the process of local people's experiment with varied livelihood, survival, and coping strategies.

The term "indigenous" is synonymous with "traditional" and "local", differentiating this knowledge from that developed by formal science in institutions, such as universities and government research centers (Walker et al. 1991; Sillitoe 2000). IK refers to local and traditional knowledge used by rural people for agriculture, natural resource management, fisheries, livestock, healthcare practices, and other activities. Sillotoe et al. (1998) maintain that IK relates to any knowledge held collectively by a population, which informs understanding of the world. It may encompass knowledge of any kind/domain including that pertaining to sociocultural and natural processes. It is culturally relative, being informed by people's sociocultural tradition and history of which it is an integral part.

Value of local knowledge cannot be ignored. Knowledge and practices of the local community cannot be termed as primitive, unscientific, or hindrance to development. Although, the local community is exposed to modern knowledge but they have not abandoned their IK acquired through many generations. It is true that local knowledge is eroding fast and much has been lost with the modernization and rapid spread of foreign technology introduced from outside. There is a clarion call to document IK before it is lost.

Coast of Bangladesh

Bangladesh is vulnerable to periodic natural disasters due to its conical shape and location on the tip of the Bay of Bengal. Bangladesh faces 710 km long coast with the Bay of Bengal containing several ecosystems of high conservation value starting from the Saint Martin's Island to the district of Satkhira (Rasheed 2008). The coastal zone is characterized by a vast network of rivers and tidal channels; erosion and accretion processes continue, siltation takes place on watercourses and river beds; and the area is prone to cyclone, storm surges, and salinity intrusion. The largest mangrove forests of the world, the Sundarbans covers a big chunk of the coast along with India. It's a land of natural disasters. Series of tropical cyclones, tornadoes, tidal bore, and floods attack the coast every year. Threat of sea level rise due to climate change is also looming large.

A coast is an area where (a) tidal water movements determine agricultural practices, movement of river transport, estuarine commercial activities, and everyday life; (b) water is under salinity intrusion; and (c) the area runs a tremendous risk of cyclones and storm surges. Based on the above three criteria: tidal water movements; salinity intrusion; and risk of cyclones, an area of 47,201 sq. km (32% of the country) of Bangladesh within 19 districts in 147 coastal *Upazilas* (subdistrict) could be termed as coastal area (Rasheed 2008). The area is populated by 35 million people (2005) representing 29% of the total population. The people of the coast are generally known for their resilience, as they fiercely face periodic natural disaster and climate change. In fact, inaccessibility and remoteness of the coast made the people fierce and self-dependent.

The people of the coast in particular and the country, in general, have developed through a process of innovation and adaptation, a variety of coping strategies and techniques that are fine-tuned to the local environment, economy and sociocultural system. The people inhabiting a disaster-prone country have their localized knowledge and practices, developed through cumulative experience, that constitute a survival strategy in the face of natural disasters. Bangladesh possesses a rich heritage of IK and practices, much of which has been lost due to their non-documentation. Nevertheless, people in disaster-prone areas still nurture such knowledge in their myths, beliefs, and traditions.

Varied Faces of Disaster

The major natural disasters that periodically visit Bangladesh include cyclones, tidal surges, tornadoes, floods, drought, desertification, earthquakes, river erosion, etc. We also face man-made disasters, like climate change leading to rising sea level. There are many mitigation measures, which are mainly scientific in nature. However, these natural disasters are nothing new. They are age-old problems and people of this deltaic region, especially in the coastal areas have developed their IK and practices over the years in combating the disaster. We are to remember that perhaps, we cannot prevent all natural disasters, but we can reduce the number of casualties. Some of the disasters can be managed, thus the question to be resolved is: how can we better manage them? As we have no choice but to live with some of the disasters, we need to find ways and means to tackle them. The coping strategies of people in disaster-hit areas should be documented for their potential utilization in other parts of the country. The policymakers and technocrats are to be made aware that there are many nonstructural solutions to these disasters, practiced by people of the locality for generations.

Cyclone and Tidal Surges

Cyclones originate when the combination of heat and moisture air forms a low pressure over the oceans. In the Northern Hemisphere, it is anticlockwise and in the Southern Hemisphere, it rotates clockwise. Bay of Bengal is the most important cyclone-prone area because of its funnel shape and most cyclones occur in the month of October and November period. In the last 150 years, 35 devastating cyclones have hit the Bangladesh Coast. Cyclone and tidal bore of April 12, 1970, killed 500,000 people. Cyclone and associated water surges of April 29, 1991, killed 140,000 people in the coast. Bangladesh recently faced two devastating cyclonic storms and associated tidal bore named *Sidr* in 2007 and *Aila* in 2009 causing loss of lives and damages to standing crops and property.

It is evident, that disasters like cyclones, tornadoes, and tidal bores will continue to hit Bangladesh and they cannot be controlled. Maybe through better management practices, damages could be reduced to a large extent. The people of the southern coastal belt use the word *tufan*, to describe a cyclone, which apparently seems close to "hurricane" in Atlantic Ocean;

"typhoon" in Pacific Ocean; "baggio" in Philippines Sea; and "willie willie" in the Pacific. Regarding low or high intensity of cyclones, they put a prefix like *choto* (small) *tufan* and *boro* (big) *tufan*, respectively. In a separate study, it was revealed that the coastal people identified five major symptoms in anticipating and predicting cyclones. They are: (a) wind direction; (b) temperature and salinity of seawater; (c) color and shape of a cloud; (d) appearance of the rainbow; and (e) behavior of certain bird species. Regarding the direction of wind, the coastal people believed that a wind blowing from *Agni-con* (southeast) is more likely to create a storm, while the wind direction from *Ishan-con* (northeast) has the potential to generate a cyclones but not to that extent in case of severity. The wind direction is also associated with other attributes, i.e., a rise in seawater temperature, red colored cloud, and the appearance of a rainbow (if it is daytime) implying the formation of deep depression in the sea. In most cases, such depressions are formed near the Andaman Islands of India. Abnormal behavior of the birds residing in trees is regarded as a signal of rapid storm approach. Also, cloud in the shape of an elephant's trunk is considered to be a symptom of tidal surge (Hassan 2000).

The cyclone warning system helps people to take refuge in cyclone shelters. Previously, we had 11 warning signals for seaports and 4 warning signals for inland ports, which were a bit misleading and created confusion among the people as they thought danger signal 8 is less severe than that of danger signal 10. In fact, the message was that the storm was passing either by the eastern or western side of the seaport. Based on the consultation held with the coastal people, the meteorological department is in the process of revising the signals to make the warnings easily understood. There is a storm warning center at the Met Office and it is responsible for issuing warnings for the tropical cyclones. There are Standing Orders on Disaster (SOD), 2010 issued by the Disaster Management Bureau, Ministry of Food and Disaster Management spelling out the responsibilities and duties of various agencies at different levels. It was prepared based on the intensive consultation involving all Ministries, Divisions, Agencies, and other stakeholders, including the coastal people. Pre-disaster and post-disaster measures/activities are laid down in the Standing Orders.

How do they survive during and after a cyclonic disaster? What are the indigenous survival strategies? In a study conducted on the people of the coastal islands, it was revealed that the islanders have developed certain short-term survival strategies of their own (Hassan 2000).

Simple tactics like holding onto and binding themselves to trees; looking for comparatively more dependable places like embankments and polders; using floating items, such as timber, thatched roof, straw piles, and bunches of coconuts represent spontaneous survival strategies. People of the locality have also devised appropriate methods for food preservation during cyclones and tidal surges. Emergency dry foods like *Chira* (rice) and molasses are wrapped in polythene are kept in earthen pots buried under the earth. They become handy when the water is gone. Self-protection instinct dominated the indigenous survival strategy. The women prefer putting on *Salwar* and *Kamiz* dress instead of *Saris* (traditional female dress in South Asia), as the latter makes it difficult to swim. During the aftermath of a disaster, coping strategies work well with the help of the community members to their mutual benefit. Generally, outside help and relief goods appear 2/3 days after the disaster. What do they do during this intervening period? People eat stems and roots of edible plants. For drinking purposes, they drink rainwater, as cyclones are always followed by rain for several hours. In the absence of rainwater, they share coconut water. Due to nonavailability of medicines, generally, the victims depend on herbs and other local substances for treating minor injuries, fever, and diarrheal diseases.

Among structural measures, the people on the coast are seen using roofing materials and design (sloping the wind direction). They plant local varieties of plants surrounding their homesteads. On Sandwip island, people plant *Hurma* (*Persimmon*) trees, which are strong and can withstand tidal waves. The lives of many people were saved during the April' 1991 cyclone, as they held on to trees of mangrove species, like *Keora* (*Sonneratia apetala*) and *Sundari* (*Heritiera fomes*). Another interesting phenomenon was that during tidal surges, people tied rafts to coconut trees so that they rose and fell with the level of the water (Haque 2000).

Bangladesh has been lauded for its good management capabilities of natural disasters. Considering the two very severe cyclones of April 12, 1970 (500,000 died), and April 29, 1991 (140,000 died), the recent cyclones *Sidr* in 2007 and *Aila* in 2009, although were very deadly, casualties were at minimum. It was possible due to the application of IK and practices of the local coastal community. The deadly super storm "Sandy" that hit New York City, Long Island, New Jersey, and six other states of USA on October 29, 2012, killing 32 people with the streets and subways flooded, airports remained inoperative for days is fresh in our memory. Most catastrophic cyclones were Hurricane *Jeanne* in Florida (2004), *Ivan* also in Florida (2004), *Catrina* in New Orleans (August 2005), and *Emily* in the Caribbean (2005). During *Catrina* in 2005, dead bodies were found floating on the water for days due to the poor management of such calamities. In Bangladesh, SOD 2010 played an important role in streamlining post-cyclone activities in various tiers of the government. A glaring example is the recent cyclone, *Roanu* that hit the coasts of Bangladesh on May 21, 2016, with less casualties.

Floods

It is a flood-prone country and floods visit us every year, often twice, and thrice. People of Bangladesh never consider all floods as bad floods. We have been living with floods for generations. There are many songs and dances on the monsoon flood. Floods are a blessing as well, as they bring alluvial soil. In the northeastern *Haor* (a bowl-shaped naturally depressed water basin) areas, all social and cultural festivities, like marriage, visiting father's house, and religious meetings are generally held during the floods. We had severe floods considering their severity and damages of human lives and properties. They are the floods of 1987, 1988, 1998, 2004, and 2007. The 1998 flood was unprecedented in the living memory, as it inundated two-thirds of Bangladesh for two long months causing damages to lives and properties. There are four main types of floods, frequently visiting Bangladesh. They are: (a) Short duration (2/3 days) flash floods, characterized by a sharp rise and drop in water level with high velocity damaging crops and property (northeastern wetlands of *Haors* and the southeastern hills); (b) Rain floods due to heavy rainfall; (c) Monsoon floods resulting from heavy monsoon rainfall over the Himalayas; and (d) Coastal floods arising from storm surges in the coastal areas. Flood severity is generally measured by the area of inundation. When the flood inundates 25% of the locality, it is considered as small flood (*Chotto Bannya* in local term); when 35% of the area is engulfed by floodwater, it is known as severe flood; and in the case of above 35% of area inundation, it is termed as catastrophic flood. In local terms, the later two floods are known as high flood (*Baro Bannya*). The floods of 1987 and 1988 are known as *Jora Bannya* (floods in pair). The floods of 1998, 2004, and 2007 are termed as *Baro Bannya*. When referring to a period, like the date of birth or marriage, etc., the villagers refer to such an event instead of date or year.

The islanders undertake human safety measures during the floods. These include making rafts from banana trunks, keeping vigil at night when the water is high or rising. They remain awake at night to keep away snakes and prevent children from drowning. The most common type of human safety measure is the construction of a high raised platform above floodwater level on bamboo poles. Often beds are raised by placing something under the legs.

Flood Forecasting and Warning Centre of Bangladesh Water Development Board has been playing an important role in flood forecasting through print and electronic media. They also have a vibrant Website on flood forecasting. Bangladesh has proven expertise in managing the floods unlike many other developed countries. The SOD, 2010 published by the Disaster Management Bureau has spelt out responsibilities of the various tiers of the government starting from the local government to the district, divisional, and central levels. It has elaborated the emergency response during the warning period, disaster, and post-disaster period. For long-term risk reduction, responsibilities of the Disaster Management Committees at various tiers are also elaborated. The stakeholders are involved in these committees. For example, in the *Upazila* Disaster Management Committee (UzDMC) at subdistrict level, representatives of local women's groups, NGOs, Red Crescent, and socially reputed persons have been inducted. SOD, 2010 have been well-acknowledged locally and globally.

In a study conducted on the assessment of the 1998 flood on Dhaka city, the respondents instead of a structural solution, suggested age-old IK and practices to tackle such floods (Haque et al. 2000). Use of sandbags, frozen cements, and bricks as walls against the onrush of floods was a common sight. Extending the silencer pipes of motor vehicles up above the floodwater level with an L-shaped pipe was an innovative coping strategy. Daily activities of marooned people did not stop with the floods. They continued developing new strategies to cope with the situation. In case of transportation in the capital, they introduced country boats, rickshaw vans, and rafts made of empty barrels. New and alternative routes were discovered everyday as floodwater continued to rise. Railway communication between Dhaka and Narayanganj gave relief to the weary commuters. People also discovered riverine communication and new routes. Small steamers started to ferry passengers to nearby locality, not accessed by boats before.

Rich "social capital" of the people of Bangladesh was found to be very handy during a disaster. The community came forward with the needed help in cash and kind (dry food, drinking water, essential medicines, etc.). During the 1998 flood, the community took collective efforts in tackling water leakages of Dhaka-Narayanganj-Demra embankments. Local youths worked day and night to protect the embankment or else there could have been a major disaster. In a disaster like a flood, the low-income and middle-income people, mostly being fatalists, try to cope with the disaster themselves and don't complain much. The low-income people help each other on a community basis. The middle-income people extend help to the lower-income neighbors and take help from friends and relatives (Islam 2000).

River Erosion and *Charland* People

In the coastal areas of Bangladesh, many rivers are changing their courses frequently giving rise to erosion and bringing miseries to riverbank people. *Nodi Sikosti* (river erosion) is a silent disaster. There is no relief and often no compensation for the victims is offered. After *Nodi Poisti* (Char formation, when a strip of land rises out of a river bed), the victims of river erosion are rarely rehabilitated (Haque 2013b). People living in the *charlands* (raised shoals in the river) are vulnerable to the vagaries of nature. They are the most desperate and vulnerable people living on these newly formed lands on the coast. Fragile riverbanks and *charlands* have never been abandoned because of flooding. On the contrary, these erosion-prone areas are accommodating a rapidly expanding population. Their survival strategy in the hazard-prone areas largely depends on how they put to use their experience and they are likely to differ according to the perception and economic ability of the persons in question. The people of the *charlands* understand their climate very well as they risk living in such a vulnerable situation. They have keen power of observation and can predict the nature of floods likely to hit them. If the monsoon rain is abundant and the water flow is relatively clean and has a current, they become alert to the onslaught of high flood (Hasan et al. 2000). However, not all of their observations are found correct. Different coping strategies are adopted by the *charland* people depending on the severity of the flood.

IK in structural measures includes those activities or practices intended to reduce damage to dwelling structures and household goods. Normal flooding does not call for repair of houses (re-thatching the straw) every year. What they do every year for their homestead is strengthening the bamboo poles. With the level of floodwater, they raise their six-legged wooden cots over bamboo platforms and store their seeds and dry food in large earthen pots. During the monsoon, the basic foundation of the house (plinth) is usually plastered with a paste of mud, jute fiber, and husks to protect the plinth from the direct impact of floodwater. On a similar such situation, people living in erosion-hit *haor* (saucer-shaped water bodies) areas in the northeast, plant *Chailla* (*Hemarthria protensa*) grass in and around their homesteads in order to halt wave action that continually erode their houses. They plant rows of *Hijol* (*Barringtonia acutangual*), *Koroch* (*Pongamia pinnata*) and *Barun* (*Crataeva nurvala*), typical *Haor* plants in front of their homesteads to halt *Afaals* (wave actions) of *haor* waters.

People living in the *charlands* are subject to the whims of the river and portray a typical man-environment interaction. IK in agricultural cropping is the adjustment with respect to crops before and after flooding. Selection of crops is very vital for the *charland* people. Usually, groundnuts and sweet potatoes are sown at the highest level of the land where the soil is slightly sandy. At the waterfront, the people plant *Aman* paddy, which is adaptable to high flooding. The deepwater *Aman* can grow at the rate of 6 inches per day and reaches a height of 15 feet to keep pace with the rising waters. There are also practices like intercropping to accommodate the risk of crop failure. The short-term flood sensitive *Aus* is sown together with the long-stemmed flood-tolerant *Aman* in the same field. Normal flooding would give two crops, while a dry year will give a good *Aus* but no *Aman* crop; abnormal floods will favor *Aman* but will affect *Aus* production. Often flood sensitive crops like *Aus* and jute are harvested before peak floods hit the *Charlands*. During post-flooding period, the farmers go for seedlings. Sometimes, they prepare floating seedbeds by horizontally placing banana trunks on the water with water-hyacinth and mud on them, locally known as "Baira" cultivation or floating gardens. As the hyacinth starts rotting, seedlings are raised on these floating gardens. Boats are a common mode of transport in the *charlands*. The poor man's boat is the raft made of banana trunks.

Population displacement due to riverbank erosion is widespread in the coastal areas of the country. A family often shifts home 8–10 times during its lifetime as riverbank erodes. IK of the people in the *charlands* and riverbanks is an important survival means for them during erosion and floods. They shift their homes only when forced to. They consider river erosion is a natural process and it should be remedied by natural measures (Hasan et al. 2000). Building embankments and repairing and maintaining them every year is a burdensome and expensive task. They remain neglected because of the inaccessibility and instability of the area far away from the growth centers.

Drought

With just 1% tree cover in the *Barind* area of Rajshahi division (in western side of Bangladesh) there are reports of creeping desertification. Despite years of irrigation using groundwater by the *Barind* Multipurpose Development Authority, the soil remains dry, dark, dusty, and without any moisture. Groundwater level is falling down due to excessive extraction of water. In order to address such a situation, people of the area have adopted many coping strategies. They dig a pond at the corner of their land and fill it with water round the year. This water reservoir changes humidity of the area and is used for irrigation and fisheries. Banks of the little pond are also afforested. There are many *Kharis* (canal) in the vicinity. Farmers fill them up with rainwater and create a reservoir by erecting cross dams. This reservoir helps irrigation during dry season. Because of dryness and aridity of the area, people plant less water-dependent trees like, *Babla* (*Vachellia nilotica*), *Shishu* (*Dalbergia sissoo*), *Pipul* (*Ficus religiosa*), Tamarind (*Tamarindus indica*) etc. (Haque 2013b).

Earthquake

Earthquakes result from seismic waves generated by the sudden release of energy from within the earth due to tectonic movement or volcanic activity. Bangladesh is mainly an alluvial plain with some strips of hills on the northeast and southeastern region. The country is surrounded by high seismic regions. It is not really an earthquake-prone country, but there are some large earthquakes occurred in the past in and around the country. Major earthquakes took place in 1869 (Kachhar), 1885 (Bengal), 1897 (Gr Indian), 1918 (Sri Mongal) and 1930 (Dhubri). Intensity of an earthquake is measured in "Richter Scale". There are 10 units. We also recently experienced "Tsunami" the earthquake in the oceans. Although, there is no scientific method of forewarning people of an earthquake, it is known that the birds and insects become restless and move haphazardly before an earthquake. People of an earthquake zone can read their movements and take the preparation of an impending disaster.

Sendai Framework for DRR

The international community recognized contribution of IK and practices in disaster risk reduction, climate change, and natural resource management. "Sendai Framework for Disaster Risk Reduction (2015–2030)" adopted at Sendai, Miyagi, Japan in March 2015 emphasizes on the IK and practices and coping strategy of the local community in facing challenges of a natural disaster. Highlighting on the role of the stakeholders, it said that the indigenous peoples, through their experience and traditional knowledge, provide an important contribution to the development and implementation of plans and mechanisms, including early warning. It further said, policies and practices for disaster risk management should be based on an understanding of disaster risk at national and local level. The Framework stated that traditional, indigenous, and local knowledge and practices would complement scientific knowledge in disaster risk assessment and the development and implementation of policies, strategies, plans, and programs. It suggested that all plans and policies, which are basically cross-sectoral in nature, should be tailored to localities and to the context.

Paris UN Climate Conference 2015

Twenty-first Conference of the Parties to UN Framework Convention on Climate Change (UNFCCC) held in Paris, France, in December 2015 adopted the "Paris Agreement", which put emphasis on the involvement of local community in disaster risk reduction. The Agreement urged upon the parties to acknowledge and adapt action, which should follow a country-driven, gender-responsive, participatory and fully transparent approach, taking into consideration of vulnerable groups, communities and ecosystems, and should be based on and guided by the best available science and, as appropriate, traditional knowledge, knowledge of indigenous peoples, and local knowledge systems, with a view to integrating adaptation into relevant socioeconomic and environmental policies and actions, where appropriate.

Conclusion

IK is now recognized as an under-utilized resource in the rural Bangladesh. All measures are to be taken to document them. Policymakers and disaster management experts are stressing on the need to create a storehouse of knowledge available at various places on the management of disasters. Local community through generations has developed many IK and practices as they have been facing an onslaught of natural disasters. Moreover, there are ethnic communities in many parts of the world, who have developed a unique system to address natural disaster. There are many NGOs engaged in field research and studies on disaster and adaptation of the local community. Community-based adaptation strategy developed by these people over many generations, remain scattered, and uncared for. Such knowledge runs the risk of disappearance in the event of a major disaster causing wiping out of the whole community.

Time and again, the people of the disaster-prone country like Bangladesh have proved that instead of structural methods, local IK, and practices have solved many problems related to the natural disaster, natural resource management, waterlogging, and river bank erosion. The people of the waterlogged area in the south didn't accept structural solution for Beel Dakatia waterlogging problem, rather they suggested a continuation of the natural tidal flow in the form of tidal river management and in fact it did work. Unfortunately, due to the arrogance of the technical experts and their reluctance to consult the local community, many development projects in Bangladesh are turning into "development disaster". Functional knowledge of the local people gained over the generations has proved to be very effective when technical solutions failed. Such knowledge is transmitted via oral traditions and need to be documented as it is fast disappearing with the decline of elderly people. For effective management of natural disaster in a deltaic land like Bangladesh, we are to bank on this traditional knowledge and practices and document them for our posterity.

Bibliography

Haque, M. (2000). "Indigenous Knowledge and Practices in Disaster Management in Bangladesh". In N. A. Khan and S. Sen (Eds.), *Of Popular Wisdom: Indigenous Knowledge and Practices in Bangladesh.* Dhaka: Bangladesh Resource Centre for Indigenous Knowledge.

Haque, M., et al. (2000). "Impact on Transport Sector". In A. Nishat, M. Reazuddin, R. Amin, and A. R. Khan (Eds.), *The 1998 Flood: Impact on Environment of Dhaka City.* Dhaka: Department of Environment and IUCN.

Haque, M. (2013a). *Environmental Governance: Emerging Challenges for Bangladesh.* Dhaka: AH Development Publishing House.

Haque, M. (2013b). *Indigenous Knowledge and Practices in Disaster Managements, ELCOP Yearbook of Human Rights.* Dhaka: ELCOP.

Hasan, M., et al. (2000). "Indigenous Knowledge and Perception of the Charland People in Coping with Natural Disasters in Bangladesh". *Grassroots Voice* III(I and II): 34–44.

Hassan, S. (2000). "Indigenous Perceptions, Predictions and Survival Strategies Concerning Cyclones in Bangladesh". In N. A. Khan and S. Sen (Eds.), *Of Popular Wisdom: Indigenous Knowledge and Practices in Bangladesh.* Dhaka: Bangladesh Resource Centre for Indigenous Knowledge.

Islam, Nazrul, et al. (2000). "Coping with the Flood". In A. Nishat, M. Reazuddin, R. Amin, and A. R. Khan (Eds.), *The 1998 Flood: Impact on Environment of Dhaka City.* Dhaka: Department of Environment and IUCN.

Khan, N. A., et al. (2000). "A Premier on the Documentation of Indigenous Knowledge in Bangladesh: The BARCIK's Experience". In N. A. Khan and S. Sen (Eds.), *Of Popular Wisdom: Indigenous Knowledge and Practices in Bangladesh.* Dhaka: Bangladesh Resource Centre for Indigenous Knowledge.

Rasheed, K. B. S. (2008). *Bangladesh, Resources and Environmental Profile.* Dhaka: AH Development Publishing House.

Sillitoe, P. (Ed.). (2000). *Indigenous Knowledge Development in Bangladesh: Present and Future.* Dhaka: The University Press Limited.

Sillitoe, P., Dixon, P., and Barr, J. (1998). "IK Research on Floodplains of Bangladesh: The Search for a Methodology". *Grassroots Voice* 1(1): 5–15.

Walker, D. H., Sinclair, F. L., and Muetzelfeldt, R. (1991). *Formal Representation and Use of Indigenous Ecological Knowledge About Agroforestry Practices: A Pilot Phase Report.* Bangor: School of Agricultural and Forests Sciences, University of Wales.

Warner, K. (1991). *Shifting Cultivators: Local Technical Knowledge and Natural Resource Management in the Humid Topics.* Rome: Food and Agriculture Organization of the United Nations.

PART II

Disaster: Identification and Use of Technology for Risk-Reduction

Mapping Vulnerable Earthquake Disaster Class-I Cities for Disaster Risk-Reduction and Community Resilience in India

Bupinder Zutshi, Homolata Borah and Priya Bhakat

Introduction

Sendai Framework of Action has rightly endorsed Disaster risk-reduction management (DRM) through community resilience measures as a key to reduce the disaster-related affects. DRM is, therefore, the key to minimize the human fatalities/causalities, loss of building and other construction and infrastructure structures, agricultural crops, loss of animal and livestock. Mainstreaming DRM within the policies and programs of different sectors ensures, that the effects of disasters are minimized. At the same time, it enables governments to ensure that these policies and programs do not put people at risk. According to the UNISDR-2009, DRM is

B. Zutshi (✉) · H. Borah · P. Bhakat
Centre for the Study of Regional Development, JNU, Delhi, India

B. Zutshi et al. (eds.), *Disaster Risk Reduction*,
https://doi.org/10.1007/978-981-10-8845-2_5

BOX 1: UNISDR-2009, DRM
The systematic process of using administrative directives, organizations and operational skills and capacities to implement strategies, policies and improved coping capacities in order to lessen the adverse impacts of hazards and the possibility of disaster.

The Sendai framework of action 2015–2030 (SFDRR) is now accepted as a global policy framework to represent the focused and coherent directions, undertaking the complexities of the present world vulnerabilities to disasters. The plan entails mostly to reduce disaster risk and losses incurred. This was considered a breakthrough on a global scale after the Hygo framework of Action encompassing securing livelihoods, and reducing losses of the vulnerable communities. The Sendai framework involves adoption of the model of integration of the role of institution and investment to lead to the decrease in loss caused during an event of disaster and to increase preparedness and build a resilient community.

While it is clear that disaster-related shocks and stresses undermine economic growth and development, there are many actions that governments and other agencies can take to reduce the risks to lives, livelihoods, and economies (Mitchell et al. 2013). Poor people suffer the most from disasters, as they lack the capacity and resources to effectively cope with disasters. Risk management in policies and programs, to reduce disaster risk reduction is vital for helping to ensure that the most vulnerable people can access the benefits of development (Mitchell and Tanner 2006). Some people see mainstreaming as a way to realize certain human rights, including the right to safety (Kent 2001). The impetus for mainstreaming risk in development can also be linked to a government's fiduciary responsibility; Jackson makes this point when he notes that mainstreaming risk is a government's "duty to their citizens to maximise the utility of the public resources disposable to them, similar to a private company's fiduciary duty to maximise value to shareholders" (Jackson 2011).

Government of India launched a comprehensive National Disaster Management Plan (NDMP), prepared by the National Disaster Management Authority of India (NDMA) in June 2016. The document was released by Prime Minister of India and it endorses disaster risk reduction, a major necessity toward disaster management by identifying vulnerable spots and vulnerable population groups. Before the onset of the NDMP, the DM Act 2005, was believed to be a milestone, owing to the Paradigm shift in the management of Disasters. The shift was introduced in response to the disasters, which initially remained relief centric and later the trajectory led to the implementation and focus toward preparatory measures of mitigation, preparedness to cover under its umbrella protection of the livelihoods, losses, and development of the community. The new NDMP focuses to reduce disaster risk, disaster damage and losses, and to create a disaster-resilient society focusing on the well-being of the people.

Sendai Framework states four priority areas for immediate implementation by the countries. The four areas are,

- Understanding disaster risk
- Strengthening disaster risk governance to effectively manage disaster risks
- Investing in disaster risk-reduction resilience measures
- Enhancing disaster preparedness for effective response and to build capacities for immediate and effective recovery, rehabilitation, and reconstruction.

India has been traditionally vulnerable to natural disasters on account of its unique geo-climatic conditions. Floods, droughts, cyclones, earthquakes, and landslides have been a recurrent phenomenon in India. About 60% of the landmass is prone to earthquakes of various intensities; over 40 million hectares is prone to floods; about 8% of the total area is prone to cyclones; and 68% of the area is susceptible to drought.

NDMA with support from academic and other research institutes have identified disaster hazard zones of India (refer Map 1—Natural Hazard Zones of India). The map depicts areas of all major hazards like; earthquakes, floods, and droughts. However, the map does not depict the magnitude of vulnerable population which are prone to such natural hazards and disasters, as all these areas are not inhabited by population. Therefore, understanding the disaster risk vulnerability from the map is not possible, unless magnitude and size of vulnerable population from these hazards and disasters are identified.

Thus, identifying actual magnitude of vulnerable population dwelling in these hazard regions/areas is an important step to understand the disaster risk from these identified hazards. The severity of earthquakes cause huge damages to buildings and other infrastructure. Fatalities/casualties due

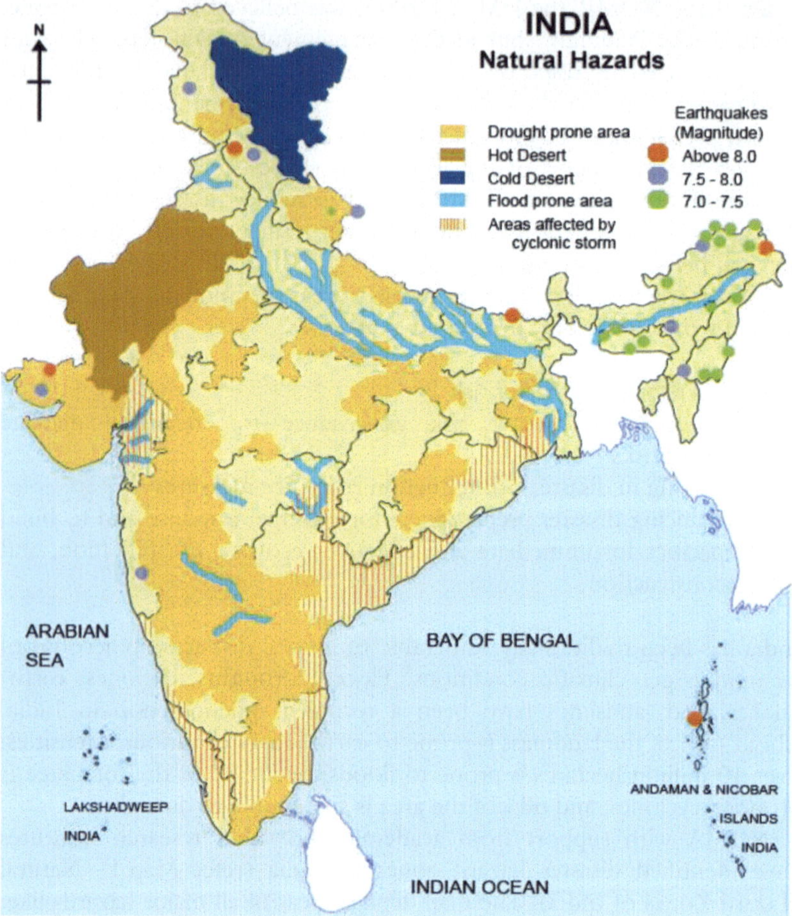

Map 1 Natural hazards in India (*Source* Poorest Areas Civil Society (PACS) Programme 2001–2008)

to earthquakes are usually high in those areas where buildings are multi-storied and congested. Situation for creating disasters from earthquake is most conducive in urban areas especially in Class-I cities and towns, where both densities of houses and density of population are very high. An attempt has been made in this paper to identify magnitude of vulnerable population and their locations prone to earthquake disaster in India.

Earthquakes in India 1991–2015

The northern and north-eastern regions, being mountainous Himalayan regions are more prone to frequent earthquakes due to the plate tectonic movements. These two regions along with Kutch region fall in Zone V (Very High Damage Zone) of the seismic activity. The northern and north-eastern part of the country is not only located at the center of one of the most active earthquake-prone regions of the country, but is also exposed to very high damage given the nature of its terrain. Relief operations in these areas become slow due to the great landslides that are triggered by earthquakes. Much of this region is also covered by thick forests adding on to the woes of rescue units.

Earthquakes hazard and its disasters have become more recurrent during last two decades (refer Table 1).

The spatial and temporal distribution of earthquakes occurred in India, reveal the highest magnitude of earthquake has occurred in the year 2009 with a magnitude of 7.7 followed by 2001 with a magnitude of 7.7–7.6 on the Richter scale. Such magnitude has caused serious damage to larger areas. Most earthquakes recorded are destructive causing threat to life and property. Major fatalities/causalities from these earthquakes have been recorded due to building collapse especially in congested areas with high densities of houses and population.

Earthquake Disaster Vulnerability: Class-I Cities in India

It is very important to understand "what emergency response" should be done in such cases where forecasting earthquakes location, time, and magnitude is complex and difficult. Although we have a network of 56 seismograph stations under the jurisdiction and coordination of Indian Meteorological department, yet, earthquake forecasting has been difficult. Research has revealed that only way to mitigate the effects of earthquake is to improve technology or engineering standards to build appropriate resilient structures and strengthen disaster risk-reduction governance measures in the seismic zones.

Earthquakes being the most common natural disaster, people have, out of experience, devised many ways of coping with them. On account of frequent occurrence of earthquakes since times immemorial, people have learnt to live with them. They have generally set up earthquake resilient buildings, which can withstand earthquake severity.

Table 1 Distribution of earthquakes in India (1991–2015)

Date	*Location*	*Comments*	*Magnitude*
October 20, 1991	Uttarkashi, Uttarakhand	Gharwal regions of Uttarakhand and occurred within the main thrust system of the Himalayas	6.8
September 30, 1993	Latur, Maharashtra	Districts of Latur and Osmanabad, including the Ausa block of Latur and Umerga of Osmanabad	6.2
May 22, 1997	Jabalpur, Madhya Pradesh	The epicenter of the earthquake was near Koshamghat village. Geologist said the quake was caused due to the presence of Narmada Fault.	6.0
March 29, 1999	Chamoli district, Uttarakhand	The earthquake was the strongest to hit the foothills of the Himalayas in more than ninety years	6.8
January 26, 2001	Gujarat	Indian Republic Day Gujarat earthquake, thousands killed	7.6–7.7
December 26, 2004	Off west coast northern Sumatra India, SriLanka, Maldives	Third deadliest earthquake in the history of the world, the tsunami generated killed 15,000 people in India	9.1
October 8, 2005	Kashmir	95 km (59 mi) NE of Islamabad, Pakistan, 125 km (78 mi) WNW of Srinagar, Kangra, Jammu and Kashmir, India (pop 894,000)	7.6
August 10, 2009	Andaman Islands	Tsunami Warning issued	7.7
September 18, 2011	Gangtok, Sikkim	Strong earthquake in NE India, tremors felt in Delhi, Kolkata, Lucknow and Jaipur	6.9
March 5, 2012	New Delhi	Moderate earthquake in national capital, CBSE Physics board exam disrupted in Delhi	5.2
April 25, 2012	Andaman and Nicobar Islands	Big earthquake in Andaman and Nicobar Islands	6.2
March 21, 2014	Andaman and Nicobar Islands	Moderate earthquake in Andaman Islands	6.7

(continued)

Table 1 (continued)

Date	*Location*	*Comments*	*Magnitude*
April 25, 2015	Northern India, North East India	Epicenter 34 km ESE of Lamjung, Nepal. Felt in eastern, northern, north-eastern India and parts of Gujarat[4]	7.8
April 25, 2015	Northern India	Aftershock (Epicenter 49 km east of Lamjung, Nepal)	6.6
April 26, 2015	Northern India, North East India	Aftershock (Epicenter 17 km S of Kodari, Nepal)	6.7
May 12, 2015	Northern India, North East India	Epicenter 17 km S of Kodari, Nepal; Felt in Delhi, West Bengal, Bihar, U.P.; 44 killed in India	7.3
June 28, 2015	Dibrugarh, Assam	3 injured in Assam earthquake, tremors felt in West Bengal, Meghalaya, and Bhutan	5.6

Source Earthquake reports, Indian meteorological department

The assumption, that it is more expensive to invest in disaster-resilient development, such as earthquake disaster-resilient shelters has not been corroborated with evidences. In fact, a cost-benefit analysis for the housing sector shows, this is not always the case. The cost of building homes that can withstand earthquakes is far lower than the cost of repairing homes that cannot stand when earthquakes occur. Results have shown that costs of building earthquake resilient structures, houses can be minimized with use of modern technology.

It is therefore pertinent to identify sites of earthquake zones, where people are most vulnerable from earthquake disasters. The study on nature of buildings, their density, and alignment at these sites would be helpful for strengthening governance and other services for disaster risk reduction in case of earthquakes. Efforts are required to create appropriate disaster preparedness responses and to build capacities of stakeholders for immediate and effective recovery, rehabilitation, and reconstruction at these identified vulnerable sites. The identification of size and density of population with density of houses at these vulnerable sites would also help in assessing the requirements of emergency operation services.

In view of the above, an attempt has been made in this paper to identify Class-I towns (Towns with a population of 100,000 and above) located in the earthquake Zone V, IV, and III, which are most vulnerable zones in terms of severity of earthquake impact.

The increased earthquake vulnerability of the identified cities is due to their locations in the earthquake severity zones and other related components like vertical height of buildings, density of houses, density of population, and issues related to unplanned urbanization like congestion and haphazard growth of buildings and other structures without following appropriate building codes, and use of substandard construction material, haphazard distribution of electric wires, and poor municipal governance.

It is also very crucial to understand the argument of why majority of Class-I cities especially their built environment is vulnerable to an elevated risk of earthquakes. The buildings in these cities are of older non-engineered masonry that do not have resistance to withstand earthquake shocks, and the new structures are not updated with the modern codes of building, abiding lifeline standards, building bye laws, and land use ordinances. Buildings in these cities are in close proximity, highly congested, and constructed with faulty structures. These cities have haphazard electricity wirings that can cause huge damages. In these cities, city planning procedures have not been adopted especially appropriate availability of services like drainage and medical infrastructure care. In case of earthquake eventuality these cities are prone to suffer heavy fatalities. Electricity and water supply lines pose threats due to liquefaction and lateral spreads. Huge spreads of settlement in large numbers are prone to be affected in the seismic zones.

Earthquakes affect India with multifarious issues owing to the unprecedented growth of population in the Class-I cities due to large-scale in-migration of population without matching housing and other infrastructure and basic urban services. In case of earthquakes, these cities are vulnerable to cause injuries, deaths, homelessness, and also the conditions prevailing after the earthquake disaster can give rise to incidences of crime like human trafficking, mental health issues, and nutritional related concerns.

In case of earthquake eventuality, these identified cities are vulnerable to large-scale fatalities/causalities and economic loses in terms of job opportunities to the surviving populations. Loss of infrastructure and basic services like healthcare, education in these cities also increases susceptibility to disruptions in the normal course of life sustenance after the disasters. Since majority of these Class-I cities have huge slum populations, the poor people are more susceptible to diseases, human trafficking, and other economic loses.

Identification of Class-I Cities in Seismic High Severity Zones

Seismic zonation map portrays the susceptibility of the regions coinciding with the high-risk zones. India has been divided into five earthquake-prone zones (refer Map 2). All states of Northeast regions (except Sikkim) and parts of Uttrakhand, Himachal Pradesh, Jammu and Kashmir, Gujarat, and West Bengal fall in very High Damage Zone (Zone V). The main cause of Earthquake in this Zone is movement of plates. A total of six Class-I cities with population of 100,000 and above are in Earthquake Zone V (the most vulnerable to earthquakes and susceptible to losses). These six cities are Jorhat, Guwahati, Imphal, Darbhanga, Srinagar, and Bhuj with a total population of 2.93 million and density of population of 4600 persons per square kilometer. These six cities also recorded 911 density of households per square kilometer (refer Map 3 and Table 2 and Annexure I and II).

High Damage Zone (Zone IV) constitutes parts of Jammu and Kashmir, Himachal Pradesh, Uttrakhand, Uttar Pradesh Punjab, Haryana, Delhi, Gujarat, Bihar, west Bengal, and Maharashtra. This Zone has 16 Class-I cities as per Census 2011, with a total population of 22.24 million having a very high density of population of 13,640 persons per square kilometer. These 16 cities also have very high density of houses (2767 houses per square kilometer). Delhi urban agglomeration with a population of 13.40 million and density of more than 20,000 persons per kilometer also falls in this Zone. Other major cities located in this Zone are Shimla, Amritsar, Chandigarh, Darjeeling, Dehradun, Patna, Gorakhpur, Ludhiana, Roorkee, Moradabad, Gantok, Pilibhit, Bulandshar, and Bairiech (refer Map 4 and Table 1 and Annexure I and II).

Moderate Damage Zone (Zone III) constitutes parts of Punjab, Haryana, Uttar Pradesh, Rajasthan, Gujarat, Madhya Pradesh, Maharashtra, Karnataka, Kerala, Tamil Nadu, Andhra Pradesh, Orissa, Jharkhand, Chhattisgarh, and Bihar. A total of 40 Class-I cities are located in this Zone. These 40 cities, recorded a population of 59.16 million in the Census 2011. These cities have 9575 density of population

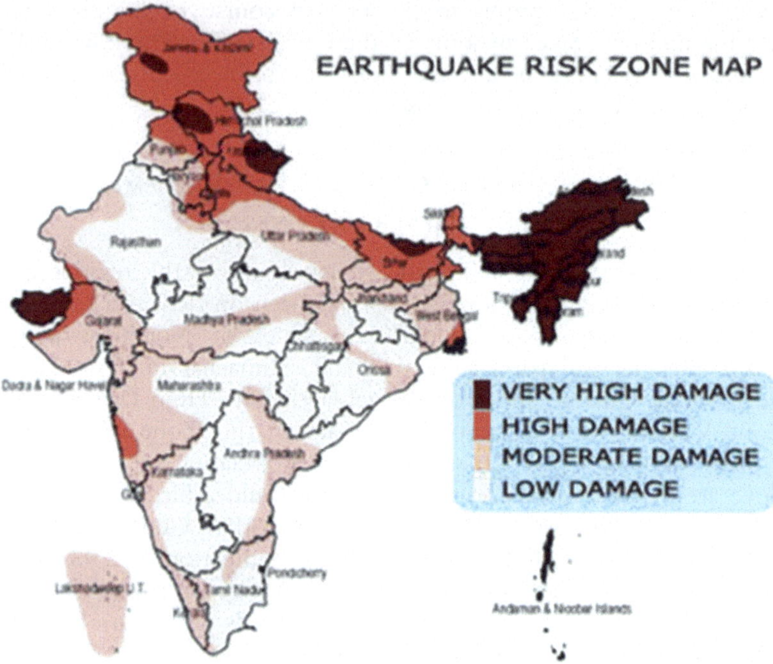

Map 2 Earthquake risk zones of India (*Source* Disaster Resistant Architecture www.slideshare.net)

and 2130 density of households. Some of the major cities falling in this Zone are Kolkata, Mumbai, Chennai, Ahmedabad, Coimbatore, Bhubaneswar, Agra, Pune, Nashik, Varanasi, Vijayawada, Vadodara, Thiruvanthapuram, Salem, Patiala, Lucknow, and others (refer Map 5 and Table 1 and Annexure I and II).

A total of 62 Class-I cities fall in the earthquake zone of V, IV, and III, which are highly vulnerable to earthquake disasters. These 61 cities have 84.43 million populations and 21.61 million houses. Thus a significant number of population is highly vulnerable to earthquakes. Both government and civil society organizations must prepare appropriate disaster management policies capacities of communities and individuals in these identified cities to reduce the disaster risks. This can be achieved by making buildings and other infrastructure, earthquake resilient and also create resilient societies to face the eventualities of future earthquakes.

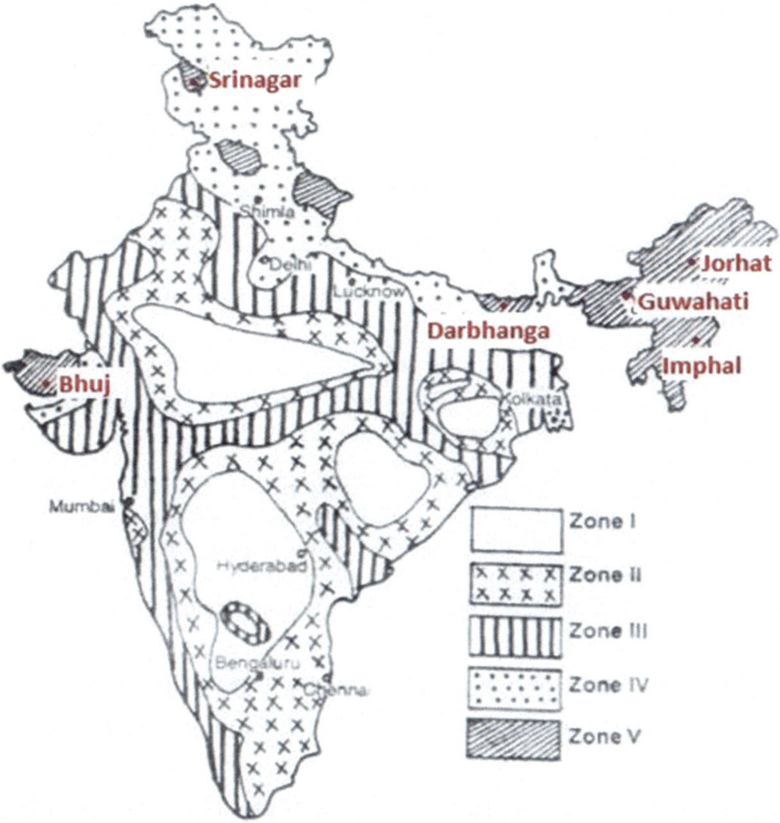

Map 3 Class-I cities falling in very high damage earthquake zone (*Source* Research Work on Census of India, General Population Tables 2011 and NDMA Map)

The argument largely revolves around how can we implement good practices; strengthen early warning system, generate community resilience practices, and save huge number of population in these cities from multiple losses of lives and assets. Disaster management in these high risk zones requires enabling urban planning in retrofitting and reconstruction, encapsulating efficient implementation of preventive measures to reduce vulnerabilities and susceptibility, reducing exposure to risk and increasing adaptive capacity toward good governance. Appropriate urban planning

Table 2 Earthquake zone wise total number of class-I towns in India

Earthquake zones	*Class-I towns*	*Total household*	*Density of household*	*Total population*	*Density of population*
Very high damage V	6	580,803	911	2,933,821	4600
High damage IV	16	4,512,154	2767	22,240,323	13,640
Moderate damage III	40	13,162,410	2130	59,169,319	9575
All (V, IV, and III zones)	**62**	**18,255,367**	**2161**	**84,343,463**	**9984**

Source Research work on Census of India, General Population Tables 2011 and NDMA Map

and its crucial role are embedded in preparing cities to face earthquakes, reduce disaster risks, and prepare community for disaster resilience measures. Urban disaster planning calls for the involvement of framing of expeditious, effective planning within which the affected populace can reconstruct, re-establish livelihood, and formulate sustainable solutions.

Policies and Plans for Earthquake Disaster Preparedness and Community Resilience in Vulnerable Class-I Cities in India

The initiatives of the nodal organization, Ministry of Home Affairs should aim to achieve sustainable earthquake risk reduction for all the Class-I cities falling within the earthquake very high, high, and moderate zones. The aim should be to decrease the casualties and fatalities due to the collapse of buildings and other structures in the congested parts of these identified cities. The earthquake reduction program calls for new constructions to be adhered to the Bureau of Indian Standards (BIS) codes and techno legal framework to be adopted for city planning with adequate services and infrastructure. For the purpose of implementation and enforcement of building bye laws regulatory bodies of Municipal Corporation, development Authorities have been established with responsibilities, but have not been able to percolate the implementation due to limitations of capacity building, lack of knowledge. Disaster mapping of these identified cities should be prepared and accordingly

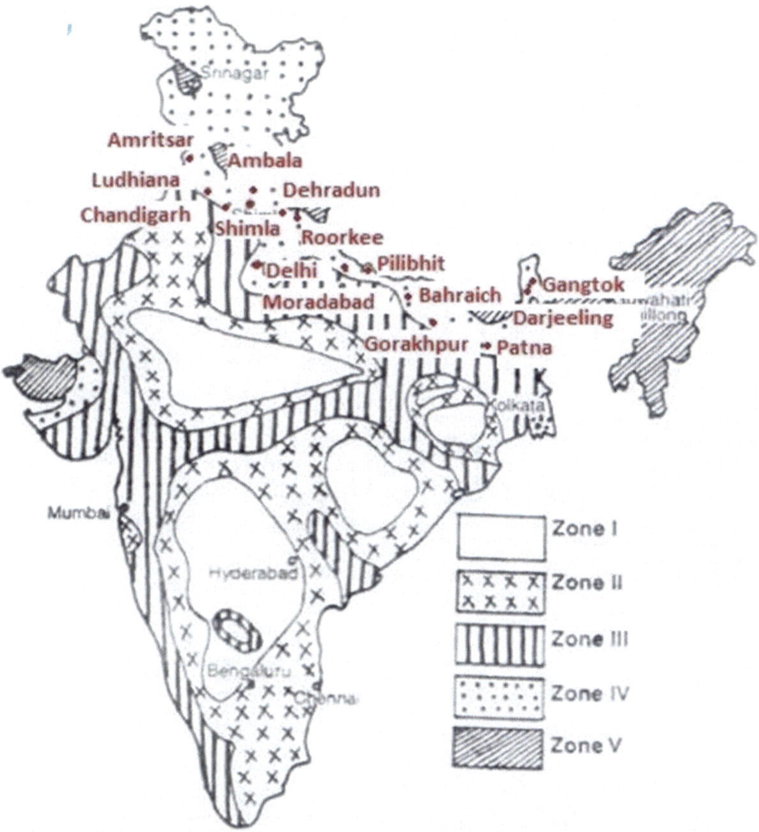

Map 4 Class-I cities falling in high damage earthquake zone (*Source* Research Work on Census of India, General Population Tables 2011 and NDMA Map)

capacities of stakeholders should be built to face the eventuality of earthquake shocks. Some of the measures required are

- To review and ensure that the government constructions (as a beginning followed by other constructions) have adhered to the building bye laws as laid by the BIS.
- To mandate for the Municipal corporations, the buildings to be prepared by an architect and engineers with due approval of buildings as per the BIS norms.

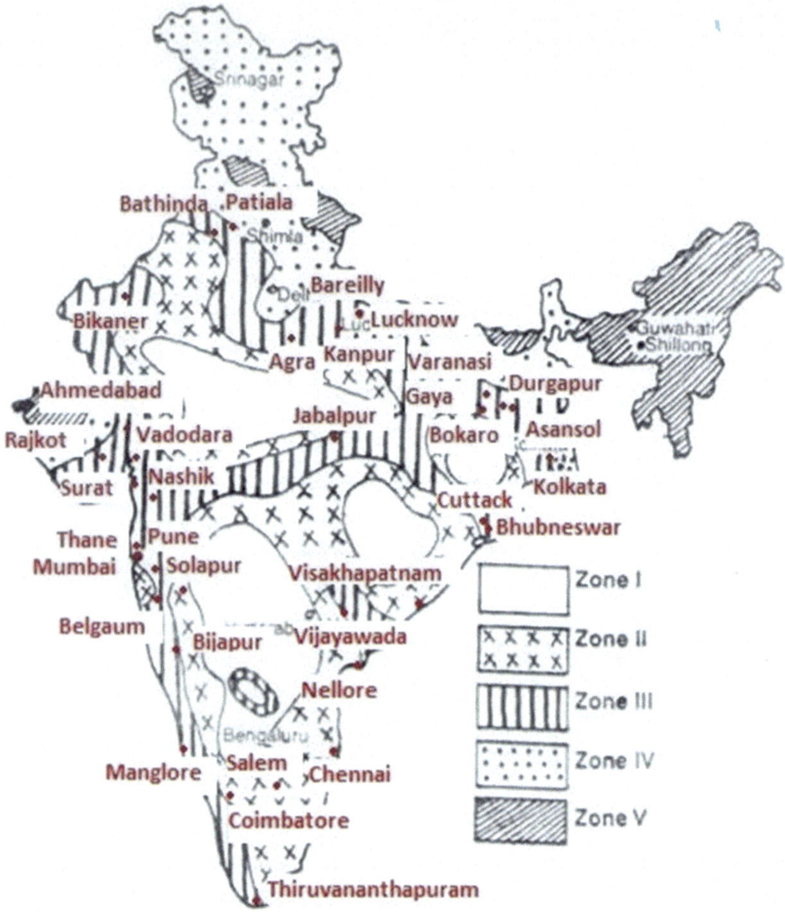

Map 5 Class-I cities falling in moderate damage earthquake zone (*Source* Research Work on Census of India, General Population Tables 2011 and NDMA Map)

- The complete protection and safety of lifeline buildings like schools, water suppliers, fire station, hospitals, etc. to be prepared in accordance to the BIS norms.
- Mandates for private builders to cultivate the culture of adopting and implementation of the bye laws.

- To generate awareness by various awareness campaigns to popularize the practices of Dos and Don'ts during earthquakes.
- Develop risk-reduction capacities of government, non-governmental staff and communities, and individuals during earthquakes as well as after the disaster during recovery and rehabilitation stage.
- The situation demands increasing accountability of government and public offices.

Annexure-I

Earthquake zone-wise distribution of class-I towns in India

Earthquake zones	*Name of the class-I towns*	*Household*	*Density of household*	*Population*	*Density of population*
V	Bhuj (M+OG)	33,402	839.46	148,834	3740
V	Darbhanga (M Corp.)	56,492	2945.36	296,039	15,435
V	Guwahati (M Corp.+OG)	230,769	1053.45	962,334	4393
V	Imphal (MCI+OG) (major part)	42,820	1548.64	193,459	6997
V	Jorhat (MB+OG)	30,119	562.97	126,736	2369
V	Srinagar (M Corp.+OG)	187,201	671.93	1,206,419	4237
IV	Ambala (M Cl)	42,027	1231.0	195,153	5716
IV	Amritsar (M Corp.+OG) (Part)	239,078	1757.9	1,183,549	8326
IV	Bahraich (NPP)	30,460	2290.2	186,223	14,001
IV	Bulandshahr (NPP+OG)	41,505	1121.8	230,024	6216
IV	Chandigarh (M Corp.+OG)	220,520	2086.7	970,602	9184
IV	Darjiling (M)	21,782	2931.6	118,805	15,990
IV	Dehradun (M Corp.+OG)	125,271	1749.1	574,840	8354
IV	Delhi Agglomeration	2,761,755	4124.6	13,400,582	20,013
IV	Gangtok (M Corp.)	23,773	1238.2	100,286	5223
IV	Gorakhpur (M Corp.)	112,237	795.9	673,446	4775
IV	Ludhiana (M Corp.)	344,333	2160.6	1,618,879	10,158
IV	Moradabad (M Corp.)	161,329	2151.1	887,871	11,838

(continued)

Earthquake zones	*Name of the class-I towns*	*Household*	*Density of household*	*Population*	*Density of population*
IV	Patna (M Corp. + OG)	294,631	2735.9	1,684,297	15,640
IV	Pilibhit (NPP)	24,341	2441.4	127,988	12,837
IV	Roorkee (MB)	22,806	2812.1	118,200	14,575
IV	Shimla (M Corp.)	46,306	1310.3	169,578	4798
III	Agra (M Corp.)	267,945	2222.3	1,585,704	13,151
III	Ahmadabad (M Corp.)	1,179,823	2516.0	5,577,940	11,895
III	Asansol (M Corp.)	113,739	908.2	563,917	4503
III	Bareilly (M Corp. + OG)	166,447	1563.9	904,797	8501
III	Bathinda (M Corp.)	60,301	880.3	285,788	4172
III	Belgaum (M Corp. + OG)	111,874	1123.1	490,045	4920
III	Bhubaneswar (M + OG)	260,000	1925.9	900,000	6667
III	Bijapur (CMC)	65,543	663.9	327,427	3316
III	Bikaner (M Corp.)	115,380	744.2	644,406	4157
III	Bokaro steel city (CT)	82,473	506.2	414,820	2546
III	Chennai (M Corp.)	1,154,982	6599.9	4,646,732	26,553
III	Coimbatore (M Corp.)	282,839	2678.4	1,050,721	9950
III	Cuddalore (M)	42,174	1523.1	173,636	6271
III	Cuttack (M + OG)	125,000	837.8	600,000	4021
III	Durgapur (M Corp.)	130,944	849.2	566,517	3674
III	Gaya (M Corp. + OG)	72,978	1454.6	474,093	9450
III	Jabalpur (M Corp. + OG)	225,340	1477.3	1,081,677	7092
III	Kanchipuram (M)	41,807	3567.2	164,384	14,026
III	Kanpur (CB)	20,553	1209.0	108,534	6384
III	Kolkata (M Corp.)	1,024,928	5540.2	4,496,694	24,306
III	Lucknow (M Corp.)	538,149	1542.9	2,817,105	8076
III	Mangalore (M Corp. + OG)	115,036	757.5	499,487	3289
III	Mumbai	2,105,604	3491.8	9,356,962	15,517
III	Nashik (M Corp.)	336,333	1297.9	1,486,053	5735
III	Navi Mumbai (M Corp.)	273,626	2517.7	1,120,547	10,311
III	Nellore (M Corp. + OG)	130,192	1335.0	547,621	5615
III	Osmanabad (M CL)	21,866	232.4	111,825	1189
III	Patiala (M Corp. + OG)	93,805	1340.1	446,246	5803

(continued)

Earthquake zones	*Name of the class-I towns*	*Household*	*Density of household*	*Population*	*Density of population*
III	Pune (M Corp.)	742,602	2686.7	3,124,458	11,304
III	Rajkot (M Corp. + OG)	286,838	1821.8	1,323,363	8405
III	Salem (M Corp.)	215,747	2362.0	829,267	9079
III	Solapur (M Corp.)	188,503	1055.6	951,558	5329
III	Surat (M Corp.)	975,797	2905.7	4,467,797	13,304
III	Thane (M Corp.)	435,341	3395.0	1,841,488	14,361
III	Thiruvananthapuram	191,446	1272.7	762,535	5069
III	Tiruvannamalai (M)	33,514	2457.0	145,278	10,651
III	Vadodara (M Corp. + OG)	393,887	1457.7	1,752,371	6485
III	Varanasi (M Corp.)	190,835	2324.4	1,198,491	14,597
III	Vellore (M. Corp)	42,598	3656.5	185,803	15,949
III	Vijayawada (M. Corp.)	305,621	1688.1	1,143,232	6315

Annexure-II

State-wise total number of cities prone to earthquake

State	*Total no. of class-I towns*				*Total*
	Very high risk zone	*High risk zone*	*Moderate risk zone*	*Low risk zone*	
A & N Islands	–	–	–	–	–
Andhra Pradesh	–	–	2	1	3
Arunachal Pradesh	–	–	–	–	–
Assam	2	–	–	–	2
Bihar	1	1	1	1	4
Chandigarh	–	1	–	–	1
Chhattisgarh	–		–	2	2
D & N Haveli	–	–	–	–	–
Daman and Diu	–	–	–	–	–
Delhi	–	1	–	–	1
Goa	–	–	–	–	–
Gujarat	1	–	4	–	5
Haryana	–	1	–	–	1
Himachal Pradesh	–	1	–	–	–
Jammu and Kashmir	1	–	–	–	1

(continued)

State	*Total no. of class-I towns*				*Total*
	Very high risk zone	*High risk zone*	*Moderate risk zone*	*Low risk zone*	
Jharkhand	–	–	1	2	3
Karnataka	–	–	3	3	6
Kerala	–	–	1	–	1
Lakshadweep	–	–	–	–	–
Madhya Pradesh	–	–	1	1	2
Maharashtra	–	–	7	2	9
Manipur	1	–	–	–	1
Meghalaya	–	–	–	–	–
Mizoram	–	–	–	–	–
Nagaland	–	–	–	–	–
Odisha	–	–	2	1	3
Puducherry	–	–		1	1
Punjab	–	2	2		4
Rajasthan	–	–	1	5	6
Sikkim	–	1	–		1
Tamil Nadu	–	–	7	3	10
Telangana	–	–	–	–	–
Tripura	–	–	–	–	–
Uttar Pradesh	–	5	5	2	12
Uttarakhand	–	2	–	–	2
West Bengal	–	1	3	–	4

Bibliography

Bahadur, A. (2008). Accountability for DRR. IDS Research Report for Christian Aid. Brighton: Institute for Development Studies. [1].

District Census Handbook. (2011). Census of India. Ministry of Home Affairs.

Earthquake Vulnerability Reduction for Cities (EVRC-2); Module 3 Session 1(a).

Jackson, D. (2011). *Effective Financial Mechanisms at the National and Local Level for Disaster Risk Reduction*. Geneva: United Nations Office for Disaster Risk Reduction. http://www.unisdr.org/files/18197_202jackson.financial-mechanismstosup.pdf.

Jain, S. K. (1998). "Indian Earthquakes: An Overview". *The Indian Concrete Journal* 72(11): 555–561.

Kent, G. (2001). "The Human Right to Disaster Mitigation and Relief", *Environmental Hazards* 3(3): 137–138.

Khan, D. (2013). *Opinion: Integrating Climate-smart DRM in Key Sectors—What Does it Take?* London: Climate and Development Knowledge Network. http://cdkn.org/2013/06/opinion-integrating-climate-smart-drm-in-key-sectorswhat-does-it-take.

Mitchell, T., and Tanner, T. (2006). *Overcoming the Barriers: Mainstreaming Climate Change Adaptation in Developing Countries*. London: Tearfund.

Mitchell, T., Jones, L., Lovell, E., and Comba, E. (2013). *Disaster Risk Management in Post-2015 Development Goals: Potential Targets and Indicators*. London: Overseas Development Institute. http://www.odi.org.uk/sites/odi.org.uk/files/odi-assets/publications-opinion-files/8354.pdf.

National Disaster Management Plan. (2016). *A Publication of the National Disaster Management Authority*. New Delhi: Government of India.

National Guidelines on Disaster Management vis-a-vis Earthquakes, Unstarred Question No. 2793, Ministry of Home Affairs, Government of India 2015.

Sendai Framework for Disaster Risk Reduction 2015–2030, United Nation.

Sergey, T., Lothar, S. G., Rutger, W., and Jochen, Z. (2004). "Vulnerability and Risk Assessment for Earthquake Prone Cities". 13th World Conference on Earthquake Engineering, Canada, Paper No. 868.

Sinha, D. K. (2015). "Essay on Earthquakes in India". Unpublished Article on Your Article Library. http://www.yourarticlelibrary.com.

Mainstreaming Nuclear Disaster Risk Reduction in India

Rajesh Kumar

Introduction

In last one and a half decade, mainstreaming of disaster risk reduction has received Government of India's urgent attention as necessary policy intervention measure for achieving goals of sustainable development in the country. With the objective of departing from relief centric approach to risk mitigation approach, Government of India enacted Disaster Management Act in 2005. As a result of new legislation, greater emphasis of the Union, as well as State Governments, is on investion in disaster risk reduction related infrastructures which would help them in reducing economic vulnerability and promoting resilience culture in India. Since 2008, India has embarked upon a huge nuclear power generation program in the direction of achieving total energy security goals. Given the energy security requirements consistent with the demands of a growing economy, it has taken to nuclear route for power generation, apart from generation through coal, gas, hydro, wind, or other renewable sources of energy. India is of the view that nuclear energy is a cheap source of clean energy and its costs can be reduced to the level of energy through coal

R. Kumar (✉)
School of Social Sciences, Guru Nanak Dev University, Amritsar, Punjab, India

B. Zutshi et al. (eds.), *Disaster Risk Reduction*,
https://doi.org/10.1007/978-981-10-8845-2_6

and it does not affect the Climate Change thereby it is considered as a source of clean energy.

Post Indo-US Nuclear Agreement of 2008, India has set a target of generating 60,000 MWs of electricity through nuclear route by 2032. Till date, it has seven nuclear power plants with 22 nuclear reactors fully functional and another 60 new nuclear reactors as a result of India's finalization of agreements with countries like USA, Canada, UK, France, Japan, South Korea, and Russia are likely to be set up by the year 2032.[1] Scholars have been of the view, "the issue of setting up and operating of new nuclear power and reprocessing plants stands intertwined between development, environmental concerns, and areas under mega projects being struck with natural calamities' like floods, earthquakes, tsunamis, manmade accidents or breach of security, etc., all might cause nuclear accidents/disasters. Dependence on nuclear power, storage and disposal of nuclear waste, areas surrounding reprocessing plants, transportations of nuclear warheads and their safety is a highly risky proposition in the country. The risk remains very high in India and raises serious questions over the capabilities and responsibilities of Union and State Governments in India which cannot match the state preparedness and disaster planning in Japan and other developed countries of the world."[2] Reports on Fukushima nuclear accident/disaster[3] have pointed out that even advanced countries can't handle nuclear disasters because of their unpredictable consequences which were experienced in case of Fukushima, for example, injection of seawater into nuclear reactor resulted into explosion instead of bringing down the temperature inside the reactor.

The Fukushima 2011 disaster has made significant impact on issue of nuclear power generation in India apart from developed countries like Germany, South Korea, and Japan who have decided to shut down some of their nuclear reactors, thereby, reducing their dependence upon nuclear power in future. One of the fallouts of Fukushima disaster on India has been rise in anti-nuclear protests in the country, thereby, raising

[1] Accessed from www.npcil.gov.in, 2008. For details of nuclear agreements with different countries please visit www.mea.gov.in.

[2] Paper presented by the author in an International Conference organized jointly by NAPSIPAG and Dhaka University, Dhaka held in Dhaka, Bangladesh, 2014.

[3] *The Japanese Nuclear Incident: Technical Aspects* by Jonathan Medalia Specialist in Nuclear Weapons Policy, March 29, 2011, Congressional Research Service, 7-5700, www.crs.gov, R41728.

serious concerns about the safety at nuclear power plants and merits of its excessive dependence upon nuclear power. For dealing with nuclear and radiological emergencies in India, **the NDMG-NRE Guidelines** titled **"Response Actions and Standard Operating Procedures (SOPs)"** are to be followed strictly and this classified document lies with the Minstry of Home Affairs.[4] The manner in which disaster relief operations were carried out in Indian states of Uttarakhand in mid-2013 and Jammu and Kashmir in September 2014, since then, serious doubts have been raised about India's capabilities to handle nuclear disasters of Fukushima scale. The doubts get more serious as politics in India is too heavy even in cases of disaster handling agency like National Disaster Management Authority (NDMA) and its functioning.

The paper has three parts. **First part**, deals with rationales for India's ever-rising dependence on nuclear power which increases the risk of nuclear disaster becoming a reality in a country with 1.3 billion populations. **Second part**, with post-disaster handling deals with post-disaster handling capabilities as a responsible nation falling within the purview of national and international legislations. **The final part** deals with emerging policy intervention issues and important lessons learnt by India in post-Fukushima period for managing nuclear disasters in the country. The paper is based on certain **assumptions**:

(1) Govt. of India's plan of generating 60,000 megawatts of electricity by setting up of 60 new nuclear reactors by the year 2032 is likely to raise the risk of nuclear accidents/disasters in India.
(2) Fukushima international disaster besides national disasters in Jammu and Kashmir and Uttarakhand states have raised serious questions over disaster handling capabilities of Union and State Governments in India thereby exposing a huge gap between our and developed countries' handling capabilities.
(3) Nuclear Disaster Risk Reduction Management in India would need massive preparation, investments, and paradigm shift in disaster policymaking.

The **methodology** used in the paper is mainly content analysis of existing resource material available freely. Primary as well as secondary reports

[4]Accessed from www.ndma.gov.in (NDMG-NRE, 2009, p. xxvii).

of agencies/think tanks like IAEA, NDMA, Japan's Diet, UN, CRS of USA, and companies who sell reactors have also been consulted.

Paradigm Shift in India's Nuclear Energy Production in the Twenty First Century

India in recent years has taken to nuclear path in a big manner for generating electricity apart from generation through coal, gas, hydro, wind, or other renewable sources of energy. For India, nuclear energy is a cheap source of clean energy and its costs can be reduced to the level of energy through coal and it does not affect the Climate Change as it is considered as clean energy. At present India has seven nuclear plants with 22 nuclear reactors (20 operational and 2 under completion) apart from research reactors at BARC, IGCAR, and other production related establishments. They produce around 4700 MWe of electricity only. Many new nuclear power plants are coming up at places like Jaitapur in Maharashtra, Fatehabad in Haryana, Haripur in West Bengal, and in Andhra Pradesh is in line with new policy of generating around 60,000 MWe of electricity through nuclear mode by 2032 after the 123 Nuclear Agreement between India and the USA.[5] As per World Nuclear Association (WNA), "India expects to have 20,000 MWe nuclear capacity on line by 2020 and 63,000 MWe by 2032. It aims to supply 25% of electricity from nuclear power by 2050."[6] As discussed, India's international obligations with regard to reducing the carbon emissions for preventing climate change, switching over to generation of electricity through nuclear mode suits her largely. India recently announced Intentional Nationally Determined Contributions (INDC) after the culmination of Paris Summit on Climate Change.

Disaster Threats Because of Nuclear Power Plants

For handling of nuclear disasters in India, an important document titled as "***National Disaster Management Guidelines: Management of Nuclear and Radiological Emergencies (NDMG-NRE)***" which is a bulky document having ten chapters, having a lengthy preface and 134

[5] Accessed from www.npcil.gov.in, 2008.

[6] Accessed from http://www.world-nuclear.org/info/inf53.html and www.npcil.gov.in.

pages covering all technical and operational aspects of nuclear disaster. The Guidelines observe, "'**nuclear disaster**' as that dimension of emergency situation leading to mass casualties and destruction of large areas and property, unlike a nuclear emergency, the impact of a nuclear disaster is beyond the coping capability of local authorities and such a scenario calls for handling at the national level, with assistance from international agencies, if required."[7]

Scholars are of view, "despite public concerns over least possibility of onsite accidents, waste disposal and uncertainties over economics, fuel switching to nuclear power currently remains the largest, proven, carbon-free generation option. One tonne of uranium produces the equivalent amount of electricity as 16,000 tonnes of coal and 80,000 barrels of oil. The spent fuel from the reactor still contains 235U, so it can be recycled. Reprocessing the spent fuel produces uranium, plutonium and waste. Safe disposal/storage of waste from the nuclear fuel cycle presents a challenge."[8] To some other nuclear experts, "the risk of major disaster is negligible, a nuclear power station is typically a system where dangerous and destructive processes can be set in motion because it is thought that all the safety backup systems in place will guarantee our safety besides views of skeptics, main threat came from the nuclear power stations also."[9] Fukushima nuclear accident has compelled countries and experts over the world to ponder over such issues again. India's records so far have been fair barring few small accidents at certain nuclear power plants, specially, Madras Nuclear Power Plant at Kalpakkam when it was struck with Tsunami and the nuclear reactor could be stopped successfully averting any disasters.

However, several lessons were thrown up in the aftermath of the Fukushima March 2011 disaster. One of the Reports strongly emphasized, "no country, developed or developing, is immune to such disasters. Advanced democracies are not necessarily more resistant or better prepared than developing countries to deal with such events. The March 11, 2011 disaster shook the very foundation of Japanese

[7]Accessed from www.NDMG-NRE, February 2009, p. xxiv. Highlights of **NDMG-NRE.**

[8]Whittington R. et al. (2002). Health Technology Assessment, vol. 17, no. 50, NIHR Journal, U.K., pp. 1653–1668.

[9]Markku Wilenius. (1996). From Science to Politics: The Menace of Global Environmental Change. *Acta Sociologica,* vol. 39, no. 1, Sociology and the Environment (1996), pp. 5–30, Sage. Accessed from Stable URL: http://www.jstor.org/stable/4194803. Accessed: 03/08/2011, 00:03.

society... Though Japan certainly had both the technical and financial capacities to rebuild the towns affected by the disaster... The Japanese experience thus offered many unique lessons for other democracies in terms of dealing with future disasters."[10] In light of above, India also learnt several lessons from Japan which are discussed in later sections.

Disaster Threats Because of the Risks of Accidental Nuclear War and Nuclear Proliferation

It is visualized that nuclear disasters may occur because of the outbreak of nuclear war between countries or because of reasons like; accident, error, or malfunction or system failure; The actions of a "rogue general", Miscalculations, The continuing military buildups, A "bolt-from-the-blue" preemptive first strike, Technologically advanced nuclear weapons, Role of third parties, and Nuclear proliferation.[11] A study by Natural Resources Defence Council (NRDC) says, "a hypothetical nuclear exchange between India and Pakistan, in which each country targeted major cities through dozen, 25-kiloton warheads, 22.1 million people in India and Pakistan would be exposed to lethal radiation doses of 600 rem or more and 8 million people would receive a radiation dose of 100–600 rem, causing severe radiation sickness and potentially death, especially for the very young, old or infirm and as many as 30 million people would be threatened by the nuclear attack. NRDC estimates that 8.1 million people live within this radius of destruction."[12] Similarly, another report of International Commission on Nuclear Non-Proliferation and Disarmament (ICNND) presented before the **United Nations**, said, "A nuclear war between India and Pakistan could cause severe 'climate cooling' and may have a devastating impact on agriculture worldwide."[13]

[10] Reiko Hasegawa. (2013) (IDDRI). Disaster Evacuation from Japan's 2011 Tsunami Disaster and the Fukushima Nuclear Accident accessed from www.iddri.org.com, p. 7.

[11] James Petras and Morris Morley. (January 23, 1988). Nuclear War and US-Third World Relations: The Neglected Dimension. *Economic and Political Weekly*, vol. 23, no. 4, pp. 151–153 and 155–158: Accessed from http://www.jstor.org/stable/4378015 on 03/08/2011, 07:03.

[12] NRDC Report and Louis Ren Beres. (1998). In a Dark Time: The Expected Consequences of An India-Pakistan Nuclear War. *American University of International Review*, vol. 14, no. 1, pp. 497–510.

[13] *The Times of India*, January 26, 2010 and www.icnnd.com.

Concerns regarding the proliferation of nuclear technologies have always dominated the proliferation-related debates at international level for controlling of such proliferations on part of nuclear have not's countries. The present stalemate in case of Iran's nuclear program and its differences with U.S. over nuclear deal has allegedly led to the beginning of nuclear programs in many of Middle-east countries like Saudi Arabia, Qatar, Bahrain, Oman, the UAE, and Jordan.[14] Several studies completed after Fukushima nuclear accident have pointed out that threats of nuclear proliferation have increased despite the fact that serious steps have been taken by international organisations like the UN for declaring the world free from nuclear weapons and a world free from nuclear reactors. The commitments made by former US President Barack Obama for making the World free from nuclear weapons in past years has been worth noting. Still, all such new measures have not lowered down the security/disaster threats arising out of cases of nuclear proliferation.

The New Challenge of Terrorism

World today is facing a very serious threat of use of nuclear weapons by non-state actors like terrorist organizations. The danger of terrorists gaining access to nuclear weapon is heightened during crises. Scholars say, "Though, nuclear weapons cannot be manufactured directly from the key raw material found in nature, uranium. For this reason, *a terrorist organization can acquire a nuclear explosive only (1) by obtaining an intact nuclear weapon from a national stockpile or (2) by obtaining fissile material from stocks that were produced in highly advanced industrial facilities and then making the fissile material into a nuclear explosive.* The most important and effective steps for reducing the threat of nuclear terrorism are therefore to secure, consolidate, reduce, and, where possible, eliminate nuclear weapons and fissile material. Programs to implement such measures are under way in many countries but are far from reaching their goals."[15] Another report points out serious hazards getting posed to nations across the world as, "There are an estimated 7,700 nuclear weapons deployed throughout the world as on today,

[14] Guzansky, Asculai, and Lindenstrauss. Civilian Nuclear Programs in the Middle East. *Strategic Assessment*, volume 15, no. 1, April 2012, pp. 100–101.

[15] Nuclear Posture Review, April 2010.

plus more than 14,000 nuclear weapons which are inactive, in reserve status, or awaiting dismantlement."[16]

Scholars are of view, "The international community urgently needs to expand its efforts to secure existing stockpiles of nuclear weapons and materials, particularly in Russia, Pakistan, and India. The elimination of nuclear weapons should be high on the global public health agenda deaths and billions of dollars in property damage if a cask of spent fuel rods were dispersed anywhere in the world."[17] In order to prohibit proliferation of nuclear technologies at the hands of non-state actors from India, additional measures were taken by Govt. of India like raising of more Battalions of Paramilitary forces, setting up of National Investigation Agency (NIA), National Intelligence Grid like institution or amendment of the Prevention of Unlawful Activities Act and making it more stringent, pointed in this direction that how serious such threats have become for India too. Countries need to pool their energies for preventing non-state actors gaining control of nuclear assets of the country.

Climate Change Catastrophe Can't Be Replaced with Nuclear Disaster Catastrophe

The Report of Women in Europe for Common Future (WECF) says, "it is erroneous to consider nuclear energy as source of clean energy and this group is campaigning for a complete ban on production of nuclear energy in Europe. As per the report of IEA and IPCC, tripling the nuclear power output by 2050 would save five billion tonnes of CO2 compared to a reduction of 25–40 billion tonnes by conventional methods by 2050. Since uranium is also a limited resource and, may last for about 70 more years, then switch to thorium for fuel, which is also finite, or to the Fast Breeder Technology together with fuel reprocessing—a polluting and dangerous production system that generates even more toxic nuclear waste."[18] A sense of prudence is needed on part of all countries, including India. "Given the health hazards of nuclear material, nuclear power has repeatedly been shown to be toxic to human health at every step of production, right from uranium mining, to fuel production,

[16] Hans M. Kristensen. (2010). Status of World Nuclear Forces. *Federation of American Scientists*. website, www.fas.org.

[17] Ibid. and Markku Wilenius no. ix.

[18] Accessed from www.wecf.com.

from power generation to storage of nuclear wastes. Climate change itself can put the nuclear power energy supply at risk: nuclear power plants need great amount of cooling water, which is why they are located along the coast or rivers. The group says, countries do not need nuclear power to avoid a climate catastrophe."[19] As discussed, nuclear disasters might occur because of manmade accidents taking place at nuclear power plants or climate change induced earthquake, tsunami, floods, storms, or any other natural calamity taking place or any use of nuclear weapon in and around the nuclear plant.

Management of Nuclear Disasters in India: National and International Obligations

India, in past one decade alone, has been struck with several natural disasters like Bhuj earthquake (January 26, 2001), the Tsunami (December 26, 2004), the Kosi flood disaster (2009), earthquake in Sikkim (2011) which resulted in total deaths of more than 50,000 persons, displacement of 6.5 lac people on account of Tsunami and 2 million people got displaced during Kosi floods and the most recent Uttarakhand natural disaster in mid 2013. The idea of setting up of NDMA was first time reflected in country's *Tenth Five Year Plan (2002–07) under Chapter "Disaster Management: The Development Perspective Document"* under the Ministry of Home Affairs and a statutory body known as **"National Disaster Management Authority of India (NDMA)" in 2005** came up and similar bodies have been created in all Indian States barring a few.[20] As of today, disaster management departments are set up in more than half of Indian states and UTs.

For avoiding nuclear disasters and implementing safety regulations, India is party to the Convention on Nuclear Safety (CNS) 1994, and ratified it on March 31, 2005, which is a legally binding international convention to govern the safety of civilian nuclear power plants. In 2007, it signed ISSA with the IAEA and brought all civilian nuclear reactors under the safeguards. The Nuclear Safety Standards (NUSS) also apply to nuclear power plants in India.[21] A National Report was also prepared

[19] Accessed from www.wecf.com.

[20] Accessed from www.ndma.gov.in, 2005.

[21] Arun Shull. (November, 2008). *Nuclear Energy Futures Paper No. 2*. The Centre for International Governance Innovation, Ottawa.

in accordance with the "Guidelines Regarding National Reports under the Convention on Nuclear Safety."[22] Thus, India's preparedness for handling nuclear disasters and the guidelines issued by NDMA is largely inspired by the IAEA and other international norms. In January 2003, the Indian Government had also established the Nuclear Command Authority (NCA) to manage its nuclear and missile arsenals and prevent its misuse. The complex system of control may be seen as a barrier against accidental or unauthorized use.[23] In order to prevent proliferation of nuclear technology, ensuring that it is not stolen or leaked in any manner to non-state actors, Govt. of India also got Nuclear Non-Proliferation Law enacted in 2007. India also got the process of setting up a Nuclear Safety Regulatory Authority started in August 2011. Even in Japan's Fukushima disaster, the operator TEPCO lacked proper training and expertise in handling disasters of Fukushima scale.

Safety Policies for Nuclear Power Plants (NPPs) Implemented by NPCIL

All nuclear power plants in India are run by Nuclear Power Corporation of India Limited (NPCIL). Under October 1996 Headquarter Instruction HQI-7003, the NPCIL accords utmost importance to Nuclear, Radiological, Industrial, and Environmental Safety, overriding the demands of production or project schedules. "Its objectives and various steps under implementation are:

- To maintain high standards for safety within the plant as well as in the surrounding areas.
- To ensure that health, safety, and environmental factors are properly assessed for all NPPs.
- To ensure that all employees, contractors, transporters working for NPPs adhere to safety requirements while carrying out their responsibilities.

[22] Govt. of India Report, 2008, pp. 82–83, accessed from www.npcil.gov.in, p. ii.

[23] Accessed from www.nca.gov.in and Hans Born. (2008). *Geneva Centre for the Democratic Control of Armed Forces (Daft) Policy Paper*—no. 15. National Governance of Nuclear Weapons: Opportunities and Constraints.

- To keep the public at large informed about the safety standards and regulatory practices that are being adopted at NPPs.
- Setting up of targets of safety performance parameters and their periodic monitoring.
- Carrying out of different levels of safety audits and reviews viz. Internal, Corporate, Regulatory and International like WANO Peer review.
- Assessment and enhancement of safety culture."[24]

Govt. of India from time to time has been taking number measures for preventing nuclear-related materials falling in the hands of the terrorists that can be used in the form of Radiological Dispersal Device (RDD), and development of crude form of "dirty bomb". It has been organising workshops on Nuclear Disaster Management through by its National Disaster Management Authority with greater frequency for prevention, mitigation, and preparedness and response at site and at hospital, rehabilitation, recovery, and research purposes. Input of specialists to delineate the threats, solutions, the technological procedures, and to gain useful insights on counterterrorist operations, decontamination, early detection, critical infrastructure protection, reconnaissance, protection, crisis management, and emergency monitoring system is also in place.[25]

As discussed earlier, highlights of NDMG-NRE discuss, "the elements of rescue, medical care, transportation, evacuation, providing food and shelter, etc. The National Crisis Management Committee (NCMC) assisted by the National Executive Committee (NEC), Department of Atomic Energy (DAE), MHA and National Technical Research Organization (NTRO) shall handle such emergencies. Specially trained NDRF, fire service personnel, civil defence, medical, transport, civil supplies, civil engineering departments, etc., are to have the radiation emergency response component as part of their response system to ensure large scale national capability in this regard. Response actions and Standard Operating Procedures (SOPs) are to be followed as per classified document with MHA."[26]

[24] Govt. of India Report, 2008, pp. 82–83, accessed from www.npcil.gov.in.

[25] Accessed from www.ndma.gov.in, 2005.

[26] Accessed from (**NDMG-NRE**, 2009, p. xxvii). The details of NDMG-NRE can be accessed from www.ndma.gov.in, 2005.

Measures Undertaken by then Govt. of India in the Aftermath of Fukushima Nuclear Accident

Immediately after the accident at Fukushima in Japan, NPCIL Chairman assured Indians that there would not be any slowdown in the country's nuclear energy program and assured that the DAE and NPCIL entailed, "undertaking an immediate technical review of all safety systems of our nuclear power plants in case of large natural disasters such as tsunamis and earthquakes.[27] Scholars in India had serious doubts about the secrecy in DAE, public health systems, their skills, skill development process, and needs urgent attention."[28] The response expected from state governments in case of nuclear disasters of the magnitude of Fukushima disaster has been very doubtful. The **NDMG-NRE**, 2009 guidelines would remain on paper if proper evacuation plans in a densely populated country are not worked out and practiced in a country like India.[29] Japan in its 750 page report on the Fukushima accident prepared by its Nuclear Emergency task force to the IAEA, emphasized upon findings like, "Japan was ill prepared, reactor design were old, lack of facilities and equipments on sharing basis, poor information and decision-making and lack of protection facilities were reasons for the disaster."[30] It is assumed that India must have adapted crucial contents of Japan's report.

It is suggested, "governments world over need to allow an informed debate on nuclear energy to take place and the exclusiveness of the nuclear energy establishments should give way to exchange of dialogues among communities, community leaders, scientists, sociologists, environmentalists, economists, health scientists, political leaderships and nuclear scientists with a view to recommend strategies to harness this source of energy balancing with safety, health and environmental concerns."[31] The NPCIL in post-Fukushima also carried out the safety evaluation of

[27] Accessed from *The Hindu*, March 22, 2011.

[28] NPCIL Report, 2011 accessed from www.npcil.org.

[29] M.M.K. Sardana. (2011). Impact of Accident at Fukushima on Nuclear Energy Programmes of India and China. ISID Discussion Note DN1109, July. *ISID Discussion Notes Public Health Issues and Disaster Management of Nuclear Fuel Cycles in India*, pp. 1–4.

[30] Ibid.

[31] *The Times of India*, July 14, 2011.

20 operating power plants and nuclear power plants under construction. The report titled, "***Safety Evaluation of Indian Nuclear Power Plants Post Fukushima Incident*** suggested a series of safety measures which pertained to strengthening technical and power systems, automatic reactor shutdown on sensing seismic activity, enhancement of tsunami bunds at all coastal stations, etc."[32]

India's Response to Fukushima

The then Prime Minister Manmohan Singh assured the country that all nuclear programs are safe shall not be slowed down and in his speech emphasized, "there would be no looking back on nuclear energy", while on a visit to West Bengal on August 21, 2011. He added, "we are in the process of expanding our civil nuclear energy program. Even as we do so, we have to ensure that the use of nuclear energy meets the highest safety standards. This is a matter on which there can be no compromise."[33] It is expected that present government of Prime Minister Modi would continue to follow the policies of previous government and avoid politicization of disaster management issues. Even the officials of AEC had the responsibility to assure the country men by stating that "Our record of nuclear safety has so far been impeccable and we have taken steps after Fukushima to ensure that it remains so," Dr. Srikumar Banerjee, Chairman, Atomic Energy Commission (AEC), said in an interview given to the Tribune Newspaper.[34] But the NDMA Chief was a pessimist about India's abilities to handle such post-nuclear disasters when he said, "India is not prepared to deal with nuke disaster."[35] Indeed, the message from India's political as well as nuclear bureaucracy is clear: despite the disaster, India's nuclear power projects will proceed unimpeded.[36]

[32]NPCIL Report, 2011 accessed from www.npcil.org.

[33]*The Hindu*, August 22, 2011.

[34]*The Tribune*, June 19, 2011.

[35]NDMA Chief's interview, Wednesday, June 1, 2011, PTI.

[36]Gaurav Kampani, How India will Respond to Japan's Nuclear Crisis, March 21, 2011. Kampani is a Stanton Nuclear Security Fellow at the Center for International Security and Cooperation. Also see, "Indian nuclear plants are safe: Scientists," *Hindu*, March 15, 2011, http://www.thehindu.com/todays-paper/tp-national/article1538642.ece, March 19, 2011.

Emerging Issues: Post-Fukushima and Lessons for India

According to the Reconstruction Agency, "The combined disaster of the earthquake, tsunami, and the nuclear accident caused nearly 16,000 deaths, over 1.2 million destroyed or damaged buildings, temporary evacuation of over 380,000 people from their home, most of whom were residents of Iwate, Miyagi, and Fukushima prefectures on the northeast coast of the Pacific Ocean. It also disrupted water supply, power distribution, and train, highway and air transport systems in a wide area of eastern Japan. Reconstruction of infrastructures has been partly hindered by radioactive contamination around the nuclear power plant, and as of spring 2013, some key infrastructures, such as a major train line and a major highway (Joban Line and Joban Expressway), have not been recovered yet. The WHO 2013 report had certified that after the nuclear accident, no deaths from radiation exposure were reported, and long-term radioactivity-related health risks for the Fukushima residents were considered to be low, still, radioactivity added a special dimension to the problem. To reduce radiation exposure, all residents approximately within a 20 km radius of the Fukushima Daiichi power plant were forced to leave their homes."[37]

According to another report, "The Nuclear Agenda: Prospects for U.S.–Japan Cooperation" edited by Yuki Tatsumi, published by STIMSON CENTER in February 2012, the Fukushima accident triggered a number of questions on how the government should be equipped to respond to nuclear emergencies. The accident at Fukushima showcased to the world that regardless of the cause, the consequences of nuclear emergencies are dire. And the discussion did not stop there. The accident also triggered public debate over the safety of nuclear energy technologies, leading many people beyond Japan to question the wisdom of relying on nuclear power as an energy source. It also revitalized the anti-nuclear movements by activist groups.

The report further said, "Nations are learning from Japan's experience of the nuclear alternative through the Fukushima Daiichi disaster. Germany, Switzerland, Italy, Israel and other critical players have joined Japan in announcing they will build no more reactors is being considered—a major blow to any prospect of curbing global carbon emissions. Some will start shutting the ones they have presently. Switzerland's

[37] http://www.reconstruction.go.jp/topics/000046.html, accessed on April 5, 2013.

cabinet has voted to phase out atomic energy by 2034, and Germany has declared to close all its nuclear power plants by 2022. Germany is planning alternatives for generating electricity through natural sources including sun, wind and water. In May 2011, G-8 leaders also took lessons from the Fukushima reactor to seek more stringent international rules on nuclear safety. Japan's nuclear disaster registered as a wakeup call by indicating the importance of a national regulatory body's independence from both government and the corporate sector."[38] It further added, "The political and social dimensions of massive shifts in environment and population are difficult to predict, but the likelihood is that over time large groups of people will become ecologically displaced persons or "environmental refugees," forced from their historic homelands and needing relocation to more hospitable places within or beyond national boundaries and such transitions will present large political and economic challenges, both for long-term humanitarian support and for immigration laws and enforcement specially, if these movements involve millions of desperate people, geographic and political boundaries will become increasingly problematic."[39] A country like India which has so much of pressure on land because of high density of population needs to put into place a very effective mechanism for dealing with such eventualities.

Safety Concerns Regarding Spent Fuel Storage

Safely securing the spent fuel that is currently in crowded pools at reactors should be a public safety priority of the highest degree. Robert Alvarej is of view, "As in Japan, U.S. spent-fuel pools are not required to have defense-in-depth nuclear safety features. They are not covered by the types of heavy containment structures that cover reactor vessels. Reactor operators are not required have backup power supplies to circulate water in the pools and keep them cool in the event of onsite power failures. Reactor control rooms rarely have instrumentation keeping track of the pools' water levels and chemistry. (In one incident at

[38] Purnendra Jain. (2011). *Japan's post Fukushima diplomacy*. Sage.

[39] Elizabeth Ferris. (February 2013). *The Politics of Protection: The Limits of Humanitarian Action* (Washington, DC: Brookings Institution, 2011), Chapter 7. Cited in Frederick S. Tipson, Natural Disasters as Threats to Peace, *Special Report 324* the United States Institute of Peace. Accessed from www.usip.org.

a U.S. reactor, water levels dropped to a potentially dangerous level after operators simply failed to look into the pool area.) Some reactors may not have the necessary capabilities to restore water to pools when needed. Quite simply, spent-fuel pools at nuclear reactors are not required to have the same level of nuclear safety protection as required for reactors, because the assumption was that they would be used only for short-term storage before the rods were removed for reprocessing or permanent storage."[40]

The NRC of the United States ordered reactor operators to:

- "... provide sufficient safety-related instrumentation, able to withstand design-basis natural phenomena, to monitor key spent fuel pool parameters (i.e., water level, temperature, and area radiation levels) from the control room."
- "... revise their technical specifications to address requirements to have one train of onsite emergency electrical power operable for spent fuel pool makeup and spent fuel pool instrumentation when there is irradiated fuel in the spent fuel pool, regardless of the operational mode of the reactor."
- "... have an installed seismically qualified means to spray water into the spent fuel pools, including an easily accessible connection to supply the water... A severe pool fire could render about 188 square miles around the nuclear reactor uninhabitable, cause as many as 28,000 cancer fatalities, and cause $59 billion in damage, according to a 1997 report for the NRC by Brookhaven National Laboratory."[41]

Finally, according to a research project titled "***Assessment: Japan-US Response to the Fukushima Crisis***," sponsored by the Sasakawa Peace Foundation in July 2011 reported, "the accident at the Fukushima Daiichi power plant has raised serious questions not only for Japan but for the entire international community about safety management at nuclear power plants and about nuclear security. In Japan, it has had a major impact on industry and energy, but also on Japan's society and

[40] Robert Alvarej (WINTER 2012). Improving Spent-Fuel Storage at Nuclear Reactors. *Issues In Science and Technology*. Accessed from www.ips-dc.org.

[41] Ibid.

lifestyle as the people have been forced to change their fundamental conception of safety management for nuclear power."[42] It further added, "The issues are far-reaching and many have yet to be resolved, but the key issues include (1) the response to nuclear accidents and safety management; (2) the revision of energy policy; (3) the handling of radioactive contamination and reparation issues; (4) concerns related to electrical power production and supply (PPS); (5) the restructuring of nuclear power safety management (the launch of a new Nuclear Regulation Authority and nuclear regulation agency in September 2012); (6) onsite nuclear power safety measures, and the strengthening of nuclear security; and (7) the question of maintaining nuclear power–related technology and personnel and the question of nuclear exports; and so on."[43]

Conclusion

In India, there have been several occasions when serious doubts about the functioning of DAE and its sister agencies have been raised specially, that their functioning is neither transparent nor much information is shared with the public. In case of enquiries, officials would not point out safety and design lapses due to fear of action being taken against their brother officials only. GOI needs to get passed the much awaited independent Nuclear Safety Regulatory Authority Bill (NSRA), 2011/2014 for bringing about much-needed independence and transparency in administering the safety oversight of nuclear operations in India. Overall it can be said that India needs to take a cautious path for implementation of its ambitious nuclear power program in light of its need as well as its realistic capabilities of managing nuclear disasters if at all they were to become a reality. A lot of education and training of citizens at massive scale need to be undertaken. There is an urgent need for developing a new culture of resilience toward such disasters on part of average citizens of the country. Union government need to encourage the state governments for preparing themselves in the light of current realities. Schools, Colleges, and Universities of the country need to introduce a course

[42] Assessment: Japan–US Response to the Fukushima Crisis, Report sponsored by the Sasakawa Peace Foundation in July 2011 accessed from www.spf.org.com.

[43] Ibid.

on Disaster Management at every level of education. The youth of the country need to be attracted toward creation of exclusive wing of volunteers who would ever remain ready to offer their services in the hours of crisis on the patterns of N.S.S. The private sector of the country will also have to assume responsibility for such causes in a very big manner. A "disaster cess" on patterns of education or petrol cess can be imposed for creating a financial reserve for training of youth and human resource exclusively for disaster-related needs.

Bridging the Gap Between Academicians and Bureaucrats: The NYSAF Way

Nivedita Haran

Introduction

It was the summer of 2013 and the author was attending a conference at a quaint location on the outskirts of Dehradun on the protection of public lands, river banks, and water bodies. There were some very eminent participants, academicians, researchers, law experts, and activists from South and Southeast Asia presenting well-researched papers. The audience consisted of scholars interested and working on this subject. But where were the administrators, the civil servants from the top, middle-, or the grass-root levels who were supposed to be doing the hard core job of enforcing the law, protecting the public lands and rivers, and representing the citizens' interest in the court of law, one asked? One could not see any of them that really got one worried and extremely uncomfortable. When this discomfort was mentioned to a few of the senior academics present at the conference, it gave rise to a trail of discussions that started off on the sidelines of the Conference at

Dr. Nivedita Haran is a Retired IAS officer, Government of India.

N. Haran (✉)
New Delhi, India

B. Zutshi et al. (eds.), *Disaster Risk Reduction*,
https://doi.org/10.1007/978-981-10-8845-2_7

Dehradun but which continued way beyond it. One could gauge that a serious spark had been ignited.

Coordination and good orchestration form the basis for good governance. At the aforementioned NAPSIPAG annual conference, whose focus was the management of land and water, the need for better communication, coordination, and interaction between academicians and civil servants was raised and the need for a forum to bring this about was underlined. From the subject of Land Governance in Dehradun, the focus homed in on disasters: Community-based activities, involvement of citizens and corporates, and the need to lay emphasis on resilience-building. In the keynote speech, the urgent need to protect public lands, to care for our rivers and water bodies, and the need to use modern technology to go about doing the same was stressed again and again. Having reasonably good contacts with the academic institutions and the work that goes on as part of research, the inability of administrators to get hold of such research had been a pet peeve of the author. Interesting and well-researched papers are presented by scholars at seminars and workshops that contain innovative ideas and possible solutions to problems on the issues of the day. Alas, these papers remain within the confines of the four walls of the seminar halls or gather dust as part of journals or books on the shelves of the college and university libraries. Academicians rue the fact that no administrator even cares to read and get acquainted with them, leave alone implement any of their suggestions. Administrators on their part struggle to find solutions for day-to-day issues and major administrative conundrums and draft major policy papers without having the benefit of the research base. Academicians and researchers struggle to find topics for research that would be relevant to society; while administrators find most research repetitive and often irrelevant. What a loss it is for the entire community that the two never meet. Due to this existing chasm, funds spent on research get wasted as many good hypotheses never find the fertile soil where they can bloom. The idea was, therefore, mooted that as part of NAPSIPAG, a standing group of members be set up who would constantly work toward bringing academicians and bureaucrats together in different fora and in different manners, thereby enabling them to interact with each other and learn from each other. This was enthusiastically accepted by the participants at the 2012 Conference that encouraged one to return from Dehradun singing "all is well!"

The positive response of the academics present at the Conference, from India and from other South and Southeast Asian countries, came as a pleasant surprise and a motivation to take this idea further. At a seminar held in JNU soon thereafter, the setting up of the Napsipag Young Scholars and Administrators Forum (NYSAF) was once again mooted. Under the able leadership of Prof Amita Singh heading the team of academicians and researchers and in the presence of the author, one of the very few administrators present, NYSAF became a reality. The aims and objectives of NYSAF were then and there drafted as a Charter and before the seminar concluded, the Charter was approved.

NYSAF: Objectives and Scope

The aims and objectives of NYSAF are as follows:

- Share knowledge within the Forum
- Synergize the functioning of scholars and administrators
- Share experiences and learn from each other
- Make research action-oriented and make administration research-based
- Limit their work on the focus area which will be land governance, disaster-resilience and subjects directly linked to it

It was agreed that young research scholars and civil servants would henceforth be invited to participate in the activities of NYSAF on a regular basis. It was also agreed that NYSAF would take up the following activities:

- Bring together administrators and research scholars to generate a core group of experts with robust academic base and practical field-experience;
- Set up a network of such experts to enable them to learn from each other and to supplement each other's knowledge and experience base making research action-oriented and administration research-based;
- Provide the environment and locale to work in synergy to implement innovative ideas impinging on the focus areas;
- Bring out occasional papers based on joint contribution and circulate these widely;
- Hold workshops, at least two every year, of select groups of scholars and administrators to expose them to innovative ideas in public service and to enable them to come up with innovative ideas on their own;

- Develop NYSAF into an active, thinking, and thought-provoking body led by a group of experts that would find novel ways to bring about interaction and cross-learning between administrators and academicians.

Any civil servant or a research scholar could be a member of NYSAF. A civil servant implies any member of the All India Service. A research scholar implies an M.Phil. or Ph.D. student or faculty who is part of a recognized University or Institution. The mode of communication within NYSAF would be the electronic media that includes the social network or any other mode found to be mutually convenient. The aim would be to ensure there is a constant open dialogue between both groups. In addition, there would be at least two Workshops of NYSAF every year where active and interested members from both groups would spend time with each other to discuss ideas, concepts, and innovative practices and work out their mode of implementation. Where possible they would together work on an idea hands-on to experience camaraderie and to face problems as found in the field and find practical solutions to them. The participants would be exposed to good practices, modern techniques, and challenging dialogues that would enable them to leave the workshop rejuvenated and even after leaving the workshop, the groups would remain in constant touch with each other and continue to share with and learn from each other. This would be a relationship for years if not for life.

Concepts like "public interest" and "public trust" that are so much a part of bureaucratic functioning and so often quoted in government files, especially in the context of land governance, need closer analysis. Alas, there are public servants who complete their careers neither being exposed to these concepts nor realizing their efficacy. The problem of corruption that has been eating into the foundation of our systems can be effectively handled only through bringing the thinkers (academicians) and the doers (administrators) face-to-face. This is only an example.

Conclusion

In conclusion, it can be safely stated that the setting up of NYSAF is one of the best and most apt developments that has taken place in the field of cooperation between thinkers and policymakers. Such interaction, cross-learning, and building up of synergy is relevant in every field, including law and governance, land management, disaster management, application of computer technology, and bringing about probity in public life. This idea now needs to be taken forward and made into a reality. Based on its success, NYSAF-like bodies need to be set up in every academic institution involved in research, engaging administrators from relevant disciplines at the level of every State government.

Sustainability of Community Managed Disaster Risk Reduction Projects

Neena Joseph

Introduction

Disaster risk reduction management (DRRM) initiatives attract political and financial commitments from government compared to rescue; relief and reconstruction and rehabilitation components of disaster management. The **importance** of DRRM is eclipsed by the **urgency** of the other components. Further, the invisibility of DRRM makes it less expedient politically. For the robustness and sustainability of risk management initiatives at microlevel, gradual building up of community's risk reduction culture and ultimate owning up and management by the community were crucial and hence the new paradigm is Community Managed Disaster Risk Reduction (CMDRR). CMDRR requires protracted and persistent effort and can be undertaken best by NGOs, who has to withdraw once the project is realized and stabilized and the community is empowered to carry out the work.

But the phenomenal efforts of community mobilization and skill building will be wasted if the DRRM initiatives are not sustainably continued and not integrated into the sustainable development endeavors

N. Joseph (✉)
Institute of Management in Government, Kochi, India

B. Zutshi et al. (eds.), *Disaster Risk Reduction*,
https://doi.org/10.1007/978-981-10-8845-2_8

of the local self-government and with that of the government. Huge opportunity cost will be entailed by way of the missed opportunity of NGO-government synergy. A credible and committed NGO can come up with good models, mobilize community, develop capabilities and build up lateral and vertical synergies with bureaucracy and local self-governments. Such initiatives get maximum support from all directions, especially during the aftermath of a major disaster. But the sustainability of such projects are beset with problems such as intersectoral rivalries, ego clashes, habit of working in silos, dearth of ideas related to relevant activities during the non disaster periods, change of leadership in the relevant sectors, absence of political and financial prioritization of CMDRR, lack of proactive measures to forge linkages from both sides, etc. Further, the knowledge and skills generated have to be melded and blended at microlevel with the total plans and projects of the local self-governments and also with those of the grassroots level activities of the government. To make this happen, there need to be consciously and deliberately crafted legal and administrative structures and mechanisms. In the absence of this, the community will not have the mandate to utilize these skills legitimately and sustainably. Another problem is the relevance of a CMDRR project in the context of a community where major disasters are infrequent. In this context, a study on the sustainability of CMDRR project is of high significance. A microlevel study is undertaken to explore the sustainability issues of a CMDRR initiative of an NGO Ernakulam Social Service Society (ESSS), which was implemented in a tsunami affected coastal ward 20 of Elamkunnapuzha gram panchayat in Kerala, India. A microlevel exploratory study can capture the dynamics of sustainability of a CMDRR project and can subsequently be scaled up by other researchers.

This project is chosen because the NGO, ESSS is a well federated one (federated with Kerala Social Service Forum at regional level, with Caritas India at National level and with Caritas International at international level) privileged with knowledge partnership of UNDP whose professionalism undergirds this. The Asian tsunami 2004 had wreaked havoc in Kerala which boasts a coastline of 580 km along its western coast. Elamkunnapuzha Grama Panchayat (EGP) in southwest Kerala had affected the vulnerable sections of poor people whose compounds were unprecedently flooded with water and resulted in considerable damages to property. ESSS implemented the project in 3 contiguous grama panchayats, but Elamkunnapuzh has the longest coastline among these and the most affected among the panchayats.

ESSS started its intervention with Community-Based Disaster Preparedness (CBDP) Project in 2005 and the project continued till 2009. In 2009, the project took a paradigm shift and was relaunched as CMDRR project. During the implementation of the project under ESSS, there were 20 wards in EGP. ESSS had already moorings in the panchayat through the secular self-groups (Self Help Groups—SHGs) of women which were federated at ESSS level. This social capital was leveraged to constitute 11 member Ward Disaster Management Committees in each ward of the panchayat. Besides this, in each ward, 5 member teams were constituted which are theme based, the themes being Early Warning Team, Health Team, Search and Rescue Team, Relief and Rehabilitation Team and Damage Assessment Team, The Ward Disaster Management Committees in each ward has the ward member as the patron. In EGP, 761 members of the community were all together trained and equipped to handle disaster situations and to take up preparedness initiatives. The panchayat level committee has Panchayat President as the patron. The teams were given generic training on disaster management as well as thematic training in their specific areas, backed up by a well-researched training module enriched with multi-sectoral partnerships including health proessionals, police, fire and rescue force. Mock drills were performed. The vibrancy of the teams was maintained through professionally organized meetings and empowering actions such as the usage of Right to Information Act to make the governance accountable for the disaster management operations. Once a certain level of capacitation was built up, the project was handed over to the panchayat in 2013 along with a directory of panchayat-wise and ward-wise contact details of all the management committees and team members and more importantly a comprehensive Disaster Management Plan, with escape routes and vulnerability mapping.[1] First Aid Box was handed over to grama panchayat who in turn entrusted the same with a committed Health Inspector who was very committed. ESSS did not hand over megaphones, because there was not enough to give to all the three panchayats studied. The assets including megaphones, sirens, and fogging machines were handed over to village office.

[1]Vulnerability mapping—the isolated places, bed tied persons, pregnant women, aged person, etc., were carefully mapped. Actually this has to be periodically updated to be effective.

The study attempts to answer the following questions.

1. Whether ingredients of sustainability were considered, when CBDP was launched and managed?
2. What is the present state of the project?
3. What can be the institutional mechanism to ensure the sustainability of the CMDRR initiatives?

This paper is divided into 6 sections. The first section covers the introduction, the relevance of the problem, area of study, background of the project, scope of the study, research problem, and the total scheme. Literature Review is given in the second section. Theoretical framework is discussed in the third section. Fourth section contains Research Methodology. The fifth section contains findings and discussion. The paper ends with the sixth section which contains conclusion of the study.

Literature Review

Community in the context of disaster management is a group of people living in proximity and sharing the same disaster (Shaw and Okazaki 2004). It is globally recognized that community partnership is of paramount significance (UN 2005, 2015). The significance of the involvement of community is highlighted in the Indian legislation (GOI 2005), in the policy (GOI 2009), and in the draft policy guidelines (NDMA 2014) and also in the literature. In the policy guidelines, community is recognized as the bedrock of disaster management. Experience from Philippines (Victoria), shows that labor of the community became a critical input in reconstructing a bridge in the flood ravaged locality (Victoria). The compelling reasons for community partnership is underlined in the legislations, documents, and literature based on global experience. Community is the primary stakeholder and recipient of direct impact (Rajeev 2014). It is the local community which bears the brunt or survives from the devastating effects of the hazards, whatever be the scale of the hazard (Salajegheh and Piramodi 2013). Communities constitute disaster fronts and being in the disaster fronts, they are able to respond to the threats themselves (Pandey and Okazaki 2015). Eighty five percent of people in a disaster either self-evacuate or are rescued by neighbors (Shaw and Okazaki 2004). Community-Based Disaster Management is cost-effective for disaster-prone developing countries who cannot sustain without external aid (Shaw and Okazaki 2004). The time-tested indigenous knowledge of the community can be utilized effectively for disaster management (Rajeev 2014). The degree of participation ranges from being passive recipients of aid and doles placing themselves in victim mode to being partners in policy formulation in the different phases of disaster management. Participation, partnership, empowerment, and ultimate ownership is needed (Pandey and Okazaki 2015). Again, the current thinking is hinged on building partnerships between different actors, multistakeholder participation; multisectoral, and multidisciplinary teaming up and creation of convergence platforms. The importance of multilevel and multistakeholder coordinating platform cannot be over emphasized (UNDP 2010). The National Policy incorporates Corporate Social Responsibility and Public Private Participation (GOI 2009). The distilled global experience is the basis of this. Philippines' experience of constructing a bridge in a flood-ravaged locality within 5 months with the labor of the community, technical expertise in the local body, and materials from Red Cross is a case in point (Victoria). The significance of convergence of grassroots level functionaries from the community bodies; nurses and health activists from health department, bureaucrats from local bodies, teachers in educational institutions, members of clubs are emphasized in the draft policy guidelines (NDMA 2014). Disaster risk is defined as the product of hazard and vulnerability, divided by coping capacity. Hence the method to decrease disaster risk is to decrease the vulnerability and to increase the coping capacity of the community. The shift is to be from emergency management framework to risk reduction framework. There are structural interventions such as hazard resistant protective engineering and structures and nonstructural interventions like policies, awareness, and education (UNSIDR 2005) to decrease vulnerability. High risk communities are to be transformed into disaster resilient communities (Victoria). These activities are not stand alone ones. The new paradigm is to make risk endogenous to development process which in turn has its own risk where gains get privatized and losses are socialized. Disasters are manifestations of development problems and hence development process must incorporate risks and strengthen resilience (Asthana 2014). A new paradigm of risk governance is required that addresses risk internalized in and sometimes generated by development process (Bazzrragchha 2012). Disasters undermine sustainable development (UN 2015). So without risk reduction sustainable development is not possible. So when community engages in disaster risk reduction, their participation in the development process is imperative. This is possible by having participation in governance.

The role and importance of the community under leadership of the local authorities is underlined by India's national policy (GOI 2009). Even if communities are initially successful in creating the project, they may lack the material resources and connections to sustain their efforts (Olekotan 2013). The national level legislation and policy in India are for mainstreaming community-based disaster management into government programs. Structural and functional convergence with government is required by community-based disaster initiatives to be sustainable. There are many studies on various aspects of disaster management. But there is a dearth of studies probing into the sustainability aspects of community-based disaster risk reduction.

Theoretical Framework

First, the project is examined for the presence of ingredients for sustainability using the UNCRD framework which gives 8 ingredients Shaw and Okazaki (2004). In the second part of the discussion, an organizational structure is discussed for the sustainability of the project using Pomeroy's framework adopted for disaster risk reduction context.

Ingredients for Sustainability in the Project

The framework is furnished below.

Existence of a culture of crisis	Well-delivered training inputs aligning with the objectives
Participative risk assessment process involving vulnerable sections	Wide stakeholders involvement and participation
Supporting agencies for sharing common motivation	Accumulation of physical, technological, and economic assets to reduce vulnerability
Genuine peoples participation involving women, elderly and children	Legislation and incorporation CBDM in development planning and budgeting

Comanagement of the Project

Robert Pomeroy had enunciated the concept of comanagement or cooperative management in the context of fisheries resources. This concept can be adapted to the management of a project for disaster risk reduction at grassroots level. Cooperative management or comanagement is a partnership arrangement in which the comprehensive set of stakeholders share responsibility and authority for management through consultations and negotiations of roles, rights, rules, and resources. Formal and informal agreements are reached on the power sharing and the output is negotiated power for various agencies. Comanagement arrangement is dynamic and responsive to changes in the environment inside and outside the system. Decentralization, social empowerment, power sharing, and decentralization constitute its core. It endeavors to overcome distrust, corruption, fragmentation, and inefficiency in the existing arrangements through collaboration. Costs and benefits of successes and failures are shared by the parties. The process could be time-consuming and complex. Three to five years will be taken by parties to address concerns of legitimacy, trust, accountability, and transparency (Pomeroy 2006). Regarding the CMDRR, the stakeholders constitute local self-governments, bureaucracy, and authorities from the new projects which hold potential threats of hazards to the community. The diagrammatic representation of the model is furnished below.

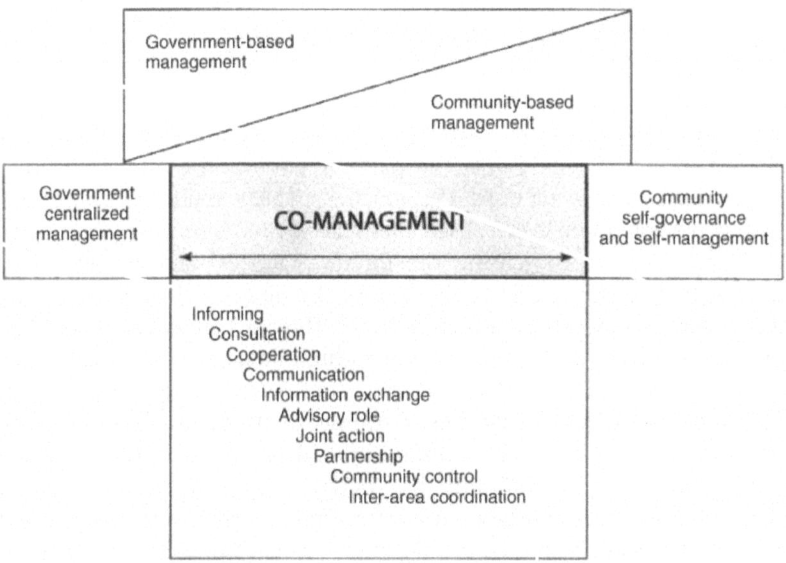

Pomeroy et al. (2015) presents additional points for the effectiveness of a comanagement model clearly defined boundaries, group cohesion, benefits outweighing costs, participation of all the affected persons, enforcement of management rules, legal rights to organize, cooperative leadership at community level, coordination between government and community, diversity of incentive to diverse groups of stakeholders, understanding of one another's needs and apprehensions, conflict management, communication, local political support, networking and advocacy, enabling policies and legislations.

Research Methodology

Interview was conducted with the ex and current directors of NGO, previous and current Project Officers of ESSS, community leaders, Panchayat President, Project Officer, State Disaster Management Authority, and key informants in the community. Two focus group discussions were held with 12 panchayat members and 15 persons in the community.

Discussion and Findings

Ingredients of Sustainability

Existence of a Culture of Crisis Consciousness

The EGP project leaders have become very keenly aware of disaster risk and have agitated against the pathetic condition of the Pallithodu road in their locality of ward 19, which had been rendered unmotorable and potholed and had caused many accidents. The community had developed acumen for sensing the unsafe practices such as the inadequate tests done for checking the leak of the pipe which was laid in the 4 km stretch of Goshree Junction to Puthuvype Junction road. The pipe has now leaks at 3 spots, leading to high damage to the road. They are also aware that many roads constructed with inadequate materials, become potholed and dangerous. Cuttings in the roads also pose serious threat to the safety. The community had protested and given many press releases. The community has become sensitive to such issues. They are able to establish a cause-effect relationship between cracked wall and the heavy traffic of giant vehicles which ply the proximal road built by giants like IOC, KRL, and LNG. They have become sensitive to the potential threats caused by an LPG storage facility which is just 30 meters away from thickly populated area. This is making the community upset, thinking about the eventuality of gas leakage. The community is actively aware of the eco impact of the new "big" developments and has many unanswered queries hinging on eco concerns: (1) is a safety audit done when the LPG is bulkily bunkered in the terminal which is just 30 meters away from the place where poor people live? (2) Has anybody done any study on whether when the tons of water from other countries which import LNG (used in the ship to balance it) is emptied into our sea, the impact it has on the habitat of the marine life in our sea? (3) Will the chemical emitted in the process of converting LNG into natural gas, will fall back on our land as chemically polluted rain? (5) What are the safety measures taken by the LNG and LPG bunkering terminals? (4) Why should we have an LPG bunkering terminal when our extra demand (27,000 tons of LPG) for the same can be met from the Mangalore (5) Will the SPM (Single Point Mooring) of KRL situated about 18–20 km away from the seashore prevents fishermen to catch the fish and thus threaten their livelihood? These queries rose in the field show that the community is no more just helpless mute spectators, but are aware of everything happening around them and are able to intervene to the extent possible for holding the big guns responsible for safety assurance. They are conscious about the fact that there is no bridge across Pallithode area for people to escape if a flood hazard occurs. Their organizing capacity and advocacy skills have heightened. During the project phase they were trained to wield Right to Information Act and question the system and hold them accountable to the people. Dr. Thummarukudy's of the opinion that risk reduction needs to become a cross-cutting theme and has to be mainstreamed into all institutions, work sites, and professions. Safety audits of schools are very important. The community has become very sensitive to the need of risk reduction in their locality. Egg the fact of lighthouse sitting down into the ground as is easily evidenced by the wall of the lighthouse which has submerged into the ground by 22 inches within 33 years. The quarters of the lighthouse has submerged by 16 inches within 9 years. The community has developed understanding about the concept of risk reduction. The houses of poor people situated across the road (jointly built by IOC, LNG, and KRL), is slowly getting submerged partly due to the natural phenomenon and partly due to the raising up of the land to build the road. There is water logging in the area and this is a sure recipe for mosquito menace and water-borne diseases. They are able to link the hazard of dog bite to careless dumping of food waste on the beach which in turn invites stray dogs. In waterlogged area, mosquitoes multiply causing diseases. Mr. Johnson a former member was explaining how the unavailability of extra oxygen cylinder in the local hospital caused the death of a woman who had to wait at the railway gate on the way to being taken to the hospital. The interviewees were very critical about governance aspects of disaster management. The community had a lot of apprehension at the upcoming IOC Bunker terminal. The daily wage members of community lost one day's wage and had to reach the Collector's chamber. It could have been better conducted at Puthuvype.

Participative Risk-Assessment Process and Incorporating Vulnerability Perceptions and Capacity
Prior to 2013, ESSS had assessed risks and vulnerabilities through PRA exercises and beneficiaries were selected accordingly. But this learnings were not utilized post-2013 period. These exercises could have been used in gram Sabhas for beneficiary selection. Special attention was made to include women in the team and in the committees. Vulnerability mapping was done and at that time all data was available for preparing a plan. But the vulnerability map would be of no use if it is not updated periodically. In the project leadership, majority were women. Like SHGs of women ESSS has SHGs of children. They are strengthened by involving them also in the project. Since risk reduction is integral to sustainable development which in turn is the mandate of a local body, the panchayat projects can be given a risk reduction angle such as projects to teach swimming, to teach how to save a drowning person, to give first aids, etc. There could be projects for enhancing safety of fisherfolk in the sea in terms of capacity building and new technologies, projects for structural safety of schools, projects to teach safety practices to people including safety practices in transportation, handling of cooking gas, usage of electronic devices healthy postures while working, etc. Risk reduction can become a cross-cutting theme.

Common Sharing of Motivation and Ownership for the Initiation and Sustainability of the Project
If the project has to remain live, it has to intervene in the live issues of the community. There is no incident of a Tsunami occurrence recorded in the history of Kerala. A community cannot be made to stay prepared for a major disaster which might happen, may be once in 500 years. The ideal thing would be to utilize the knowledge of the trained individuals for dealing with minor day-to-day small-scale disasters. This approach

will help to sustain the interest in the project and bring relevance of the project to their lives. EGP faces acute water shortage. The trained force at EGP has intervened on such occasions. Being a coastal area, fishing is a major means of livelihood. During off season, providing alternate means of livelihood would be beneficial in sustaining the project. During the project period, i.e., from 2005 to 2013, ESSS intervened and supported the community to start microenterprises under the Theera Mythri scheme of Fisheries Department. Some of these enterprises continue to date. There is scope for addressing localized issues such as waterlogging, drug and alcohol peddling, etc. Such interventions make the project relevant. The involvement in the project familiarized them to the government offices especially the panchayat office which stands them in good stead. Smt. Shyla Attippetty who was one of the leaders of the project narrated a recent experience how Pallithodu road was in a deplorable condition and how the panchayat procrastinating the Rs. 2 crore and 20 lakh which LNG had agreed to hand over in a phased manner to the panchayat. Finally, Shyla and team were able to lobby for handing over the road construction project to Construction Corporation.

But the real issue is that even without the functioning of such giant force of 1810 persons, a few trained people joining together on interested issues could do the same. The project lacks a central binding and coordinating force.

Genuine People's Participation and Capacity Building and Focus on Sectoral Groups like Women, Elderly, Children, and Ethnic Minorities

There is no incident of a Tsunami occurrence recorded in the history of Kerala. A community cannot be made to stay prepared for a major disaster which might happen, may be once in 500 years. The ideal thing would be to utilize the knowledge of the trained individuals for dealing with minor day-to-day small-scale disasters. And much more important is to make this trained force to make risk assessment and work toward risk reduction. This has happened in the case of EGP. The EGP group has become very keenly aware of disaster risk and has agitated against the pathetic condition of the Pallithodu road in their locality. The community had developed acumen for sensing the unsafe practices such as the inadequate tests done for checking the leak of the pipe which was laid in the 24 km stretch of Vypin to Munambam road. The pipe leaked causing heavy damage to the road. The community had protested and given many press releases.

Again, involvement in the project had nurtured leadership qualities in the members. For example, one member of the team Smt. Aleyamma Isaac contested grama panchayat election and won. She is now Block president. Another lady has become the chairperson of tailor's welfare board.

Well-Delivered Training Inputs

"The project had woken up sleeping persons like us and elevated us to a very high level of awareness and we will always stay there". There were many training programs of generic nature as well as specialized programs for the five thematic teams. During the interviews, many members of the community said that if any hazard occurs they would be in a much better position to manage it. They specifically mentioned that they now know which is the escape route, where to congregate in the eventuality of a disaster, how to rescue the vulnerable sections such as differently abled, pregnant women, bed tied individuals and aged persons. They know that valuable documents need to be kept well packed and in an easily accessible and known place so that in case of emergency, it can be readily picked up. There were many training programs including mock drills with fire force rescue teams. Many community members mentioned that they now know how to give first aid, and hold and transport a wounded person to the hospital. Unless the project is sustained, the benefits of such leanings will slowly diminish. Panchayat members told that the intervention power of the people has increased. Mr. Johnson, the then panchayat member said that his leadership qualities were enhanced by involving with the project and now many leadership positions in the community are gravitating to him and the latest in the list is as joint Secretary, Labor Commission of one diocese. He takes lead in risk reduction initiatives such as safe construction of roads, safety measures demanded from IOC, LPG storing facility, etc. The ward level coordination committees were asked to file at least two RTI cases each. Such training had enhanced the intervention capabilities of the community. The coordination committee detected fraudulent cement mixing and had raised this issue of irregularity. Children who were terrified by the tsunami were trained back into normalcy through play therapy.

Wide Stakeholder Involvement and Participation

The ward management committees had ward member as the patron. But the perception is that of an invitee presiding over a meeting. During the project phase, there was wide multistakeholder participation especially for the training. There were interconnections to Health personnel, police, fire force, etc. There was tie ups to certain hospitals with which ES already had affiliations.

Accumulation of Technological and Economic Assets Against Hazards

The project accumulated assets such as sirens, megaphones, and fogging machines and tools for retrieving people and animals from debris. All these were handed over to village office.

The project was not structurally or functionally integrated to EGP or village office. The siren is not functional. The fogging machine if maintained could have been very useful for eradication of mosquito menace. EGP which is a coastal area often plagued with waterlogging.

Legislation and Incorporation of the Project in Development Planning and Budgeting

This has not happened. ESSS operates on a project to project basis. The project was handed over to the panchayat along with the Directory containing the contact details of 1810 trained leaders in the three panchayats together. A full report of the project too was handed over. The assets were handed over to the village office. The project structure is intact, the team meetings and meetings of the ward management committees continue. But now since the last 6 months they are raising doubts regarding the relevance of such meetings. ESSS has hold on the community through SHGs. ESSS continues to have linkages with community. Many risk-reduction issues are discussed informally. But the structure is fast losing its robustness. Although it was handed over to the panchayat, it might be 'resting peacefully in the shelves of the panchayat'.

What eroded its sustainability is that the project was not institutionalized as a part of the panchayat. There was no attempt to make a bye law. Community standing on its own cannot sustain a structure beyond a particular time. An authority has to be legitimizing it. Thanks to the national level Act, We have Disaster Management Authority at national, state, district, and local level. A linkage could have made the project legitimate. Again, being a coastal area with security issues, Police Department constituted jagratha samithies. When it came to selecting members, selection was not made out of this pool. Out of the 23 members who were interviewed, 3 remember this project only vaguely! All these point out to the eroding sustainability. All the members interviewed including those who do not "remember" the initiative told that such an initiative.

Possibilities of Comanagement

First the project is examined using the framework, to assess the presence of the essential ingredients required for its sustainability. The main point which emerged was that left alone, the project cannot sustain. It needs multistakeholder participation. It needs legitimacy. Hence it needs a legitimate platform to survive and thrive. With a bye law, it can find its legitimate position in a local body. The technical know-how of NGOs and the authority and funds of local body can be synergistically blended. The Disaster Management Act, 2005 envisages Authority at national, state, district, and local level. A bye law can be purposefully crafted so that the trained Community Disaster Risk Reduction set up gets a say in ensuring the safety standards of constructions and development plans, livelihood protection (and thus reducing vulnerabilities), legitimately negotiating with mega projects which threaten safety, gaining better access to services of health, police, water authority, PWD, etc., in the eventuality of a hazard. A structural and functional convergence with panchayat will keep it connected to local body and they would not be "forgotten" to be included in the disaster-related functions and the painstakingly acquired training will not go untapped. Since risk reduction is an inevitable component of sustainable development, the team can facilitate waste management, micro planning, strengthening grama sabhas, protecting water bodies, micro planning. Thinking along these lines, early cancer detection and kidney problems can be taken into its ambit, etc. The authority of the NDMA will thus cascade down to the implementation level. This will be a synergistic blending of the technical competence of trained and organized human force and the power and funds of the panchayat. The trained human force if managed effectively could ease the workload of the ward members leaving them more room for pursuits of strategic significance and thus bring about good governance, provided they know how to utilize the organized forces to tackle the issues in the local body. In this context, training/academic/research organizations have a big role to play in training the local bodies to utilize the wealth of knowledge and experience of the groups. There are many such organized groups within the panchayat. They should be able to coordinate and use these groups so that they can multiply their effectiveness. The disaster management team needs to expand its scope as function as community task force or multitask force, since risk reduction is inherent to sustainable development. Again, bringing in such forces and linking them to governance will augment good governance. Good Governance is crucial for risk reduction. Thus risk reduction can be mainstreamed into good governance.

ESSS could remain as a knowledge partner. NGOs can make models. Government has money and NGOs have certain unique skills and expertise. Synergistic operations are needed to produce results. NGOs need lobbying skills. The Government and panchayat officials need to be given training to utilize the expertise of these skilled groups.

Conclusion

Well-conceived and effectively launched and run grassroots level CBDM initiatives containing majority of the ingredients for sustainability, will fail to become sustainable if the initiatives are not integrated legally into the local bodies and then into the local plans, programs, and budgets of the local government. A multistakeholder comanagement with clearly demarcated boundaries and well-spelt out duties and responsibilities will keep it alive and vibrant.

Consultations

1. Dr. Muralee Thummarukudi, Disaster Risk Reduction Chief, United Nations Environment Programme.
2. Mr. Saran, Project Officer, District Emergency Centre, Ernakulam, Civil Station.
3. Lighthouse Man.
4. D. Suseendran, Navigational Assistant Grade I (Head Lighthouse Keeper), Puthuvype, Elamkkumnnapuzha Panchayat.
   Fr. Mathew Kallinkal, the former Director, Ernakulam Social Service Society.
5. Fr. Antony Raphael, Director, Ernakulam Social Service Society.
6. Fr. Romance Antony, the then Director Kerala Social Welfare Forum.
7. Mr. Titson, the then Project Officer, Ernakulam Social Service Society.
8. Mr. Vipin, the current Project Officer, Ernakulam Social Service Officer.
9. Mr. Benny Bernard, then Panchayat Member and current Vice President, Elamkkunnathupuzha Grama Panchayat.

10. Mr. Johnson, the then Panchayat Member, Elamkkunnathupuzha Grama Panchayat and current associate of Ernakulam Social Service Society.
11. Smt. Shyla Attippetty.
12. Smt. Shybi Peter, one of the 5 Health Team Member.
13. Smt. Sindhu Sabu, one of the then 11, Ward Disaster Management Committee Member.
14. Smt. Sheela Shaji, one of the then 11, Ward Disaster Management Committee Member and currently a candidate contesting Grama Panchayat election.
15. Smt. Beatrice Joseph, President, Elamkunnapuzha Panchayat.
16. Mr. Benedict, Chairman, Standing Committee, Welfare, Elamkunnapuzha Panchayat.
17. Smt. Sajini Joy, Standing Committee, Development, Elamkunnapuzha Panchayat.
18. Smt. Devayani Radhakrishnan, Chairman, Standing Committee, Health, and Education.
19. Mr. Radhakrishnan, Member, Member, Elamkunnapuzha Panchayat.
20. Smt. Rosily, former President and now member at Elamkunnapuzha Panchayat.
21. Shaji, Member, Elamkunnapuzha Panchayat.
22. Mr. Madan Mohan, Coordinator, Association of Grama Panchayat Association of Kerala.

Abbreviations

CBDM	Community-Based Disaster Management
CMDRR	Community Managed Disaster Risk Reduction Disaster
DDMA	Disaster Management Authority
EGP	Elamkunnapuzha Grama Panchayat
ESSS	Ernakulam Social Service Society
IOC	Indian Oil Corporation
KRL	Kerala Refineries Ltd.
LDMA	Local Disaster Management Authority
LNG	Liquefied Natural Gas
LPG	Liquefied Petroleum Gas
PRA	Participatory Rural Appraisal
NDMA	National Disaster Management Authority
SDMA	State Disaster Management Authority

Bibliography

A Guide To Community Based Disaster Risk Reduction in Central Asia (2006).

Annex A: Community-Based Disaster Risk Reduction, Community Resilience and HFA1 (UN, 2005). Retrieved October 23, 2015, from https://www.google.co.in/?gfe_rd=cr&ei=Tb8pVtZMpu_zB8fYkcgD#q=Government+accountability+through+comm.

Asthana, Vinay. (2014, September 27). "Disaster Risk Management, Shifting Paradigm". *Economic and Political Weekly* XLIX(39): 25–29.

Bazarragchaa, Sodnam. (2012). "Community Participation in Disaster Risk Mitigation: A Comparative Study of Mongolia and Japan". Retrieved October 23, 2015, from http://www.adrc.asia/aboutus/vrdata/finalreport/2011B_MNG_Final_Report.pdf.

Ernakulam Social Service Society. (2012a). "Disaster Management Plan".

Ernakulam Social Service Society. (2012b). "Directory of Task Force".

Ernakulam Social Service Society. (2012c). "Care, Community Acumen and Risk Elimination".

GOI. (2005). "NDMA, Government of India, The Disaster Management Act, 2005". http://atingl.nic.in/Downloads/THE%20DISASTER%20MANAGEMENT%20ACT%202005.pdf.

GOI. (2009). "Government of India, Ministry of Home Affairs, National Policy on Disaster Management, 2009". Retrieved October 2015, from http://www.ndma.gov.in/images/guidelines/national-dm-Policy2009.pdf.

Infed. (2010). "The Learning Organization: Principles, Theory and Practice". Retrieved October 27, 2015, from http://infed.org/mobi/the-learning-organization/acessed.

International Institute of Rural 1. Reconstruction and Cordaid. (2013). "CMDRR Training, Design and Implementation, Building Resilient Communities, Philippines".

Local Self Government and Disaster Risk Reduction. Retrieved October 23, 2015, from http://www.unisdr.org/files/13627_LocalGovernmentsandDisasterRiskRedu.pdf.

Nambiar, Malini. (2015, January 31). "A Decade of Disaster Risk Management in India". *Economic and Political Weekly* L(52): 36–40.

NDMA. (2014). National Disaster Management Authority of India, National Policy Guidelines on Community Based Disaster Management, February 2014. Retrieved October 25, 2015, from www.ndma.gov.in/pdf/draftindianpolicyguidelinesonCBDM.pdf.

Olekotan, Adimole. (2013). "Achieving Project Sustainability Through Community Participation". Retrieved October 23, 2015, from http://www.krepublishers.com/02-Journals/JSS/JSS-17-0-000-000-2008-Web/JSS-17-1-001-08-Abst-Text/JSS-17-1-021-08.

Pandey, Bishnu, and Okazaki, Kenji. (2015). "Community Based Disaster Management: Empowering Communities to Cope with Disaster Risks".

Retrieved October 21, 2015, from http://unpan1.un.org/intradoc/groups/public/documents/UN/UNPAN020698.pdf.

Petra, and Komal. (2014, December 27). "Addressing the Emergent Disaster Riskscape in Urban India". *Economic and Political Weekly* XLIX(52): 38–42.

Planning Commission, Ministry of Panchayati Raj, Report of the Working Group on Democratic Decentralization and PRIs, 2006. Retrieved October 27, 2015, from http://planningcommission.nic.in/aboutus/committee/wrkgrp11/wg11_demo.pdf.

Pomeroy, Robert. (2006). *Fisheries Co. Management: A Practical Handbook (2006)*. Ottawa, Canada: International Development Research Centre.

Pomeroy et al. "Fisheries Co-management: Key Conditions and Principles Drawn from Asian Experiences". International Center for Living Aquatic Resources Management, Philippines. Retrieved October 27, 2015, from http://dlc.dlib.indiana.edu/dlc/bitstream/handle/10535/1220/pomeroy.pdf?sequence=1.

Rajeev, M. M. (2014, December). "Sustainability and Community Empowerment in Disaster Management". *International Journal of Social Work and Human Services Practice* 2(6): 207–212 (Horizon Research Publishing).

Salajegheh, S., and Piramodi, N. (2013). "Community Based Disaster Risk Management in Iran". Retrieved October 22, 2015, from http://www.ijerd.com/paper/vol7-issue9/G0709060069.pdf.

Shaw, Rajib, and Okazaki, Kenji. (2004). *Sustainable Community Based Disaster Management (CBDM) Practices in Asia: A User's Guide*. Hyogo: United Nations Centre for Regional Development (UNCRD), Hyogo Office.

Thummarkudy, Muralee. (2015). "Community-Based Safety Management". Retrieved October 23, 2015, from http://mathrubhumi.com/english/columns/be-on-guard/community-based-safety-management-97684.html.

UNDP. (2010). "The Real Weath of Nations, Pathway to Development".

United Nations. (2015). "Sendai Framework for Disaster Risk Reduction 2015–2030". Retrieved October 25, 2015, from http://www.preventionweb.net/files/43291_sendaiframeworkfordrren.pdf.

UNSIDR. (2005). "Hyogo Framework for Action (2005–2015), United Nations International Disaster Reduction". Retrieved October 31, 2015, from http://www.unisdr.org/2005/wcdr/intergover/official-doc/L-docs/Hyogo-framework-for-action-english.pdf.

Victoria, Lorna. "Community Based Disaster Management in Philippines". Retrieved October 21, 2015, from http://www.preventionweb.net/files/733_8363.pdf.

What Is Community Managed Disaster Risk Reduction. Retrieved October 21, 2015, from http://www.gfdrr.org/sites/gfdrr.org/files/20120119_Cordaid_SUMMARY_What_is_CMDRR.pdf.

Innovative Use of Technology (GIS) for Disaster Risk Reduction: A Case Study

Namrata Agrawal

Introduction

The world is facing an increasing frequency and intensity of disasters either natural or man-made. India is prone to natural calamities due to distinctive geo-climatic conditions. The country has to face frequent occurrence of floods, droughts, cyclones, earthquakes, and landslides. Approximately, 60% of Indian landform is vulnerable to earthquakes of various intensities; around 40 million hectares is susceptible to floods; around 8% of the total area is vulnerable to cyclones; and 68% of the area is prone to drought (Ahmad 2003).[1] The disastrous cyclone of Orissa (October 1999), Bhuj earthquake of Gujarat (January 2001), and North Kashmir earthquake (October 2005), emphasize the necessity to embrace a multidimensional effort encompassing varied scientific, technological, financial, and social processes. Further, there is an urgent requirement to adopt and implement broad disciplinary and

[1] Centre for Research on the Epidemiology of Disasters (CRED) "The International Disaster Database" http://www.em-dat.net/.

N. Agrawal (✉)
National Institute of Financial Management, An Institute of Ministry of Finance, GoI, Faridabad, India

B. Zutshi et al. (eds.), *Disaster Risk Reduction*,
https://doi.org/10.1007/978-981-10-8845-2_9

multi-sectoral approach to manage disasters more effectively. It is also required to incorporate risk reduction approach in the developmental plans and strategies that would aid in effective and efficient disaster management.[2]

Geographic Information System (GIS), a technical tool, is very effective in disaster management but its use is still 'quite limited in the country like India' and was not available even in the highly disaster-prone State of J&K at the time of study. Drive to make the common populace aware of such technology and its virtue would help in reduced causality. It is essential to convince the decision-makers to allocate resources for appropriate investments in technologies like GIS and Remote Sensing for Disaster Management for minimal impact of disasters on human life and property. GIS would help in improved and quality assessment and analysis of natural hazards leading to effective and realistic guidance in developmental activities including facilitation of the planners in selection of mitigation measures and in the execution of emergency readiness and response action (van Oosterom et al. 2008).

Types of Emergencies[3]

Caused by Humans

These are unplanned happenings or accidents that are caused due to human activity or during the process of human development. Emergencies arising due to chemical spills, utility failures, epidemics, crashes, explosions, and urban fires are some of the examples of such exigencies.

Natural Disasters

These are unplanned happenings or accidents that occur due to natural phenomenon of the universe. Emergencies arising due to earthquakes, tornadoes, tsunami, freezes, blizzards, extreme heat or cold, drought, or insect infestation are some of the examples of such exigencies.

[2] http://www.idrn.gov.in.

[3] Types of Geodatabases (ArcGIS 9.2 Desktop Help). ESRI.

Internal Disturbances

These are the intentional activities caused by a group or individual with a motto to cause disturbance. Riots, prison breaks, and violent strikes are some of the examples of such exigencies.

Attacks

Large-scale terrorism or war using nuclear, conventional, or biological agents are some of the examples of such exigencies.

Phases of Emergency Management[4]

The emergencies as stated above can be broadly managed in phases involving different skill sets.

Planning

Planning includes all those activities that are related to analysis and documentation of the likelihood of an adversity or disaster including the probable consequences or impacts on life, property, and the environment. This includes assessing the hazards, risks, mitigation, attentiveness, response, and recovery needs.

Mitigation

Mitigation activities do away with or reduce the likelihood of a disaster (e.g., arms build-up to prevent enemy attack, or legislation that requires strict building codes in earthquake-prone areas). It also includes long-term actions designed to reduce the effects of inevitable disaster.

Preparedness (Twigg 2002)

In the preparedness phase, governments, organizations, individuals, and other stakeholders develop plans to save lives and minimize disaster damage (e.g., compiling state resource inventories, accelerating training exercises, installing early warning systems, awareness drive, action-oriented

[4] http://egsc.usgs.gov.

emergency response forces, etc.). These measures also seek to augment disaster response operations (e.g., piling of essential food items and medical supplies, extensive training exercises, and mobilization of emergency response personnel on standby).

Response

These are the set of activities which are effected after the occurrence of an emergency or disaster. These are the activities that are intended to provide emergency assistance to disaster victims. To name a few—search and rescue operations, emergency shelter, medical care, and mass feeding. Further, these activities are also meant to stabilize the emergency situation, speed up the recovery operations, and reduce the possibility of secondary damage by shutting off contaminated water supply sources, securing and patrolling areas prone to looting.

Recovery

These are the set of actions required for bringing the situation close to normal or better. They include two sets of activities:

3.5.1 Short-term recovery activities—return of essential life support systems to minimum operating standards such as cleaning up of the site, providing temporary shelter including access to food and water.
3.5.2 Long-term recovery activities—this activity may go on for several years after the occurrence of the disaster. The duration depends upon the intensity and damage caused by the calamity. This activity leads to recovery in real terms wherein the life of the effected populace returns closer to normalcy or may be of improved levels (in some cases). Some important activities such as provision of affordable redevelopment loans, legal assistance, community planning, etc., help in enhanced recovery.

Research Problem

The study has been carried out with a view to suggest ways to implement GIS techniques for Disaster Management in India with special focus to the State of Jammu and Kashmir (J&K).

The main objectives of the study are as follows:

4.1 To understand the working of a GIS
4.2 Benefits of using GIS in Disaster Management
4.3 To analyze the impact of GIS implementation for effective and realistic disaster management including mitigation, with special focus on the year 2005 earthquake of Jammu and Kashmir.

Data Source

The impact of the earthquake which occurred on October 8, 2005 in the State of J&K has been dealt thoroughly by survey of affected villages.

- Primary Data was collected by conducting focussed group discussions and informal as well as formal interviews with the stakeholders and the effected populace. Interactions were also held with the District Administration and also with the head of Disaster Management Cell of the State.
- Secondary data was also collected to understand and analyze the impact of the technology and its benefits thereof.

Survey Area

The state of Jammu and Kashmir (J&K) is situated at the northernmost part of India. It has three regions: "Jammu", "Kashmir Valley", and "Ladakh".

Based on the climatic conditions, the state has two different capitals. The summer capital is "Srinagar" whereas "Jammu" is its winter capital. The State has a total area of 222,236 km^2. The population of Jammu and Kashmir is around 10,143,700. Jammu and Kashmir is divided into 22 districts.

The survey of earthquake affected villages was carried out in two districts of Jammu and Kashmir. These two districts, namely, "Kupwara" and "Baramulla" were the worst affected areas.

Kupwara District[5]

Kupwara is the northwest border district of Kashmir Valley. It is situated at an average altitude of 5300 feet from the sea level. The geographical

[5] http://kupwara.nic.in.

Fig. 1 District Kupwara (*Source* Wikimedia)

area of the district is 2379 km^2. The total population of the district is estimated to be around 640,013 (Fig. 1).

Baramulla District[6]
Population-wise and also area-wise, it is the largest district in entire valley. The district is spread over an area of 4588 km^2 and its population is around 11.51 Lakhs (Fig. 2).

Geographic Information System[7,8]

Human beings are also known for thinking, planning, predicting, and taking decisions based on geography. This ability or skill of human beings can be improved and enhanced with the help of a scientific/technological tool, which could comprehend geography and make realistic and intelligent decisions. Such technological tool is termed as GIS, as it

[6] http://baramulla.nic.in/.

[7] Thinking About GIS: Geographic Information System Planning for Managers.

[8] http://kupwara.nic.in.

Fig. 2 District Baramulla (*Source* Wikimedia)

is able to give intelligent decision based on the geography. A GIS program has capability to process geographic data obtained from several sources and integrate it into a Map project. Such maps are interactive and are very informative. These maps can be effectively scanned in any direction, zoomed (in/out), and provide numerous important information and decision outputs. They can be further customized or configured or chosen to display the objects based on the users choice of numbers (how many roads/objects to display), and how to display the selected roads/objects. Smart GIS programs are designed to perform complex calculations for tracking storms or predicting erosion patterns.

GIS Functioning and Data Integration

GIS has an excellent integration capability. It can link any homogeneous or heterogeneous geo or special data which is otherwise difficult to relate through any other means. It uses a combination of mapped variables to build and analyze new variables. Using this technology, we may be able to integrate agricultural records with hydrograph data to determine which streams will carry certain levels of fertilizer run-off. Even, agricultural

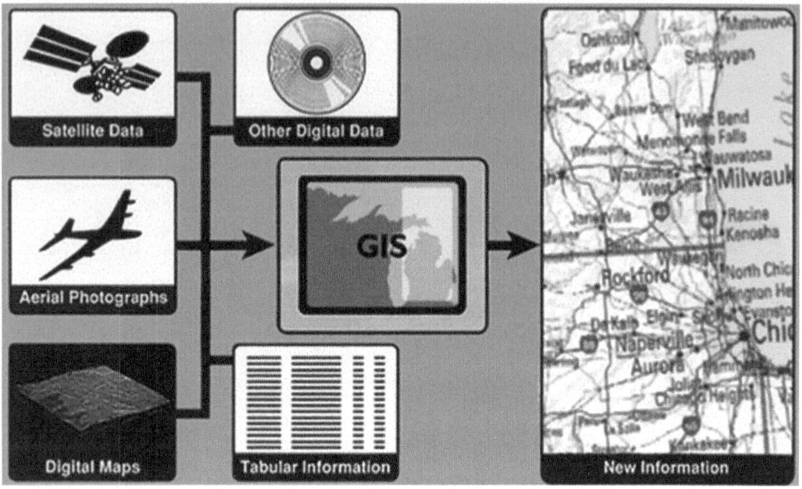

Fig. 3 Data integration is the linking of information in different forms through GIS (*Source* US geological survey)

records can be integrated to estimate the amount of pesticide used in a land/field. By locating these fields and intersecting it with streams, the GIS can be used to predict the amount of nutrient run-off in each stream. Moreover, as the streams converge, the total loads can be calculated downstream where the stream enters a lake. The following figure depicts the way a GIS integrates data for making it relevant in decision-making (Fig. 3).

GIS is based on the principles of abstraction. It represents real-world objects such as roads, land use, elevation, bridges, etc., with digital data. Real world objects can be divided into two abstractions: discrete objects (a house) and continuous fields (rainfall amount or elevation).

For abstraction purposes, GIS uses two broad methods to store data:

1. Raster: It is a digital image represented in the form of grid of pixels. Data/image is usually stored in its original form. Rasters appear to be pixelated as each pixel has its own value or class.
2. Vector: It is a simple vector map which uses various vector elements, such as a well or a city can be represented by a dot/point, rivers are represented by lines, lake by polygon. Vector data consists of coordinates and have vertices and paths.

GIS Tools (Ormsby et al. 2004)[9]

GIS software provides the functions and features required to store, analyze, and display geographical information, thus facilitates effective and intelligent decisions.

Although, there are several GIS Solutions available off the shelf, the present study has been restricted only to two popular GIS Tools which are generally used in Disaster Management applications. These tools are—ArcGIS and ITRIS (Integrated Seismic Research and Information System).

The Kashmir Earthquake of 2005[10]

On October 8, October 2005, at 8.50.38 A.M. (local time), an earthquake of severe intensity (magnitude 7.6 on the Richter scale), with epicenter at 34.432°N, 73.537°E, occurred in the Muzaffarabad Region of Pakistan Occupied Kashmir. The tremor which lasted for 6 minutes resulted in uncountable fatalities, huge property destruction, severed communication networks largely in Pakistan Occupied Kashmir, and the North West Frontier Province (NWFP) of Pakistan including the adjoining areas of India Administered Kashmir (Fig. 4).

In areas closer to the epicenter, most of the buildings made out of cement and sand collapsed. Within 25 km from the epicenter, approximately, 25% of the buildings collapsed and another 50% were severely damaged. The effected districts of Indian side include, "Poonch", "Baramulla", "Jammu", "Udhampur", "Ramban Kathua", "Srinagar", "Budgam", "Anantnag", "Pulwama", and "Kupwara". However, due to the proximity to the epicenter, the districts of "Baramulla" and "Kupwara" faced maximum damage and fatalities.

As per the official figures, 1360 persons lost their lives, 6622 were reported to be injured and 33 persons were reported to be missing after this calamity.

Survey Observations

A survey was conducted in the following six villages of the two worst effected districts, namely—"Kupwara" and "Baramulla" (Table 1):

[9]Types of Geodatabases (ArcGIS 9.2 Desktop Help). ESRI.

[10]http://www.jammuandkashmir.nic.in.

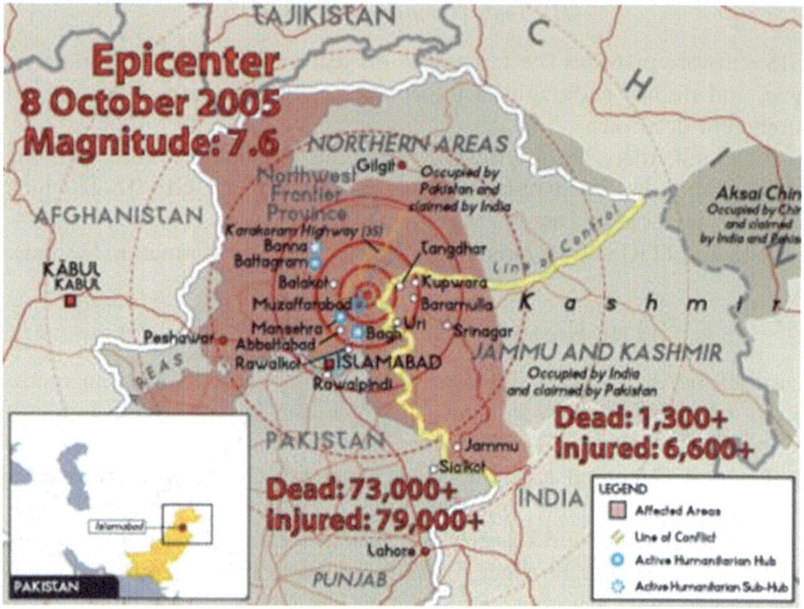

Fig. 4 Epicenter of the 2005 earthquake (*Source* Centre for Excellence in Disaster Management and Humanitarian Assistance)

Table 1 Villages visited during survey

S.No	Village	District	Approximate distance from epicenter of earthquake (km)
1	Thandipora	Kupwara	90
2	Chowkibal	Kupwara	70
3	Teetwal	Kupwara	20
4	Tanghdar	Kupwara	40
5	Panzgam	Kupwara	45
6	Uri	Baramulla	25

Casualty Figures

The 2005 Kashmir Earthquake (also known as the South Asian earthquake or the Great Pakistan earthquake) was a major earthquake which shook the Pakistan Occupied Kashmir and NWFP near the city of Muzaffarabad, including the adjoining areas of Indian Administered Kashmir on October 8, 2005 for long 6 minutes. The intensity of this earthquake (7.6 on Richter Scale) was quite close to the 1906 San Francisco earthquake, 1935 Quetta earthquake, 2001 Gujarat earthquake, 2009 Sumatra earthquakes, and 2015 Nepal earthquake (7.9 magnitude).

Table 2 Depicts the Official Casualty Figures and the Relief in Terms of Cash/Damage Compensation, Distributed to the Indian Side of Effected Persons.

Table 2 encompasses the various crucial parameters such as: District-wise, lives lost, injured, fully/partially damaged homes including amount of damage compensation based on the nature and type of fatality. It is evident from the above figures that the districts of "Kupwara" and "Baramulla" were worst affected, resulting in large relief distribution/compensation in these two districts.

Casualty Figures of Surveyed Areas

The following table depicts the casualty figures of the areas that were surveyed during the said study:

The following chart shows the casualty figures of the six villages that were surveyed (Fig. 5).

The casualty figures of "Uri" and "Teetwal" villages is toward the higher side as these are the areas closest to the epicenter of the earthquake. The total deaths that occurred in the surveyed areas were 423. Officially, the total population of the six areas covered in the survey is 125,653.

The analysis of the above-surveyed data yields an extremely crucial result. This provides information regarding the death of 0.33% of the population in the said disaster.

It was extremely difficult to approach the effected villages as the region is mountainous. If an effective GIS technology would have been in place at the time of study in the state of J&K, it would have provided a complete and realistic assessment of the damage to roads, houses, people, etc., thus facilitating the Disaster Management agencies in reaching difficult areas in less time resulting in reduced causality figures (Fig. 6).

The above chart depicts the casualty figures in the villages surveyed under the study. It is observed that only the villages—"Teetwal", "Tanghdar", and "Uri" have recorded death figures. Further, it is evident that the sample figures collected during survey are in sync with the official figures of casualties.

Table 2 O/o the divisional commissioner of Kashmir (all amounts in Rs. lakhs)

S.No	*Anantnag*	*Baramulla*	*Budgam*	*Kupwara*	*Pulwama*	*Srinagar*	*Total*
Lives lost	0	674	01	276	0	2	953
Amount paid @ Rs. 50,000	0	336.50	0.5	138	0	1	476
Injured	0	399	8	94	0	316	817
Amount paid @ Rs. 5000	0	0	0.04	5.9	0	15.8	21.74
Fully damaged homes	10	14,710	44	8994	13	11	23,782
Amount paid @ Rs. 100,000	10	14,642	44	8495	11.8	11	23,213.8
Partial Damaged Homes	112	91,334	10,560	66,887	120	5857	174,870
Amount paid @ Rs. 30,000	10.38	8346.63	144.29	12,305.32	29.15	477.32	21,313.09
Shelters Constructed	0	11,527	0	7995	0	0	19,522
Amount paid for shelter construction @ Rs. 30,000	0	3548.10	0	2392.5	0	0	5940.6
Incentives paid for on time construction @ Rs. 5000	0	448.40	0	266.80	0	0	715.2

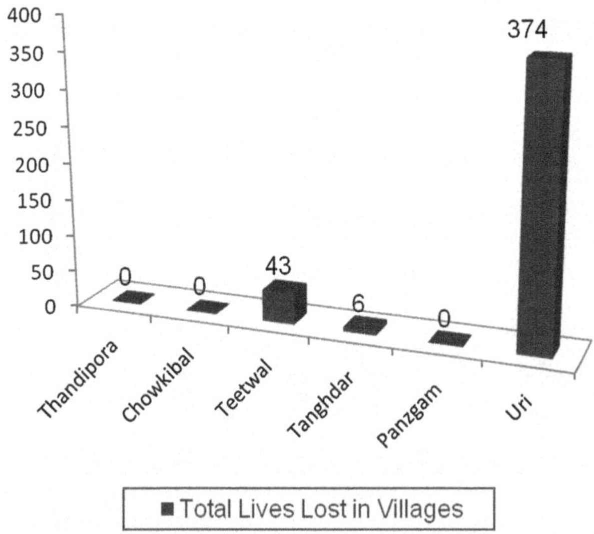

Fig. 5 Total lives lost in villages

Casualties Due to Unavailability/Delayed Medical Aid
During the survey, people of the effected areas were interacted to assess the unavailability/delay (if any) in obtaining the medical aid (individual/family members). From Table 3, it is observed that 13 deaths had occurred in the villages, namely, "Teetwal", "Tanghdar", and "Uri" due to unavailability of medical aid. The total deaths that occurred in these three areas were 23. The following chart represents the same (Fig. 7):

It is inferred that 57% of deaths that occurred in the regions were actually due to non-availability of Medical aid and other disaster-related aids.

GIS System, in place, would have provided real and much-needed relief in time. Described below is a snapshot from a GIS mapping tool where the impact of an earthquake is seen on a road route between two cities. Information on this view would be vital in rushing supplies and relief to the concerned towns as well as planning for the same (Fig. 8).[11]

[11] http://lagic.lsu.edu/gisprimer.

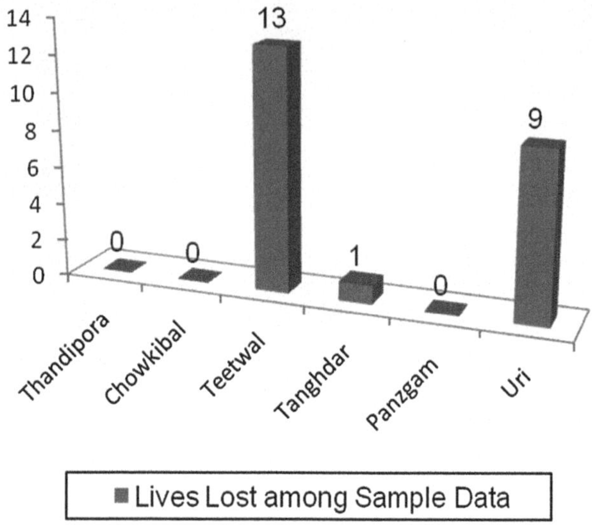

Fig. 6 Lives lost among sample data

Table 3 Figures of casualties in surveyed villages

S.No	*Name of village*	*District*	*Population*	*Total lives lost as per official data*	*Lives lost among sample data collected*	*Average no. of days taken by DMA*	*Deaths due to delay*	*% deaths due to delay*
1	Thandipora	Kupwara	2122	0	0	0	N/A	N/A
2	Chowkibal	Kupwara	3224	0	0	14	N/A	N/A
3	Teetwal	Kupwara	2986	43	13	6	8	62
4	Tanghdar	Kupwara	2665	6	1	5	1	100
5	Panzgam	Kupwara	2944	0	0	13	N/A	N/A
6	Uri	Baramulla	111,712	374	9	66	4	44
	Total		**125,653**	**423**	**23**		**13**	**57**

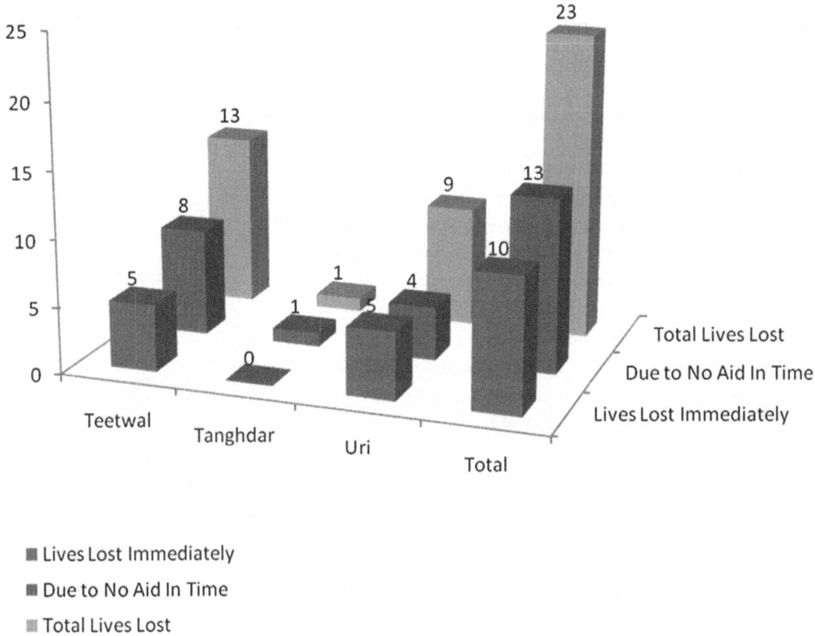

Fig. 7 Total lives lost in villages due to delayed aid

Effectiveness of GIS in Disaster Management

Reduced Causality (Erickson and Moskalev 2009; Landefeld and Seskin 1982)

From the above study and analysis, it is evident that there would have been high probability of saving many more lives provided an effective GIS System would have been in place during the time of disaster in the state of Jammu and Kashmir. Further, it is clear from the study that, 57% of deaths had occurred due to nonavailability of Medical and other disaster-related aids.

From the sample data of 80 people accurately representing the population of earthquake affected residents on the Indian side, a minimum number of 13 lives could have been saved out of the 23 lives that were lost without GIS in place.

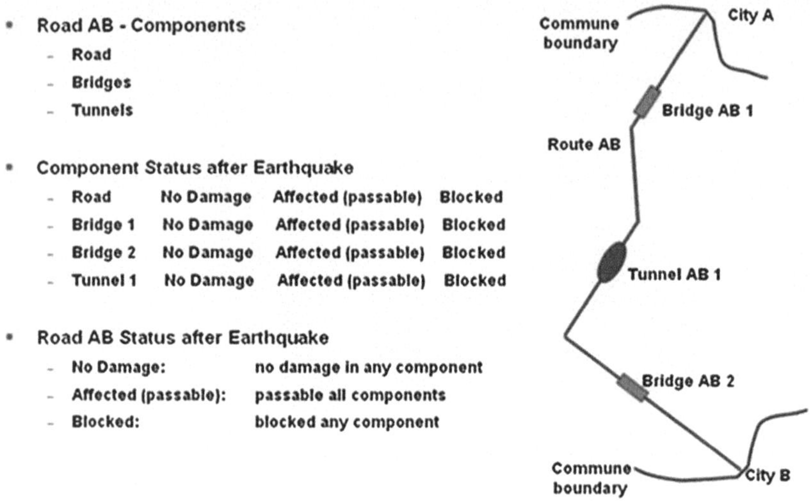

Fig. 8 Reduction in reaction time during disaster (*Source* gisdevelopment.net)

Damage to Property (Mechler 2005; Roy 2008)

The respondents were surveyed to estimate the loss due to damage of their homes and damage relief received by them.

It was observed that the amount of relief was categorized as follows:

i.	Full damage to house	Rs. 1,00,000
ii.	Partial damage to house	Rs. 30,000
iii.	Constructing shelter	Rs. 30,000
iv	Constructing shelter within stipulated time (bonus)	Rs. 5000

About 100% of the respondents claimed that they had lost much more than they were compensated for. These claims could not be verified hence this gap. Generally, after severe disasters/earthquakes, the estimate of loss is done from outside the affected areas, resulting assessment gaps. Such gaps can be minimized by implementation of GIS-based systems which can issue real-time loss estimates, using its datasets including integration of various databases and data sets. Further, availability of a satellite photograph of effected area soon after the disaster would facilitate direct damage assessment. A quantitative measure of the extent of devastation may be possible by comparing the images, before and after the devastation, if a satellite photograph of the area in question is available in the image library before the disaster.

By implementing the technique of mapping to 3D-GIS on the entire area, it would have been possible for the Disaster Management Authorities to determine the exact position of the houses and infrastructure that existed before the disaster. Similarly, with 3D-GIS, it would have been possible to know the extent of damage caused by the earthquake. Using appropriate mapping technique with strong 3D-GIS software, it would have been possible to extract the level of damage to each building within the area, thus eliminating any doubt/gap regarding claims made after disaster.

The following chart shows the difference between the claims and the amount received as relief by the respondents of the survey (Fig. 9).

The timeline in the Figs. 8 and 9, clearly depicts a vast difference in the claim (damage to property) and the subsequent relief amount distributed.

In fact, the authorities need to authenticate and verify the claims before the distribution of the relief. In the absence of a realistic and authentic assessment facility, they decided not to compensate the people but only to provide relief in a blanket fashion. However, with an appropriate GIS-enabled disaster management system, the authorities would have been in a better position to verify the claims and perhaps the actual victims would have been given their due rather than fake claimants and influential people of the villages.

Conclusion

The scope and size of the ICT implementation in Disaster Management in India is expected to be \$10 billion in 10 years. This fact was even stated in a business conclave during the 58th International Astronautically Congress held in 2007. The annual revenues of the ICT in Disaster Management/GIS market are expected to grow from an estimated \$4–150 billion in the next decade globally.

With more and more government agencies, private companies and individuals using ICT (GIS) and high-resolution imagery services, the market is growing by leap and bounds and would certainly facilitate in realistic and quality assessment and analysis of disasters by helping the stakeholders in planning, readiness and execution, and mitigation measures.

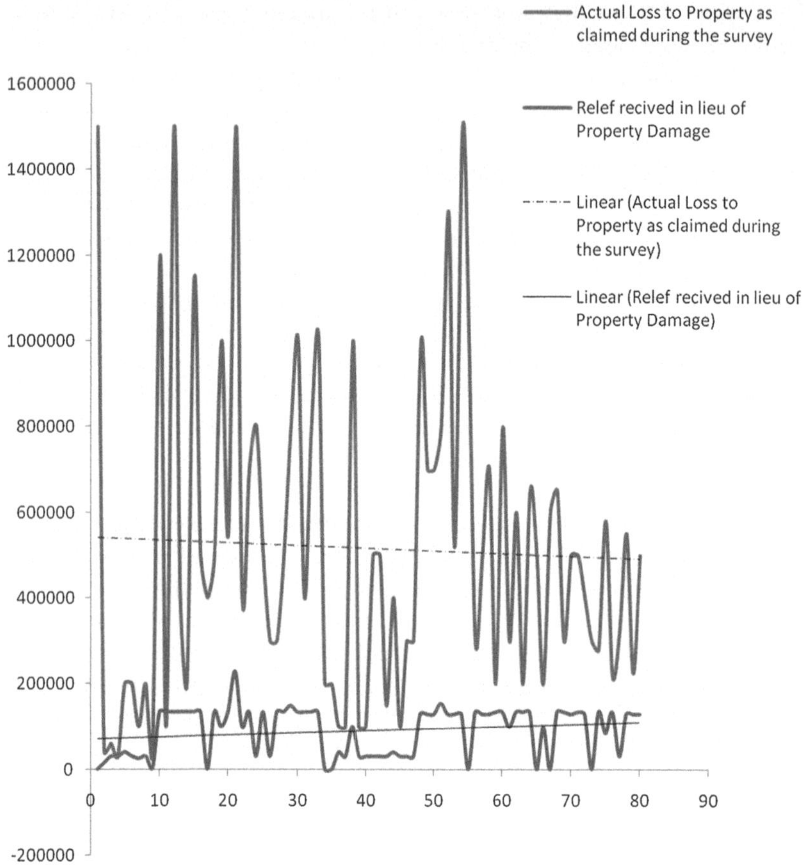

Fig. 9 Gaps in claims and relief distributed

Recommendations

The disaster mitigation programs using Innovative Governance Techniques through ICT implementations must be extensively implemented and practiced. The broad recommendations are as follows:

9.1 A multi-disciplinary and pro-active approach will result in improved disaster management and mitigation. The various stakeholders such as the community, civil society organizations, and the media have to collectively play a key role, besides, the various vital initiatives, institutional and policy framework, disaster prevention, mitigation and preparedness initiatives and measures taken by the Central and State Governments in India. We can march toward a protected and sustainable nation development only when the development projects are sensitive toward reduction of risks to human life.[12]

9.2 GIS as a tool is useful in all spheres of Civil Administration across the globe. The power of decision-making provided by these tools cannot be ignored for long. In order to deliver better services to its citizens, our country has to move fast in line with western world in implementing and harnessing the advantages of advance technology.

9.3 Use of indispensable and vital technological advancements/tools such as Internet, GIS, Remote Sensing including Satellite communication, which are central to planning and successful implementation of most Disaster Risk Reduction initiatives should be promoted. Further, the potential of most advanced technologies should be harnessed in early warning, preparedness and response systems along with adequate emphasis on building human capacities to use these tools and technologies (Kwan and Lee 2005).[13]

References

Ahmad, Ayaz. (2003). *Disaster Management Through the New Millennium*. New Delhi: Anmol Publications.

Erickson, Charles, and Moskalev, Roman. (2009). "Economic Value of a Human Life". Retrieved February 8, 2010, from http://www.ecosys.com/spec/ecosys/download/UniFr/Travaux/Life.pdf.

Kwan, M.-P., and Lee, J. (2005). "Emergency Response After 9/11: The Potential of Real-Time 3D GIS for Quick Emergency Response in Micro-Spatial Environments". *Computers, Environment and Urban Systems* 29: 93–113.

Landefeld, J. S., and Seskin, E. P. (1982). "The Economic Value of Life: Linking Theory to Practice". *American Journal of Public Health* 72(6): 555–566.

[12] http://www.esri.com.

[13] http://www.esri.com.

Mechler, R. (2005). "Cost-Benefit Analysis of Natural Disaster Risk Management in Developing Countries".

Ormsby, Tim, Napoleon, Eileen, and Burke, Robert. (2004). *Getting to Know ArcGIS Desktop: The Basics of ArcView, ArcEditor, and ArcInfo Updated for ArcGIS 9*. Redlands: ESRI Press.

Roy, Sandip. (2008, October 1). "The Future of Earthquake Disaster Management: Use GIS and Probabilistic Risk Assessment to Enhance Preparedness (GIS Analysis)". GEO World.

Twigg, J. (2002). "Lessons from Disaster Preparedness". Benfield Greig Hazard Research Centre, University College London.

United Nations Development Programme. (2004). *Reducing Disaster Risk: A Challenge for Development*. New York: UNDP.

van Oosterom, Peter, Zlatanova, Siyka, and Fendel, Elfriede M. (2008). *Geo-information for Disaster Management*. Berlin: Springer.

Urls:

http://www.censusofindia.net.

http://www.gis.com/.

A Sociological Perspective on "Institutional Coordination" for Disaster Risk Reduction in India

Sanghamitra Nath

Introduction

Malin landslide in Pune, cyclonic storm "NILOFAR" in Gujarat, Jammu and Kashmir floods, Hindu Kush earthquake, heat wave, and the list just goes on. Presently, the only solution to India's alarming vulnerability to natural disasters depends upon disaster resilience. The English word "resilient" was derived from the Latin word "resilïre" literally meaning "bounce back". It was used by risk managers referring to a return to a normal functioning state rapidly after a disturbance (Giroux and Prior 2012). The significant utilitarian connotation makes "resilience" a desirable feature in a number of fields, especially disaster management. The Hyogo Framework for Action 2005–2015 and the results of the 2005 World Conference on Disaster Reduction led to a re-thinking of disaster risk reduction (DRR) discourse and policy formulation from a new perspective (Manyena 2006: 434–435). It highlighted the need for a new culture of disaster response with greater

S. Nath (✉)
Department of Sociology, Bajkul Milani Mahavidyalaya, Vidyasagar University, Midnapore, West Bengal, India

B. Zutshi et al. (eds.), *Disaster Risk Reduction*,
https://doi.org/10.1007/978-981-10-8845-2_10

emphasis on resilience rather than vulnerability (Ibid., pp. 433, 435). The Sendai Framework for 2015–2030 further entrenched "resilience" into DRR. India, being a signatory to the Sendai Framework, adopted resilience into its disaster management legislation, plan, and policy. In 2016, Prime Minister Narendra Modi emphasized India's disaster resilience, "to maximize the ability of the country to cope with disasters at all levels" through "sincere cooperation of the center and state governments, coordination of various ministries and departments" among an array of actors.

Disasters pose a frightening range of risks making risk reduction a core concern of emergency governance which, in turn, is best actualized through coordination with related institutions. Optimal networking at the central, state, and district levels is essential for effective disaster preparedness, prevention, mitigation, rescue, rehabilitation, and relief. Institutional coordination is a vital mechanism since disasters overwhelm public administration so much so it throws disaster handling arrangements into inaction, disorder, and frenzy. As a core feature of emergency governance, institutional coordination "inspires social confidence, which is the one thing [it] must have if it is to be universally acceptable" and "[b]ecause of this, they are able to serve as links between one situation and another" (Collins 1985).

This paper is a study of institutional coordination for DRR in India. It explores the subject sociologically, particularly from a functionalist perspective. Not only it helps to understand the manifest, moral, and latent functions but also provides an insight into the element(s) which make up effective institutional coordination. For this, the paper is divided into the following sections: Institutional Coordination Problems in Current Times, Provision for Institutional Coordination in the Disaster Management Act 2005, and Sociological Perspective on Institutional Coordination in Modern Times.

Institutional Coordination Problems in Current Times

In June 2013, Uttarakhand became infamous due to the tragedy that erupted out of torrential rainfall. One of the main causes attributed was poor coordination between the state government, the central government, and the Indian Meteorological Department (IMD). While the state government did not adopt and apply the national guidelines on disaster preparedness and risk reduction, the National Disaster Management Authority (NDMA) failed to enforce the same (Upadhyaya 2013).

In September 2014, Jammu and Kashmir witnessed unseasonable and extreme rainfall that led to severe floods. Though the IMD had issued a warning about the probability of heavy to very heavy rainfall, the state government failed to comprehend the warning and take precautionary actions (Narain 2014). *The State Department of Irrigation and Flood Control did not monitor and maintain embankments or provide any warnings when they were breached. The Central Water Commission failed to provide any information on river flow in the downstream areas* (Thakkar 2014).

In December 2015, Chennai was ravaged by floods due to failures in urban governance, water and environmental governance (Pereira 2015). *Worst, post-disaster period witnessed absence of coordination between government agencies busy with rescue operations and independent relief teams. It was reported until the government was done with rescue operations, it would not be ready to receive relief and help* (Janardhanan 2015).

The aforesaid tells of the status of institutional performance as well as institutional coordination in the country. Particularly, institutional coordination is extremely important because the tasks of DRR are so large it cannot be accomplished by a lone bureaucratic organization. The central legislation, the Disaster Management Act 2005, makes institutional coordination a vital instrument in the preventive approach to disasters. Besides large-scale institutionalization of disaster management, it provides for integration among national, state, district disaster management authorities, National Disaster Response Force (NDRF) and National Institute of Disaster Management (NIDM). Coordination is also required among weather stations, emergency administration, nongovernmental organizations and other diverse but related agencies. The emphasis on institutional coordination is carried forward in the National Policy on Disaster Management 2009, and National Disaster Management Plan 2016 as "Inter-agency coordination".

Between April and May 2015, a pilot survey was conducted in New Delhi with respondents associated with disaster management. Using non-probability purposive sampling and snowball sampling, it was intended to study how core institutions in disaster management relate to each other in the form of regular meetings, common capacity trainings, consultative meetings, joint fieldwork, and paper exchange only. The respondents had to rate their answers on a scale of 1–10 whereby 1 indicated very poor and 10 indicated very high. This would help to shed light on the extent of institutional coordination for efficient and effective DRR. The results were as follows:

- Regular Meetings: The highest response corresponded to rating 7 or "High".
- Common Capacity Trainings: The highest response corresponded to rating 1 or "Very poor".
- Consultative Meetings: The highest response corresponded to rating 1 or "Very poor".
- Joint Fieldwork: The highest response corresponded to rating 1 or "Very poor".
- Paper Exchange Only: The highest response corresponded to rating 7 or "High".

In short, regular meetings and paper exchange were highly used by the core institutions to relate to one another. The data reflects an uncomfortable reality—regular mutual cooperation and coordination across line departments, ministries, and disaster management authorities remains little and ineffective. A fundamental aspect in the governance of disasters is networking or building partnerships with concerned agencies within the government and outside. Interviews of respondents from the pilot study revealed not every organization had sound rapport with other organizations leading to lack of coordination and conspicuous absence of convergence of actions.

Provision for Institutional Coordination in the Disaster Management Act 2005

The Disaster Management Act (DMA) 2005 sets out explicit guidelines for coordination among institutions at the three tiers. It presents a federal arrangement to tackle disasters in the country. "Disasters" is not particularly dealt with in the three lists of the 7th Schedule of the Indian Constitution. In the event of a natural disaster, the principal responsibility rests with the state government and the central government provides additional support in the form of financial and logistical support. The extent of central assistance depends upon seriousness of a natural disaster, scale of necessary relief operation, central assistance requirements to supplement state's financial resources, and logistical support needed by the state government (Disaster Management In India—A Status Report 2004). The Ministry of Home Affairs (MHA) is the apex body for disaster management and coordination of relief and response. The NDMA is the central organization for disaster management. The

NDMA coordinates with Lal Bahadur Shastri National Academy of Administration, Mussoorie, and Sardar Vallabhbhai Patel National Police Academy, Hyderabad, via short duration capacity building programs in disaster preparedness. The State Administrative Training Institutes (SATI) have the duty to train capacities at the state level.

The 2005 Act informs "who" may take part in institutional coordination and "for what" institutional coordination may take place. This is highlighted by the main general processes of disaster management under Section 35 (2) (DMA 2005)

- Coordinating actions with Ministries or Departments of the Government of India (GoI), State Governments, National Authority, State Authorities, governmental and nongovernmental organizations in the context of disaster management
- Ensuring consolidation of measures for prevention and mitigation of disasters with development plans and projects with Ministries or Departments of the GoI
- Cooperating and assisting State Governments when requested by them or deemed appropriate it
- Coordinating with the agencies of the United Nations, International organizations, and governments of foreign countries for the purposes of this Act
- Establishing institutions of research, training, and developmental programmes in the field of disaster management.

The focus on institutional integration is true even at the state level. Some of the chief measures of the State Governments would include under Section 38 (2) (DMA 2005: 1):

- Coordination of actions among different department of the State Government, State Authority (SDMA), District Authority (DDMA), local authorities, and other nongovernmental organizations
- Cooperation with and assistance to the National Authority and National Executive Committee, State Authority, State Executive Committee, and District Authorities in disaster management
- Cooperation with and assistance to Ministries or Departments of the GoI as requested or when deemed appropriate
- Make certain that measures for prevention and mitigation are merged with development plans and projects by the departments of State Government

- Take into account that State development plan integrates measures for reduction or mitigation of vulnerability of different parts of the State to different disasters
- Confirm that different parts of the State prepare disaster management plans in accordance with the guidelines provided by the National Authority
- Ensure that different departments of State Government and District Authorities take appropriate preparedness measures
- In the event of a threatening disaster situation or disaster, the resources of different departments of State Government should be made available to the National Executive Committee or State Executive Committee or District Authorities for the purpose of effective response, rescue, and relief.

The NDRF also works in cooperation with the MHA, NDMA, SDMAs, Block Development Officers (BDOs), Tehsildars, Panchayats, stakeholders, communities, and State Governments. When the NDRF battalions are dispatched to demarcated areas of responsibility, the Commandant establishes contact with the District Magistrate. The NDRF conducts both structured and unstructured training. The structured training on preparedness and response is held for the BDOs, Tehsildars, Panchayats, health authorities, education authorities, relief distributors, civil defense, police, fire fighters, voluntary organizations, and other stakeholders.

The unstructured training is conducted for the community to build community capacities and resilience. Communities are physically and socioeconomically vulnerable to periodic disasters occurring with varying intensities. They have "most to gain" if the impact of disasters on communities can be reduced (Shivaji 2016). The imperative for community-based disaster management arises from the recognition that the community is the actual first responder to disasters. Training of first responders makes the case for a proactive approach to disaster management. The training comprises of components viz Community Capacity Building Programme and Public Awareness Campaign within the area of responsibility (About NDRF, n.d). The NDRF holds interaction/meeting/exercises with voluntary agencies such as National Service Scheme, Civil Defense, Nehru Yuvak Kendra, School Children (School Safety Programmes), Home Guards, National Cadet Corps, and Nongovernmental Organizations. They arrange demonstrations on First Aid, a method to save oneself and others with household/local items, organize film shows on past disaster mitigation scenarios, and conduct mock drills along with representatives of the community and local authorities. Such good practices help in promoting self-reliance, ensuring emergency management plans meet local needs and circumstances, attaining cost-effective and sustainable solutions by way of harnessing traditional organizational structures and mechanism, and capacity building activities and partnership with community disaster committees and volunteers using local dialects as well as local values and cultures (Shivaji 2016). In short, community sensitization to precautions and preventive actions helps to build community resilience (Avasthy, n.d.). In March 2016, the NDRF conducted Familiarisation Training and Community Awareness/Preparedness Programme in Arunachal Pradesh, West Bengal, Chhattisgarh, Lakshadweep, Maharashtra, Gujarat, Jammu and Kashmir, Haryana, Uttar Pradesh, and Karnataka (Familiarisation Training and Community Awareness/Preparedness Programme 2016).

Natural disasters are nonroutine events which occur routinely. Of late, there has been a remarkable increase in natural disasters in India. As a result, the Act puts in place public administration which should prevent, prepare, mitigate, rescue, rehabilitate, and provide relief in times of natural threats. The legislation has the manifest function of effective governance of disasters that is contingent upon, besides other mechanisms, institutional coordination. Beneath the manifest function, there is a moral function to the law. It comes across as a "savior" of precious lives as well as resultant outcomes from cumbersome, expensive and painful processes of development. In a disaster, there is so much to protect because the fear of loss rides high. The law adopts a systematic approach whereby a range of institutions are mobilized to work together to save the threatened society at large. Kant believed that morals emerged from duty to the ethical principles. The moral duties underlying the law echoed of moral obligation to ethical principles. The ethical principles, in turn, reflected Kant's moral philosophy of "[d]uty and willing" (Swingewood 1984).

While institutions are required to coordinate dutifully and willingly within well-defined institutional frameworks, they engage in social exchange through individual representatives of those institutions. These representatives are already socialized into the law's and institutions' moral philosophy that gets reflected in role performances. During social exchange, their role performances have direct bearing on the institutions' image and credibility. The social exchange is mandatory not simply because it is legally compelling. It is mandatory because it is morally compelling—it exerts constraining and enabling power on the collective representatives thereby demanding moral obligation to ideals through which society sees itself (Swingewood 1984).

Sociological Perspective on Institutional Coordination in Modern Times

Disasters are periods of instability. It throws a society into chaos and confusion. Since human beings are not passive recipients of social realities, they enact upon them to improve their conditions. While external factors determine social reality as well as governance practice, institutional coordination becomes a well-thought-out calculated response to external vagaries that transform a "normal" society into an "abnormal" one. It acts as a mechanism to plan a course of action for the future in order to minimize the transformation and related collective losses. The frequency of disruptive events shapes experience, law, and governance. The institutionalized response may be counted as rational action because "[w]e think ourselves as rational, masters of our own destinies" but "our rationality itself is given to us by the social structure we inhabit, a structure that forms us to think in one particular way rather than another" (Collins 1985).

As mentioned earlier, institutional coordination is a form of social exchange. On the surface, items exchanged are information, knowledge, and data among bureaucratic representatives. The items are exchanged to bring about convergence of plans and actions as part of pre-, during, and post-DRR measures. According to Levi Strauss, in any social interaction it is the exchange which matters rather than the things exchanged. For Malinowski, the exchange carries a symbolic meaning—it strengthens the network of institutional relationships (Abraham 1982). In this case, the symbolic meaning indicates integration of various social organizations which become bound in a relationship of reciprocity. Bureaucratic representatives' exchange involves giving, receiving, and an obligation to repay. It is this principle of reciprocity that is carried forward in further exchanges. In recognition of the benefits of this shared value during catastrophic times, "men attribute meanings and values to the things exchanged independently of the intrinsic value of the items involved" (Abraham 1982). Two critical assumptions constitute the institutionalization of the norm of reciprocity in the views of Gouldner (Abraham 1982),

- People should help those who have helped them
- People should not injure those who have helped them

For Gouldner, internalization of the specific norm of reciprocity has many advantages such as (Abraham 1982):

1. It checks the disruptive potentialities of power differences
2. Besides the fulfillment of obligations that contributes to the stability of social systems, it caters to the period when there is a pending obligation that helps in cementing social relationships. Hence, neither the debtor nor the creditor is eager to break off mutual relations
3. Due to the sentiments of gratitude and rectitude, it serves as a second-order defence of stability by providing an additional moral sanction for conforming to the obligations
4. It acts as a "starting mechanism" initiating social interaction.

Despite high functional relevance, regular institutional coordination remains inadequate and ineffective. As early as 1954, Jawaharlal Nehru was vocal about the administrative value of institutional coordination because "Administration is meant to achieve something, and not to exist in some kind of an ivory tower, following certain rules of procedure and, Narcissus-like, looking on itself with complete satisfaction. The test, after all, is the human beings and their welfare" (Administrative Reforms Commission 2009). Even today, institutions exist as ivory towers due to functional specialization and autonomy. Much before Nehru, the French sociologist Emile Durkheim was worried about the general atomistic tendencies in modern societies. While Nehru's concern was with institutions, Durkheim's was with individuals in modern societies. Modern societies have inherently typical characteristics such as growth and spread in population over a territory, and the concomitant rise of complex division of labor. Highly developed division of labor causes social roles to become increasingly specialized and fashions individuals increasingly different from each other (Collins 1985). He observed that division of labor had emerged to be an imperative rule of conduct, a duty (Durkheim 1964). It was indeed the source of civilization and the necessary condition for development in societies (Durkheim 1964).

Durkheim's interest remained firmly entrenched in the issues of social harmony and social cohesion. He mused upon what held modern societies together given the expanding social differentiation and functional specialization. However, those very features prompted mutual inter-dependence among individuals "for the fulfilment of a wide range of social needs possible through the ongoing exchange of goods and services" (Milbrandt and Pearce 2011). Not only did mutual interdependence reinforce social bonds but also forged solidarity in modern societies. He argued social solidarity to be "the most essential condition of social life" (Durkheim 1964). Better known as organic solidarity, it appeared only where there was cooperation which itself came out of division of labor (Durkheim 1964). The concept of "organic solidarity" emerged from organic analogy wherein each organ, with its special physiognomy and autonomy, worked together to maintain the unity of the organism (Durkheim 1964).

Durkheim surmised that the most visible symbol of organic solidarity was restitutive law. This law primarily ordained a return of things as they were in the past (Durkheim 1964). Restitutive sanction was not expiatory because it did not arise out of collective conscience. If it did, it was only a feeble state of it. Unlike repressive law, it constructed specialized organs through which functionaries, who underwent special training, adjudicated between interested parties. On its own, it did not intervene; it did only in case of petitions submitted by interested parties (Durkheim 1964). Restitutive law was civil law that primarily facilitated and integrated varied organs to maintain the whole (Milbrandt and Pearce 2011).

Institutional coordination depends on organic solidarity that results in effective social exchange.

The Disaster Management Act is a form of restitutive law based on division of labor where specialized and autonomous institutions are in a relationship of mutual dependence. It aims to return normalcy and restore society to its previous state after a disaster. Corollary, the objective entails tackling changes brought about by disasters. However, the ultimate purpose of the Act and institutional coordination is more than organic solidarity—it has a latent function. It is to keep social order in society and prevent phases of anomie during times of disaster. Disasters have a high propensity to generate social instability and normlessness creating widespread anomie in the crisis-stricken society (Swingewood 1984). Social order is a function of organic solidarity externally expressed or indicated by restitutive law. A pre-requisite for social order is strict compliance to legal rules and regulations reinforced by penalties of fines and punishments (DMA 2005). The real function of institutional coordination, thus, revolves around the twin tasks of maintenance of social stability and social order as well as resumption of the central values of the old society (Swingewood 1984).

Returning to social stability, social order, and values reminiscent of the previous version tends to be problematic. While natural disasters exaggerate the existing social divisions and inequalities, restoration of social order could imply restoration of earlier hierarchy and stratification which intensifies deprivations and human rights abuse of the vulnerable in the reconstructed society. Needless to say, this type of "order" will hardly produce stability and question values hitherto held onto. The Disaster Management Act 2005 and the Disaster Management Plan 2016 do not address the complications underlying a return to "normalcy". While the latter explicitly aims at building resilience, it is not known how it will be achieved since it fails to visualize different vulnerable groups experience disasters differently and a return to normalcy is possible only when their needs are understood and met by state institutions. This glaring amiss puts at great risk millions of women, children, disabled, and elderly people as well as lower caste and tribal communities.

The United Nations Development Programme (UNDP) India found inequality persistent among the country's most marginalized groups including women, scheduled castes, tribal, and rural populations. In 2014, India ranked 127 out of 146 countries with a value of 0.563 in gender inequality. Overall, India ranked low on UNDP's Gender Inequality Index at 12%. The ranks remained low for Human Development Index for Scheduled Caste at 0.27 and Human Development Index for Scheduled Tribe at 24.01. Post-disaster, adolescent girls, and women face many problems from lack of sanitary towels, sexual violence in camps, to poor healthcare for expectant mothers. Many disabled people cannot physically access buildings or transport, relief camps, or aid distribution points. Lower-caste communities or Dalits, according to a 2013 report, were found to be excluded from access to emergency aid such as clean water, dry food rations, or shelter as their higher-caste neighbors. They were exploited for their labor to remove corpses and debris (Bhalla 2016).

Disasters may be seen as opportunities to reconstitute societies that are more equitable and enabling. Earlier, the Hyogo Framework for Action 2005–2015: Building the Resilience of Nations and Communities to Disasters (HFA) made DRR a national and a local priority via strong institutionalized implementation. It envisioned community resilience in the form of reduction of disaster losses but the losses were symbolic of vulnerabilities and associated risks already present in the community. While it proposed to build a culture of safety and resilience at all levels, it rightly recognized the need to reduce underlying societal risk factors (Hyogo Framework for Action 2005–2015). Of the four Priorities for Action listed in the Sendai Framework for DRR 2015–2030, Priorities 3 and 4 referred to resilience and "Build Back Better" in recovery, rehabilitation, and reconstruction. The former suggested public and private investments in DRR to enhance economic, social, health, and cultural resilience of persons, communities, countries and their assets as well as the environment while the latter recommended integrating DRR into development measures as part of build back better efforts (Sendai Framework for DRR 2015–2030). Together, these priorities may be seriously pursued to create stronger institutions with new moral code and practice which help to attain multidimensional resilience. Risks, in the form of vulnerability, remain entrenched in existing social structures and social order which get amply exposed during disasters. "Build Back Better" may be perceived to rebuild social structures and societies where vulnerabilities are reduced through coordination and partnership with local people, local government, private sector, and other stakeholders, and the vulnerable are protected from discrimination and injustice through easily accessible empowered institutions.

Conclusion

For a long time, the external environment was neglected and ignored in organizational studies.

Modern organization theory includes external environment that exerts grave influence over the functioning of organizations. Chester Bernard's "The Functions of the Executive" (1938) is a classic illustration that modern organizations cannot be closed off from outside influences. Institutions specifically established for DRR prove the constraining power of external environment. However, institutions on their own cannot overcome the gigantic challenges posed by natural calamities. They need to coordinate preferably regularly and effectively to realize the vision of disaster resilience.

In India, one of the core processes of DRR is institutional coordination at all levels of the government machinery. This was reinforced by the Disaster Management Act, National Policy on Disaster Management 2009, and National Disaster Management Plan 2016. However, pilot survey showed that institutional coordination was poor and ineffective despite the legal and policy provisions for the same. This paper was a sociological study of institutional coordination from the functionalist perspective. It was found beneath the instrumental purposes of institutional coordination were moral goals. Within the moral goals were embedded latent objectives. The latent objectives were recommencement of social order and the central values of the old society. However, return to old social order or status quo was highly problematic due to the oppressive nature of traditional social structure.

The paper also discussed the Disaster Management Act as a form of restitutive law. The need of the hour requires converging restitutive law with 73rd and 74th Amendments of the Constitution for making communities resilient through local governance. The Sixth Report of the Second Administrative Reforms Commission (2007) strongly advocates implementation of the 73rd and 74th Amendments of the Constitution that creates an institutional framework for self-governing local bodies in both urban and rural areas for generating grassroot democracy. It follows the principle of decentralization operationalized through the principle of subsidiarity thereby reconfiguring governance where citizen and community are the center. It promotes local decision-making to improve efficiency and self-reliance as well as replaces traditional hierarchical government with concentric circles of government with outward delegation of responsibility depending on necessity (The Sixth Report of the Second Administrative Reforms Commission 2007). Hence, decentralized governance will go a long way in helping disaster-affected communities to meet their expectations; however, the growth of self-governing local bodies has been uneven, halting and slow as per the Sixth Report of the Second Administrative Reforms Commission.

Institutional coordination facilitated adaptation to external environment. As a form of social exchange, it opened the door to multiple ideas which circulated among administrators "lodging for a while in their individual minds... use them to guide their own thinking and make their own decisions until these ideas are brought back again to other situations that charge up their ideas with further social significance" (Collins 1985). Thus, Donald F. Kettl called modern organisations "Learning Organisation" (Donald 1994). Institutions, at the national, intermediary, and local levels, become learning organizations when they incorporate lessons learnt from governance of disasters into laws, policies, guidelines, and practice. Institutions may draw upon academic research to become learning organizations. The current field study found regular meetings and paper exchange were highly used by core institutions of emergency governance to relate to one another. It is recommended that common capacity trainings, consultative meetings, and joint fieldwork be given equal importance for meaningful engagement among core institutions and local communities. In addition to the laid out mandate for core institutions, there should be scope for deepening local governance, skill up-gradation, knowledge sharing and technological innovation, and psychological counseling, to mention a few. Moreover, the range of incidents called "disasters" should be widened, for example, cold wave in North India and snake bite in Sundarbans claim human lives enough to be termed as "natural disasters" in this developing country. Therefore, what constitutes a disaster should go beyond political decision-making.

Assessment and analysis of factors leading to successes or failures in disaster management also speak of institutional capacity to handle disasters in the country. Further research should be done to examine whether the federal arrangement is serving the purpose of disaster management effectively, whether the three-tier institutional setup is really integrated facilitating institutional coordination and capacity, whether the institutions apply global frameworks such as the Sendai and Hyogo frameworks to actualize local community resilience, and how institutionalization of disaster management should be improved to meet the aspirations towards building back better and empowering local communities.

Bibliography

About India: Challenges. (n.d.). Retrieved from http://www.in.undp.org/content/india/en/home/countryinfo/challenges.html.

About NDRF. (n.d). Retrieved from http://ndrfandcd.gov.in/cms/Ndrf.aspx.

Abraham, M. Francis. (1982). *Modern Sociological Theory: An Introduction*. Delhi: Oxford University Press.

Administrative Reforms Commission, Twelfth Report, Citizen Centric Administration: The Heart of Governance. (2009, February). Retrieved from http://arc.gov.in/arc_12th_report/arc_12th_Report.pdf.

Avasthy, Alok. (n.d.). *Role of NDRF*. Retrieved from http://nidm.gov.in/idmc2/PDF/Presentations/Armed_Forces/Pres2.pdf.

Bhalla, Nita. (2016, June 8). "India's Plan to Tackle Disasters Neglects Women, Disabled and Lower Castes: Aid Workers". *Reuters*. Retrieved from http://in.reuters.com/article/india-national-disaster-plan-idINKCN0YU1BR.

Collins, Randall. (1985). *Three Sociological Traditions*. New York: Oxford University Press.

Disaster Management—A Status Report. (2004, August). Retrieved from http://www.preventionweb.net/files/2534_2534DisasterManagementIndia.pdf.

Disaster Management Act 2005. (2005, December 26). Retrieved from http://www.ndma.gov.in/images/ndma-pdf/DM_act2005.pdf.

Durkheim, Emile. (1964). *The Division of Labour in Society*. New York and London: The Free Press and Collier-Macmillan Limited.

Familiarisation Training and Community Awareness/Preparedness Programme—March 2016 by NDRF. (2016). Retrieved from http://ndrfandcd.gov.in/WriteReadData/userfiles/file/ndrf/Training/Famex%20Calender%20March%20.pdf.

Giroux, Jennifer, and Prior, Tim. (2012). Expressions of Resilience: From 'Bounce Back' to Adaptation. *3RG Report, Fact Sheet Risk and Resilience Research Group, Center for Security Studies (CSS), ETH Zurich*, pp. 1–17. Retrieved from http://www.css.ethz.ch/content/dam/ethz/special-interest/gess/cis/center-for-securities-studies/pdfs/Factsheet-Expressions-of-Resilience.pdf.

Hyogo Framework for Action (HFA). (n.d.). Retrieved from https://www.unisdr.org/we/coordinate/hfa.

Introduction, National Disaster Management Authority. (n.d.). Retrieved from http://ndma.gov.in/en/introductions.html.

Janardhanan, Arun. (2015, December 8). "Chennai Floods: Coordination Issues Hinder Relief Work". *Indian Express*. Retrieved from http://indianexpress.com/article/india/india-news-india/chennai-floods-coordination-issues-hinder-relief-work/.

Kettl, Donald F. (1994). "Managing on the Frontiers of Knowledge: The Learning Organisation". In Patricia W. Ingraham and Barbara Romzek (eds.). *New Paradigms for Government: Issues for the Changing Public Service*. San Francisco: Jossey-Bass Publishers.

Manyena, Siambabala Bernard. (2006). "The Concept of Resilience Revisited". *Disasters* 30(4): 433–450.

Milbrandt, Tara, and Pearce, Frank. (2011). "Emile Durkheim". In G. Ritzer and J. Stepnisky (Eds.), *Major Social Theorists*. West Sussex: Wiley-Blackwell, pp. 236–282.

Narain, Sunita. (2014, September 22). "Was Kashmir Flooding a Man-Made Disaster?" *Al Jazeera*. Retrieved from http://www.aljazeera.com/indepth/opinion/2014/09/was-kashmir-flooding-man-made-d-201492272058673744.html.

National Disaster Management Plan 2016. (2016, May 20). Retrieved from http://www.ndma.gov.in/images/policyplan/dmplan/National%20Disaster%20Management%20Plan%20May%202016.pdf.

National Institute of Disaster Management. (2013). Conference Proceedings of National Workshop on Uttarakhand Disaster 2013: Lessons Learnt. Retrieved from http://nidm.gov.in/pdf/pubs/proc%20ukw-13.pdf.

National Policy on Disaster Management 2009. (2009, October 22). Retrieved from http://ndma.gov.in/images/guidelines/national-dm-policy2009.pdf.

Pereira, Aaron. (2015, December 4). "Chennai Floods: Decoding the City's Worst Rains in 100 Years". *Indian Express.* Retrieved from http://indianexpress.com/article/india/india-news-india/chennai-floods-rains-jayalalithaa-imd-reasons-rescue-news-updates/.

Sendai Framework for Disaster Risk Reduction. (n.d.). Retrieved from http://www.unisdr.org/we/coordinate/sendai-framework.

Shivaji, Sl Mulgiri. (2016). "Community Awareness in Disaster Management—Role of NDRF". *The Saviour* 2(1): 20–23. Retrieved from http://ndrfandcd.gov.in/WriteReadData/userfiles/file/ndrf/NEWS%20LETTER/news12016.pdf.

Sixth Report, Second Administrative Reforms Commission. (2007). *Local Governance: An Inspiring Journey into the Future.* New Delhi: Government of India.

Swingewood, Alan. (1984). *A Short History of Sociological Thought.* London: Macmillan.

Thakkar, Himanshu. (2014, September 18). "Chronicle of a Tragedy Foretold". *The Hindu.* Retrieved from http://www.thehindu.com/opinion/op-ed/jammu-and-kashmir-floods-chronicle-of-a-tragedy-foretold/article6420015.ece.

Upadhyaya, Himanshu. (2013, June 21). "Uttarakhand Tragedy: How We Ignored the Writing on the Wall". *India Together.* Retrieved from http://indiatogether.org/disaster-government.

Yadav, Yatish. (2013, June 23). "As Floods Rage, Institutions Fail". *New Indian Express.* Retrieved from http://www.newindianexpress.com/nation/As-floods-rage-institutions-fail/2013/06/23/article1648469.ece?service=print.

From Response to Responsibility: An Academe–Industry Partnership on Solid Waste Management in the Philippines

Marlon de Luna Era

Introduction

By the late 1980s and early 1990s, a shift in environmental governance began to focus on advocates for effective environmental management by looking into communities and local institutions as important actors to involve in the process of environmental governance. There emerged a new view in environmental governance by pinpointing three (3) social actors, the states, markets, and the civil society (Lemos and Agrawal 2006).

The private business corporations are generally motivated by a return of its investment. These organizations seek to constantly garner public interest for their company or service businesses to sustain interest in their business, product or services. The way the private companies "pay back" to the community is through the Corporate Social Responsibility (CSR). CSR is a vivid example of partnership called

M. de Luna Era (✉)
Behavioral Sciences Department, College of Liberal Arts, De La Salle University, Manila, Philippines

B. Zutshi et al. (eds.), *Disaster Risk Reduction*,
https://doi.org/10.1007/978-981-10-8845-2_11

Social–Private Partnership (SPP). The SPP as a new form of partnership was created between the business sector and the academic institutions. The Industry–Academe partnership, while an emerging partnership during our present time and age, was commonly known to have carried out programs, projects, and activities through partnerships that involve provision of scholarship funds, on the job trainings, and similar endeavors. The emerging partnership of Industry and Academe through SPP added another dimension to its joint undertaking in the context of implementing a school-based ecological solid waste management.

In the implementation of the waste management program, it was noted that among the non-state actors to have that strong contribution in waste generation and waste management aside from the industrial and commercial establishments are the schools and universities. Such is the joint undertaking by the Toyota Auto Parts ***(TAP) as*** among those industries actively performing its CSR in the City of Santa Rosa. The TAP is the global production and supply base for manual transmissions of Toyota for its Innovative International Multipurpose Vehicle (IMV) project which was initiated to create an optimized global manufacturing and supply system for pickup trucks and multipurpose vehicles to satisfy market demand in more than 140 countries worldwide. TAP has a total of 1063 team members with main customers from other Toyota affiliates located in other Asian countries, Argentina, and South Africa. Empowered by teamwork and good product, TAP is committed:

- To provide products that guarantee customers' delight done in a timely and efficient operation empowered by continuous improvement and challenging spirit, assuring safe working condition, and in harmony with the environment.
- To sustain company growth and profitability by promoting efficient management of its resources and cost-effective programs for the benefit of customers, stockholders, employees and their families, business partners, and the community.
- To continuously uplift team member's self-esteem, quality of life and productivity through developmental programs, strengthening mutual trust, respect, and effective communication.

Santa Rosa City: The Detroit of South Luzon

The City of Santa Rosa is located in the Province of Laguna, in the southernmost part of Metro Manila (Fig. 1). It is linked to Metro Manila and other southern provinces by the South Luzon Expressway (SLEX) and has a total land area of 5514 hectares which represents 3.08% of the total land area of the province. It is mainly composite of land plains making up 97.89% of its total land area; having slopes ranging from 0.0 to 2.5% indicating level to nearly level lands. It has 18 barangays of which 3 are coastal Laguna Lake villages.

In the year 2005, the annual population growth is estimated at 6–8% resulting in a population density of 3417 per sq. km. Fifty percent of its population belong to the age bracket of 15–40 years of age.

Fig. 1 Map of Laguna Province highlighting the City of Santa Rosa

Santa Rosa became a city on July 10, 2004, as predicted in a study conducted by the Japanese International Development Cooperation (JICA) in 1994, Santa Rosa is now fast becoming the premier business city South of the Philippines. It hosts several industrial parks, such as (a) Laguna Technopark Inc., (b) Santa Rosa Business Park, (c) Meridian Industrial Complex, (d) Greenfield Development Park, and (e) Toyota Industrial Complex, and also considered as the "Little Detroit" of the Philippines with the presence of Ford Motors, Toyota Motors, Nissan Motors, Honda Cars, Star Motors, and Columbian Motors premiere car manufacturing companies. In terms of contribution to the economy, 10% of total electronics exports and 50% of local automotive production come from the City of Santa Rosa, Laguna. However, Santa Rosa, for some scientists and city planner, is not ready as a city because of the lack of physical infrastructure (i.e., wider roads and other infrastructures) in the province. Thus, poor solid waste management among other problems in the city of Santa Rosa became likewise a fast-rising problem in the becoming industrialized city which is Santa Rosa, Laguna. Given the reality stated above, it has become evident that a new partnership be formed in the upkeep of the city. Leading to the necessity of a Solid Waste Management project and the search for active partners to which Toyota Auto Parts along with the schools and universities of the city of Santa Rosa heeded the call in carrying out said task.

Roles of the Non-state Actors in the Academe–Industry Partnership

One of the projects initiated by the Toyota Auto Parts in the City of Santa Rosa is the implementation of the ***School-based Ecological Solid Waste Management (SBSWM) Project.*** The project is known as another type of Social Project Partnership or SPP in joint partnership between an industry and all the schools in the City of Santa Rosa focusing on the implementation of an on-campus solid waste management. The partnership also introduced a sustainable Ecological Solid Waste Management and in compliance with the provision of Republic Act 9003. Sought to define the following specific roles and participation of the non-state actors in the school-based solid waste management using the Solid Waste Management Framework. Using the SWM framework, the paper will be discussed based on the topic on Environmental organization, Engineering, Education, Equity, and Enforcement.

Environmental Organization in the Partnership

Implementation of the School-Based Solid Waste Management (SBSWM) project in most of the schools, revolve around student organizations as its active participants. This was made possible through the partnership between the TAP and schools in the City of Santa Rosa with the project aimed at implementing a comprehensive ecological solid waste management that would include formation of student environmental organization on campus. An initial result of this partnership is the establishment of a Young Environment Society Organization (YES-O) or the Environmental Science Club for each school.

In the Industry–Academe partnership, a number of non-state actors were directly involved in the project namely, the school principal or administrator, campus coordinator (usually a teacher), student leaders from the Environmental Science club or the Young Environment Society Organization (YES-O) and the utility workers of the partner school. The identification of non-state actors (NSA) in this project reveal the significant role the NSA play, particularly in the areas of capacity building, Information Education and Communication, actual waste segregation, waste collection, and composting of biodegradable wastes. Another NSA identified in the partnership are the junkshop owners wherein the different participating schools have partnered with a local junkshop creating an opportunity for them to sell their recyclables weekly. In some schools, some parents of school children were actively involved in the Solid Waste Management Program by helping their children in bringing recyclables in the schools. Individual parents are actively involved which is not the case among the officers of the Parents Teachers Association (PTA). As many of them have not been involved in the implementation of the SWM program, this non-involvement could be attributed to the lack of awareness on the importance of SWM implementation among the PTA officers.

Borne of the earlier mentioned partnerships, a particular environmental organization was formed and utilized in order to implement SBSWM project. The YES-O, proved to have a more sustainable implementation of SWM program. This was affirmed by the presence of several active student organizations in each of the top schools who implemented the SWM project by batch.

Engineering Aspect in the Partnership

Implementation of a solid waste management project includes numerous activities and structures necessary to the same which include: construction of facilities such as sanitary landfill, conversion of open dumpsite into controlled dumpsites and setting up of a Material Recovery Facility (MRF) per barangay. Initially, for the *SBSWM* Project, there was no expectation to construct a sanitary landfill in that locality. However, in all the partner schools the setting up of a MRF is required as well as the installation of a solid waste segregation bins and putting up a composting facility. Results from the study's FGD, also attest that a significant number of partner schools were able to install three waste segregation bins for recyclables, compostables, and residuals in compliance to the SBSWM project. In addition, a distinct food waste container was likewise put up in several school canteens where the putrescibles (food wastes) are collected daily and utilized as animal feed.

Education Aspect in the Partnership

An SWM project that is known and owned by the community has a great chance of sustainability; major to this act is to make the community aware of the project through the aspect of education. The education aspect includes dissemination of Information, Education and Communication on SWM, capacity building on SWM through seminar workshops and similar activities. The SWM campaign utilizes various media such as poems, music, poster, comics, and other forms of media. Supplemental to said activities are a number of Trainers' Training Workshop organized by the TAP in cooperation with the National Solid Waste Management Council (NSWMC), United Nations Development Programme (UNDP), Department of Environment and Natural Resources (DENR), Environmental Management Bureau (EMB), Community-Based Ecological Solid Waste Management Council (CBESWMC), Laguna Lake Development Authority (LLDA), and Save Silang Santa Rosa River (S3R2). The various workshops initiated with the different trainers aimed at integrating SWM in the school curriculum. These consisted of topics such as Republic Act (RA) 9003, community-based ecological solid waste management, air and water quality management, and ecological solid waste management modules for the secondary schools based on the curriculum set by the DENR. More than

20,000 students were able to participate in the trainers' training activities on waste management in a span of three years from school years 2007 to 2010. Also, the students have intensified their information dissemination campaign on SWM by putting up bulletin boards and publishing newsletters about the different initiatives on SWM.

Meanwhile, outcomes from Key Informant Interviews with the different school representatives yielded results with a number of them having introduced competitions such as poem and songwriting contests as well as beauty pageants that promote fashionable outfits made from recyclable materials. Such competitions, as school partners observe, create a significant increase in the level of environmental awareness among students and teachers. Further, the actual participation of schoolchildren in said competitions enabled the participation of even some of their parents in SWM activities. But, there was no tangible proof on the part of the Parent-Teacher Association (PTA) in almost all the school partners showing its support and participation in SWM programs of said partners. The incorporation of solid waste management in the different learning areas particularly in Science and Values Education subjects evidently was realized on paper. With the full-blown implementation of such SWM project yet to be realized, this completed module was turned over by TAP to the City Government of Santa Rosa and the Department of Education Division of Santa Rosa.

Enforcement Aspect of the Partnership

All schools in the City of Santa Rosa, (both private and public) have been part of the SWM partnership through the project. The designated project management team in each school serve as enforcers of the SWMs on various campuses. The type and level of enforcement varies from campus to campus depending on the kind of SWM projects implemented. Common strategies of enforcement include competition of best SWM project implementation based on the criteria set in RA 9003. A project monitoring team was organized by the TAP to determine the extent of SWM implementation and the level of compliance in the SWM implementation. This aspect of SWM has strengthened the partnership between TAP and the partner schools as it provided an opportunity to clarify expectations and accountability between the project implementors and the donor. Good practices were given incentives while areas for improvement were rectified.

Equity Aspect of the Partnership

Another important aspect of SWM implementation is the availability of resources that will be utilized to sustain the project. A mobilization fund was provided to all participating schools. An incentive system was introduced to encourage compliance and innovation in the project implementation during the partnership. During the period of six months, top three schools by batch. The top schools are given cash rewards and certificates of recognition. This strategy was regarded as venue for healthy competition among various schools in the City of Santa Rosa.

One of the positive outcomes of the partnership includes setting up of a more systematic way of selling recyclables. Students in the elementary levels were encouraged to bring dry and clean recyclables (pet bottles and tin can), which could be sold to the junkshop, these same collected recyclables are then sold to the junkshop. Eliciting positive reaction from respondents through expressing appreciation in having raised additional funds from selling these recyclables notably, with recyclable items gathered and sold resulted in additional school funds that were used for special projects on campus. It has also been observed that not only has said partnership contributed to the enhancement of creativity especially among students in making craft. It has also resulted to discovering more innovative fund-raising activities that could be undertaken after the SWM project life.

Performance of Accountability Roles in Industry–Academe Partnership

Accountability needs to respond to the questions such as what are the measures of accountability and accountability to whom. For the latter, Najam (1996) and Laratta (2009) in their individual studies conducted; utilized two types of accountability namely: internal (downward) accountability and external (upward) accountability. The downward (internal) accountability refers to the responsibility of the non-state actors to the needs and requirements of the intended beneficiaries. The condition being when there is a strong culture of internal participation, then downward accountability will occur naturally at least in theory but when there is a form of internal representational advocacy, downward accountability depends on those who are representing the constituency. On the other hand, upward (external) accountability refers to the responsibility of an individual or an entity with the funding agency/donor and meeting the formal requirements of regulatory provisions where they exist. In this type of accountability, the Civil Society Organizations (CSOs) are mentioned to be the strongest. In a related study, Fry (1995) as cited by Laratta (2009) noted that downward accountability is more complicated than upward accountability because it has two dimensions: a feeling of responsibility (or internal dimension) and responsible behavior (or external dimension). This is somewhat consistent with the definition of Keohane's internal and external accountability. Internal accountability is being referred to authorization and support by the principals to the agents within an institution while external accountability is being referred to individuals or groups outside the acting entity who are nevertheless affected by its activities. This dual perspective is not unlike that proposed by democratic accountability theorists, even though their focus is on public institutions and representation by elected officials rather than on nonprofit organizations (e.g., Behn 2001; Przeworski et al. 1999).

The opportunity presented to the NSA respondents whereby they were asked to indicate on a three-level scale on whether Statutory Organizations (SOs) should have more or less or about the same degree of involvement in doing their NPO's tasks (by requirement of reports and monitoring of their performance) during a formal inquiry. This provided an impetus for them to distinguish themselves from other study groups in viewing statutory accountability that demands more as enabling them rather than be given the view as enforcing the same.

Thus, posing the hypothesis that a positive relationship exists between the NSAs perception of SOs and their positive or negative experiences of statutory accountability demands. If the representatives perceived state requests as related to the "greater good" then accountability demands were viewed as enabling rather than enforcing (Fry 1995). Naidoo (2003) on the other hand, discussed the three levels of accountability namely, upward accountability, downward accountability, and horizontal accountability aside from the previously mentioned upward and downward accountability by providing a definition of each of the levels of accountability. For instance, upward accountability refers to the responsibility of an individual or an entity with the funding agency/donor and meeting the formal requirements of regulatory provisions where they exist. CSOs fall in this category and have the strongest in compliance of the upward accountability. The second is known as downward accountability which refers to the responsibility of an individual or organization to the people who are being served and benefitted from the project. The third one is horizontal accountability or peer accountability which requires much greater effort and attention for compliance. Failure to address this question could lead to unnecessary duplication and a failure to forge the appropriate synergies and the wastage of resources.

I concur with Brown and Moore (2001) in their argument that different accountability regimes are needed depending on the role of the NGO (e.g., service delivery or political advocacy) and how they structure their accountability will have a "decisive impact on the strategies they execute" (p. 585). In short, that once the organizational mission has been created would define accountability regime for environmental governance and in which all likelihood would craft the organization's mission and strategy. However, there is the caveat that the application of accountability regimes, one that is patterned after business sector to that of an NGO setting could produce negative results especially it this is done haphazardly and without much consideration where and how this was to be applied by NGO managers.

The last two decades saw non-state actors (NSAs) witnessing the application of adapted tools from the business and public sector in environmental governance. This was realized as it dealt with the challenges posed by rapid growth in turnover of staff, global reach, membership and so forth in its field. Leading to what would be inevitably become the emergence of governance and accountability issue. One of the basic issues that typically may arise from the same is how one sector may define its accountability as it shapes its own distinctive identity and role in society. Consequently, this necessitates that environmental governance devise its own tools, processes and systems or accountability regime scheme, and convince others of its merits. Although, the governance and accountability debate needs pacing so as to develop and test the concepts and necessary tools to create an accountability scheme; nonetheless, it is essential to hasten the pace to usher in a new style of environmental governance for the twenty-first century.

However, all debates move at their own pace, to cite the local management of forestry revenue in the South and East Cameroon resulting in a decentralized forestry taxation system introduced in 1994. Ideally, the various NSAs involved in the decentralized management of forestry revenue should be accountable not only upwards, toward higher levels in the administrative hierarchy, but also downwards, to the beneficiaries of the income and expenditure registered, that is the neighboring village communities. But, in actuality, such a thing does not occur, it is impossible to get a hold of accounts that are a faithful representation of all the income and expenditure registered each year in relation to the management of forestry revenue. This lack of downward accountability is one of the factors facilitating misappropriation of funds, which in turn hampers local development and restricts the effectiveness and efficiency of governance and decentralization. Logo (2003) breeding a failure in accountability.

Below are some of the initiatives which took a look into whether private companies report to the concerned communities that have social and environmental implications and likewise if legally constituted community representatives account to donor corporations resources provided for environmental projects.

In a SPPs, the business sector including transnational corporations (TNCs) and the local business have been observed to play a significant role in environmental governance. For example, the past ten years saw the rise of corporations declaring to have "greened" themselves and claiming to be environmentally and socially responsible through a number of voluntary corporate-based programs. To an extent, even promoting sustainable development in the south generated a particular interest in the field of corporations in global environmental governance by author Marian Miller (Clapp 2005). Documenting how her observation of the rise in power of TNCs, tended to drain away the sovereignty of these countries and exhausting their resources through their ability to influence policies and environmental outcomes in developing countries. The need to create a formal mechanism to influence the environmental behaviors being exerted by TNCs and to promote environmental practices among them has become necessary as argued by Clapp (2005). At present, efforts made through the issuance of Guidelines for Multinational Enterprises by the United Nation's Global Impact through the Organization for Economic Cooperation and Development (OECD) took precedence from calls for an international treaty on corporate accountability. The potential of these issuances and calls for corporate accountability while voluntary at the moment; maybe well realized if these are incorporated into legally binding contracts as some NGOs are proposing. In determining the performance of accountability roles of the TAP and the partner schools, the accountability indicators based on the concept of ***upward and downward accountability*** by (Naidoo 2003) was utilized.

Upward Accountability

The partnership between TAP and the different schools was with funding support from the Toyota Foundation in Singapore. The accountability measures were established between stakeholders through a Memorandum of Agreement (MOA). The MOA was executed between the TAP (project implementor) and the Toyota Foundation Singapore (donor) to thresh out expectations between two parties. The provisions in the MOA are the regular submission of progress reports of TAP to the donor agency. On the other hand, there is no MOA or even a Memorandum of Understanding (MOU) extant between TAP and the Partner Schools. All the 109 school partners were aware of their responsibility to complete a solid waste management project in their respective schools within the prescribed period. Since there was no formal agreement about the project, the participating schools submitted their reports only when being requested by the TAP. Representatives from some partner schools that even without a MOA, the school as guardian of values and integrity have innate responsibility of informing their donors of the project progress. Project Coordinators were designated in every participating school to ensure program implementation will be sustained and inherent to produce progress reports to donors as part of their function. The common activities undertaken by the partner schools were (1) Submission of accomplishment and/or progress reports to the School Administrator, Department of Education-City of Santa Rosa and the TAP), (2) Conduct of regular meetings with the school teachers and staff, (3) Submission of action plan to the school administrator and TAP, and (4) Conduct of visitation, inspection, and monitoring of project implementation.

The TAP organized a Project Monitoring Team, which is composed of experts on SWM from the TAP, LGU, DENR, NSWMC, and S3R2 on a voluntary basis. Resultant of the successful partnerships created through meetings conducted by the team with school participants, feedback on the progress of the project; and used of a standard assessment tool, used for their monthly activities.

The DENR—National Solid Waste Management Commission and the Environmental Management Bureau have provided for the technical assistance in the preparation of the training design and in the formulation of the monitoring and evaluation assessment. The monitoring covers only the actual progress of the implementation of the SWM in the school such as solid waste segregation bins, composting facility, and others but does not look into the record on how the funds were utilized. Financial auditing was not part of the process in the partnership while implementing the project. The reason for this was, the TAP relied on the usual accounting and auditing procedures and statutory requirements imposed by the Bureau of Internal Revenue and Security and Exchange Commission had been consistently done.

Downward Accountability

Aside from upward accountability, the partnership between TAP and the partner schools also involved downward accountability. This accountability refers to the ability of the stakeholders in the delivery of SWM projects and activities. The other aspect of downward accountability is whether TAP had conducted a systematic assessment of needs of the partners (project beneficiaries). The partnership in the implementation of the SWM project was done using top-down approach. There was no prior assessment conducted to determine the whether SWM project is the best project option available. It seems that the project implementer made an assumption that SWM program interventions are necessary and a priority considering that SWM is a vital concern of the schools (Fig. 2).

Facilitating Factors in the Accountability Role in the Academe–Industry Partnerships

The accountability as an important aspect of governance including in the implementation of SWM project through a partnership, in this case between the Academe and the Industry. Recognizing the value of accountability, the study also looked into the factors that facilitate and hinder accountability between partners.

Being compliant to the provisions of Republic Act 9003 which requires the incorporation of ecological solid waste management in the school system at all levels, and *having funds available* for SWM implementation; these were the two major considerations which could have facilitated the SWM engagement of the TAP and the Academe using a SPP model. More often than not, schools particularly the public ones, have no funds earmarked for solid waste management facility. The mobilization funds to support the engineering component of the project and the cash incentives (i.e., top school performers and selling of recyclables)

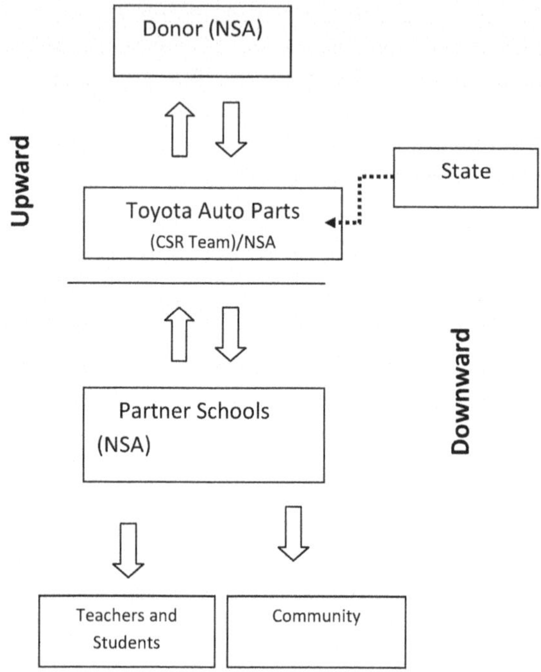

Fig. 2 Accountability of the non-state actors in Academe–Industry Partnerships

for its sustainability have definitely facilitated and motivated the schools to participate in the implementation of the SWM on campus. The financial incentive aspect on the other hand, enabled the partner schools to commit in the project and be more accountable in the partnership; having innate moral obligation to get involved in environment-related projects and to do what is right despite the absences of a formal agreement. There are some factors considered to facilitate in the performance of accountability role on SWM, such as:

- Recognition by appropriate government agencies. This enabled the NSAs to implement SWM projects and to ensure that accountability was looked into as required by the concerned government authority. The TAP has been given a formal recognition by the DENR for their active involvement and effort in the successful implementation of the School-Based SWM in the City of Santa Rosa. Active involvement of the school community, particularly the teachers and students who were given trainer's training on SWM, had contributed to the transparency effort of the project implementation.

- Intensification of SWM information and education campaign on campus and in the community at large. as well as advocacy and inspiration from the mass media provided transparency in the project implementation with the constant updates via press releases, press conferences and the like undertaken to report on the progress of the project. This has contributed in the enhancement of the level of environmental awareness not only of the stakeholders on campus but also the other stakeholders outside of the schools.
- Lessons learned and insights gained from previous experiences resulting from flooding aggravated by poor waste management. This has not only increased the level of people's environmental awareness but also has improved the waste management practices of the people in schools and in the community.

Hindering Factors in the Performance of Accountability Role in Academe–Industry Partnerships

Formal agreements and similar documents presented show an enhanced performance of accountability roles in created partnerships. This is because, in such document, basic information required of the partnership such as project objectives, expectations from both the parties, roles and responsibilities, functional structure, project output, outcome expected, and project implementing guidelines have all been enumerated and defined to ensure accountability. This facilitated the processes of accountability not only in other private partnerships but proven to be true also in a partnership between the industry and the partner schools hence, an enhanced performance of accountability role.

Earlier, a MOA was executed between Toyota Foundation (donor) and the TAP. However, there was neither MOA nor MOU between the TAP and the partner schools. The absence of a formal agreement may be considered a setback in looking into the practice of accountability but also in the actual performance of accountability role of NSAs in a partnership. The context and level of expectations have become not clear that had resulted in a differing interpretation of the provisions of the partnerships. For some schools, it was revealed during the FGD that the project had been "passed on" to the schools without prior consultation with them. But since TAP has provided mobilization funds, there is no reason not to accept the project. In other words, the project when offered with equity and funding would be hard to refuse but the downside would be that the absence of formal agreement may result in some confusion and weaken accountability performance. Other factors believed to cause a lack of accountability performance could be attributed in the perceived lack of political will from the state particularly among the local government units which were manifested in the lack of local ordinances that would improve implementation of the SBSWM.

Bibliography

Agrawal, A., and Lemos, Maria Carmen. (2007). "A Greener Revolution in the Making". *Environment* 49(5): 36–45.

Behn, R. (2001). *Rethinking Democratic Accountability.* Brooking Institution Press, p. 317.

Brown, L., and Moore, M. (2001). "Accountability and Startegy and International Non-Governmental Organization". Hauser Center for Non-Profit Organization. Working Paper 7.

Clapp, J. (2005, August). "Global Environmental Governance for Corporate Responsibility and Accountability". *Global Environmnetal Politics* 5(3): 23–34.

Fry, R. (1995). *Accountability in Organizational Life: Problem or Opportunity for Non-profits.* Wiley Periodicals Inc.

Laratta, R. (2009). "Autonomy and Accountability in Social Services Nonprofits: Japan and UK". *Social Enterprise Journal* 5(3): 259.

Lemos, M. C., and Agrawal, A. (2006). "Environmental Governance". *Annual Review of Environment and Resources* 31: 297–325.

Lemos, M. C., and Oliveira, J. L. F. (2004). "Can Water Reform Survive Politics? Institutional Change and River Basin Management in Ceara, Northern Brazil". *World Development* 32(12): 2121–2137.

Logo, P. B. (2003). *The Decentralized Forestry Taxation System in Cameroon: Local Management and State Logic. Environmental Governance in Africa.* Washington, DC: World Resources Institute.

Naidoo, K. (2003). "Civil Society Accountability: Who Guards the Guardians-GDRC". https://www.gdrciorg/ngo/accountability/ngo-accountability.pdf.

Najam, A. (1996, December). "NGO Accountability: A Conceptual Framework". *Development Policy Review.*

Przeworski, A., Stokes, S., and Manin, B. (1999). *Democracy, Accountability and Representation.* Cambridge University Press.

Republic Act 7160. The Local Government Code of 1991.

Republic Act 9003. Ecological Solid Waste Management Act of 2000.

PART III

Disaster Risk-Reduction: Structural Inequality, Women and Child Resilience

The Struggles of Women: An Experience from the Flood-Prone Area

Shishir Kumar Yadav

Introduction

Gender is one of the major categories that have been recognized in disaster studies as holding importance in any assessment of risk and vulnerability in a disaster. The vulnerability and risks that women face during a disaster differs from that of male members of the society. This vulnerability has to be understood to be primarily cultural and organizational rather than biological or physiological (Wiest et al. 1994: 3). Women are systematically marginalized from the access to major resources in the community and are discriminated in the private and public sphere. Males are favored in the access to the scarce resources such as food and nutrition, health care and so on during the crisis situation. However, all women are not universally or identically impacted by disasters. Even among the females, adolescents, pregnant women, single mothers, lactating mothers, the disabled, and the aged make up particularly vulnerable groups in emergencies.

S. K. Yadav (✉)
Centre of Social Medicine & Community Health (JNU), Delhi, India

B. Zutshi et al. (eds.), *Disaster Risk Reduction*,
https://doi.org/10.1007/978-981-10-8845-2_12

Disaster and Women

Women are one of the important indicators to compartment who explain the social realities and existing real situation in the field. As we know that women in developing counties have been called the "invisible earners" (Weist et al. 1994: 15). Women's productive work, particularly in child-rearing and other domestic work, as well as their enormous contribution to national food production requirements, is hidden in statistics (Chiu 1982, cited in Weist et al. 1994). Women are not only responsible for attending to the basic needs of their children and families, but account significantly for productive and income-generating activities in their respective communities (Weist et al. 1994). This pre-disaster condition in many societies, deny to women recognition for the work they actually carry out rendering them and their dependent children relatively more vulnerable than men (ibid.) do. Women are likely to be adversely affected by damage to economic livelihoods because basic survival strategies such as securing water, food, and wood for heating purposes, often fall on women, representing an extra burden on top of caring for and nurturing the family (Neumayer and Plumper 2007).

Females, generally have lower socioeconomic status than males, and therefore, females are more sensitive to the possibility of resource loss (e.g., monetary loss) (Ho et al. 2005). Further, women in comparison to men have restricted access to the formal and regulated labor market (Monzini 2001). In India, women's work in the household is not acknowledged as a form of work and not given credit as such. Only the work fetching cash outside the home is counted as a females' contribution to the household and given due credit. Further, sociocultural beliefs and practices often preclude women's ownership of land and other production technologies such as tractors or grinding mills (Weist et al. 1994). Statutory and/or customary laws often restrict women's property and land rights and make it difficult for them to access credit and agricultural extension services (Brody et al. 2008). Low literacy coupled with very low levels of ownership of land and other productive assets makes women prone to destitution. Along with this, patrilocal residence, patrilineal descent, and the prohibition against women inheriting property tend to enforce the social norms that women are dependent on men from birth until death (Samarasinghe 2008, cited in Locke 2010).

Methodology

In aligning with the broader objectives of the study, it deemed feasible to resort to a qualitative study as qualitative research provides richness and holism with strong potential for revealing complexity. The paper conceptualized the study in context of accessibility, affordability, and availability. The study is carried out in one of the flood-prone districts of Uttar Pradesh, Siddharthnagar. The latter has been experiencing floods on recurrent basis. The subjects for this study have been selected via purposive sampling. One of the affected villages of the district was the sample of this study. Data has been collected from the heads of the female members. The first preference has been given to older member women in the household, and if older women were not available then the next head in the family was considered. To fulfill the objective of the study, households from different castes and class were selected.

The method that was adopted for following the main study was the ethnographic method. It included in-depth interviews, group discussions, focused group discussions, observations, case studies, and narrative documentation. For the detailed and better insight in the issue, an in-depth study was undertaken by the researcher. A lot of times, it was continued through group discussions and focused group discussions. At some times, it was attempted by the researcher to convert the group discussion into focused group discussions. In group discussion and focused group discussions, the researcher was able to look at the various perspectives of the masses. Since, a lot of time, the group had people belonging to same age group; it was possible to know the perception of people belonging to a particular age group on the same event. Different women in groups also talked about the same event and other variants and shared their information with the researcher. The names of the informants mentioned have been changed for ethical purposes.

The Struggles of Women in Recurrent Floods Experience from the Field

The gender hierarchies in the society are an overwhelming reality that intervenes in all affairs. Disasters are considered to be the altering moments where natural interventions disrupt the existing social structures in the society. However, on the reverse, the people are seen to be acting even in the crucial moment, keeping in mind the social norms and concerns. Further, after the disaster, these norms are deciding factors in relief and coping strategies also.

Enarson and Morrow in an important 1996 report argued that the disaster does not unfold in isolation. Disaster "affirms, reflects, disrupts and engages with gendered social relationships" (Enarson and Morrow 1998). The gendered division of labor where women have to take up the care responsibility of the children and old is a social norm in the society. During a disaster, it has been observed that it only accentuates burdens on the women where they have to take care of the dependable members and look for selves (Peek and Fothergill 2009). The intersectional analysis of women across castes, class, age groups, and ability complicate the understanding of the experience of women during a disaster. In this head, the researcher attempts to bring out the differentiated and increased vulnerability of the women in the village. From the discussions and interviews held with women in the village, the impact of the recurrent floods on the women is highlighted.

Gendered Roles in the Village

The patriarchal setup prevalent in other parts of the state and the country has the leading edge in the village also. As noted previously, the recurrent floods in the village does away with the shock or trauma of the disaster. The recurrent floods are an important concern of the people in the village. The social structures and norms adapt themselves to the conditions of floods and afterwards instead of radically changing or altering itself. The gendered division of roles and labor in the society continues to persist unperturbed. During the situation of floods and afterwards, the gendered norms in the society govern the social life in the village. It is well known that the patriarchal structures in the society work to the bias of one gender over others. Women are in general treated as inferior beings as against the men in the society. They are denied equal opportunities in both public and private spheres. Traditionally, women are confined to the private sphere in the household and not allowed to access the public utilities and resources as freely as men. In the division of power, in a patriarchal society the structures are built and support a particular gender as against the others. The perception toward women and the roles designated to them are often governed by these patriarchal structures. Obviously, they work to the detriment of women.

In the village, while men took lead in earning the livelihoods for the households, women performed maintenance work such as cooking, cleaning, and caring for elderly and the children. Most of the households in the village struggled to make the two ends meet. In such a scenario, it was difficult to rely on the income of one person in the family. In most households, women helped the male members in the fields. They also took work under MGNREGA or in other villages as construction workers. However, this does not lighten their burden from the work inside the households. The duties in the private sphere of caring, cooking, and cleaning remain their responsibility unshared by the male members. Women wake up earlier than other members so that they can cook and clean before leaving for work. They also sleep later than other members to wind up the works while other members rest. Thus, while women share the burdens outside with the male members, the duties inside the house are not equally shared by others. It is only daughters who then help their mothers in the works inside the house. Even during the floods, these duties do not disperse. In the relief camps and sometimes under makeshift arrangements for protection against the floods, cooking is ensured by women only. When the researcher posed the question about the sharing of work, they mocked at the question. They jokingly instead asked the researcher and said,

> Or koun khana banaye? Pralay rahay na rahay, khana humi banoube.
> Who else will cook? Disaster or no disaster, we will cook.

They do not get enough or equal time to look after their personal needs and concerns. The elderly and children are also dependent of the care taken by the women in the family. Male members do not participate in the care given activity as much as women. Some of the women in a discussion among themselves hinted that during floods, because of the lack of work outside, male members treat it as a vacation where they can retreat into shelters. They did not join in them in the care work or in cooking. One of the women said,

> Pani me haman ke kaam badh jaye. Lekin eh loganke paas kaam nahi rahe, jonau karat rehat hin uho thapp ho jaat hai. Sara time chat par baeth ke araam karat rehat aur taash khelat hai. Ae logan chhuti manawat hai. Bachha budhan sabke jimmevaari humar ho jaat hai. Iho nahi hot hai ki jab kaam nahi hai toh bacchoun ke do minute dekh liye.

> In the flood, our work increases. These people have no work, whatever work is there that also stops. They just go and enjoy at the rooftop. They play cards in group there from morning to evening. They enjoy like it's a vacation. We handle the kid and the old. If you hand over the baby to them for two minutes, they never hold. Every work has to be done by us.

Clearly, for women, men are freer as compared to them during floods. Irrespective of the inconveniencies faced by all the members of the household, women are busier during floods. They do not get a "break" as others. Initially, women were hesitant to discuss these issues. They did not think they were worthy enough to be mentioned. They treated these everyday habits as obvious. However, when emphasized, they opened up and talked with details about their varying status in the family as against the male members. With reference to the work done by women outside and in the home, they hinted at the disregard the household work faces.

The unequal sharing of burdens between men and women in the society puts the latter under vulnerable conditions during the floods. Women are exhausted with the care and households work even during the floods. They do not enjoy equal freedom as men to take care of selves during the floods. Physically and mentally, they remain in a weaker condition as against men who are taken care of by the former. Their work often gets invisiblized bagging them hardly any importance or attention.

Societal Challenges

There are several other challenges that society poses to women which puts them in a vulnerable situation as against men. Women are perceived as the reservoirs of honor of the society. Women are bound by the social norms in the society where they are refrained from participating in the public sphere as openly and freely as men. They have to be cautious about time and space they are accessing. It is often not permissible for women to move out on the streets late in the evening and night. The men can loiter on the streets for a long time. There is no social norm restricting their movement. Further, women are expected to be dressed in a particular standard to be able to access a public arena. They are expected to not be out alone. They must roam around in groups or in the company of a male member. Even during floods and afterwards, these norms continue to govern the actions of most women. These norms exist in the village uncontested. In the conversation with the researcher, women expressed their helplessness against these norms but did not seem to question them. These norms vary for women belonging to different caste groups. The accessibility to the resources and utilities available in the public sphere is severely limited for women because of the social norms that control their movement and behavior. This limitation increases their vulnerability to floods and puts them in a dangerous condition.

During floods, people get together on roofs or other higher platforms to take shelter. They gather all the crucial objects that they might need for several days. As stated previously, women get together to cook, secure expensive objects, and take care of the children. Men and women often divide up into groups for most of the period during floods. Some of the women have stated the issues of being able to access proper spaces or nearby spaces such as toilets or granaries once the men have settled down there. A woman in her early twenties told the researcher about her experience where she could not access the bathroom to change her clothes because all the men were sitting in the room which had the bathroom. It is not perceived as respectable for a woman to walk into a room where men are sitting together. Women also mentioned how they had to send their children for fetching objects that they need from the spaces occupied by men. Even when they want to talk to a male member, they have to send the message through some kid to pass it on to the respective person among the group.

These everyday limitations that exist for women only aggravate the situation for them during floods. Since the floods are seasonal in the village, they gear up for the floods early on. But even then the important decisions are taken up by men only. Women wait for the instructions from male members before they can take any step. Male members decide where to move the stuff to, when to do that, and what all they take. One of the women mentioned about her experience where her husband was drunk and slept off when everybody else in the village was preparing to move to roofs of the houses. While she moved most of the things, she stayed with her husband who will not wake up. She kept sitting next to him until he woke up after a few hours. She talked about her fear and concern for her husband and her children whom she had sent off to the neighbors.

Women who belonged to the lower castes further had to keep in mind the presence of men of their community but also those belonging to the upper castes. They got together among the women of their castes only and did not mingle with the high caste women. In the village, it was strictly ruled out for women of lower castes to move in areas where upper caste men are present. During floods, they mostly stayed in their respective corners and did not go out. When the researcher asked the women about their accessibility to the relief materials and to health services in the camp, they replied that if they wanted to go, they went in groups. Mostly, they avoided by sending their male members. Most of the women did not question these hassles that they have to take care of. Some of them took pride in the security and honor hey maintain. They mentioned about these problems as consequences but did not therefore perceive that these must be challenged or overthrown. One of them said,

> E eta sabka karin padit hai. Eme khehu ka dos thodi ba. Aurat jaat na dhyan rakhi toh koun kari. Mard apan kaam kari hum aapan.
> One has to do all of this. This is not anybody's fault. If women will not care then who will. Men do their work and we ours.

The situation is even more complicated with the women whose husbands have migrated out of the village. Women take lead in the households where men have migrated. All the responsibilities of taking care of the cattle, children, elderly, and household needs fall on them. The accessibility for them is even more limited for they often do not have anybody to send for their needs. One of the women told the researcher that sometimes they just go and get the work done amid the stares and taunts but other times, they give up. Most of these women headed households are dependent on the family next door or a relative so that they can stay with the latter's family during the floods.

Economic Challenges

The position of women with regard to the control of economic resources impacts their vulnerability to floods in the village. Under a patriarchal setup, while women may join in earning, they may not be able to own the fruits of their income. The issue goes even much deeper than this. Since women are perceived to be inferior to men, the work performed by them is also obviated as not-up-to-the-mark. They are treated as lower than men in the labor market also where their labor performed by women is paid less. The work and the income of the male members are treated as primary and important while the perception toward women's work in the public sphere is seen as contingent and less worthy. Despite having a law for equal wages for men and women, the latter are paid less in reality. Because of the work crunch, women do not contest and accept whatever crumbs are thrown to them especially in poorer areas. Further, the women who have a family to take care often just have to give away whatever their income is to the head of the family. She is rarely the in-charge for spending the income of the family. This lack of control over economic resources especially puts them under vulnerable condition vis-a-vis others.

In the village, women who work for wages struggle with the unequal wages. In their conversation, they appreciated MGNREGA for some stability. When they seek work in other villages, they end up working the same hours for lesser money. Most of the women said they could not keep money they have earned from the work. Some of them gave it away to the elder members of the family and others to the male head. They did not have much say in the way money was to be spent. One of the women, Jamni Sahni, told the researcher that her husband kept all the money and spent it the way he thought. Although the money was spent on the needs of the households but there was always some money saved for bidis and liquor. She said she could not save any money because he would take it all. The other women burst into laughter and revealed that they would not give away all the money. They would hide some money for their own use.

Women also talked about the way expenditure during floods includes money for liquor and bidis. They mentioned how money was spent on liquor but not on dry wood for fuel for cooking. Dry wood is more expensive than the wet ones but the latter is more unsafe for the one who is cooking. It causes much more smoke that makes breathing difficult while cooking. One coughs and chokes their way through cooking. Since it is the women who are mostly cooking, they do not bother much to invest in better fuel. Several issues that trouble women and require investment are often sidelined when the expenditure is controlled by the male members. Women themselves begin to put the needs and demands of other over their own. Even in the relied products they receive, food, clothing, etc., chances are that they would first be spent on males and children and lastly on the women. This only aggravates their vulnerability. Floods only worsen this scenario.

Health and Sanitation Issues

One of the most important concerns that affect women the most, especially during floods, has to do with the health and sanitation related concerns. We know now that women are not rendered equal attention and respect in the society as much as men. Their inferior status in the society puts their concerns and issues at the backseat. Health and sanitation is one aspect that gets ignored amidst different issues. In general, open defaecation is a norm. Women have their separate space where they go together to defaecate. It is unsecure for lone woman to go to relieve herself. During floods, the situation gets worsened to an extreme. For men, it is not much of an issue to find a place but women have to struggle to find a secure lone place where they can defaecate especially during floods.

Initially, women were hesitant to talk about the issue with the researcher. However, with the help of a female key informant, women began talking and revealed many issues they face. One of them women, Sarita Devi, talked about how during floods she struggles to find along with other women a private, dry, and secure place where they could relieve themselves. She said that mostly these areas would be occupied by men or sometimes kids scattered and played in such areas. Dry areas are a rarity during floods. They either manage to find a less inundated area closer to the river or an abandoned wall. The situation at the shelter house or the relief camp is equally worse. Another woman, Barfi Sahni, mentioned that for the organizers or leaders at the shelter house or camp, sanitation is an issue that is better left to the villagers themselves. They do not even talk about it. The key informant, who was an educated woman in her thirties, confirmed this. There was an acceptance among all to leave the issue of sanitation on to them. The issue of menstruation was also never touched upon. The lack of female personnel made the issue invisible. Although there was an ASHA worker who was frequent to the village, floods only made her come less often. For daily needs and concerns, the local government primary health care center was inadequate. They could not rely on the subcenter for most of their ailments. The lack of female personnel at the subcenter was a major handicap. It is because of this most women preferred to go to the civil hospital in the district for medical concerns. However, the visit would not be frequent or quick enough. The traveling costs and the social hesitation in traveling alone made them wait for the longest time possible until the situation becomes absolutely in the need of a consultation. The visit to the civil hospital was treated as an event where most women preferred to travel together in groups so as to meet respective doctors. Their priority was to manage the ailment on their own as long as possible, once it went out of hand, did they go to the civil hospital.

Floods aggravate the already poor conditions of sanitation especially for women. The ASHA worker told the researcher that during and after the floods, most women suffer from urinary tract diseases. For menstruating and pregnant women, sanitation is a bigger problem during floods. Often with no outside support, women have to work out their way on their own. The ASHA worker talked about how it is very difficult to cope up for women after floods. Sometimes, it lasts long enough till the next floods. The government officer told the researcher that if the flood lasts for more than ten days, they provide for mobile toilet facility to the people. But when confirmed by the women in the field, all of them denied the availability of any such provision. Women mentioned about the difficulty for delivering women. It is very difficult for women delivering during floods to recover from the post-pregnancy effects. A lot of times the child does not survive the weather conditions. Women who are delivering often fall sick and take time to get well. The lack of privacy is a major constraint for women to prevent them from taking care of the self.

The inadequate functioning of the primary health care subcenter also puts women in a vulnerable condition. Especially for pregnant women, immediate medical health is unavailable. The center provides for minimal sickness only. It has no provisions for women struggling with urinary diseases, skin infections, sanitary products for menstruating women, and for immediate help for pregnant women. Women talked about their struggle of tending for their children also at the same time. One of the women in her late 60s, Bina Devi, told the researcher, about the major complexities in the delivery of her daughter-in-law in the floods two years back. She said how she had to struggle to find a way to take her to the hospital in the city when she complained of severe pain. It was her eighth month but during the preparation for moving to another place, she got hurt and had to deliver.

Despite several cases of this kind, the situation remains the same. The vulnerability of the women in the floods with regard to the health and sanitation concerns is a serious condition. Because of the recurrent floods and the situation not improving, women are doubly impacted by the floods.

Issues with Young Girls

Among the many concerns that women face, young girls are especially vulnerable during floods. Although the researcher could not directly establish a conversation with young girls, but through the key informant, some of the issues of the young girls could be presented. The key informant conveyed to the researcher that while older women were still aloof from sexual abuse during floods, younger girls faced it often. Floods create difficult situations in the village where elders are often busy in preparing and moving to a safer location. Children are often left out to play when the water is at a slightly lower level. Young girls are quick to be a victim of abuse from elder boys and men in the village. Although women maintain caution and care to a greater degree, but according to the key informant, every year after the flood there is at least one incident that is rumored to have happened. Such incidents are not reported to the police and are often resolved at the level of village. In a conversation to the elder women, they confirmed that they took special attention of the young girls so that they fall into any "trap". They said they kept them busy in some work so that they don't have much time to trot around. Young girls belonging to poorer households are especially vulnerable for they have to join their parents in work outside. They work as child laborers for they have no option but to support the family. Education is also a distant reality for most of them as they either take care of the siblings while their parents are out or they join out for work as well.

Recommendations and Conclusion

We see that women are particularly vulnerable after a disaster due to a combination of factors, such as economic dependency and lack of adequate financial resources, illiteracy, discriminatory cultural and social attitudes, physical infirmity, and so on. Women's and men's differential access to social and physical goods or resources is one of the key dimensions of gender inequality and an important predictor of post-disaster vulnerability of the females. Gender inequality is a major factor contributing to the increased vulnerability of women and girls in disaster situations, which is further intensified in the aftermath of disasters. Women have special needs and their needs that should be given due recognition in the post-disasters relief and welfare programs. Women experience the greatest stress during a disaster due to their multiple responsibilities.

It has been found that women and girls who are involved in income-generating projects gain economic independence, improved self-esteem, social recognition, and dignity and are quick to recover.

It is important to recognize the specificities of the concerns and requirements of women in a post-disaster context because of their place in the society. Emerging from the narratives presented in the article, it becomes very crucial to include women from the field in the decision making process. This inclusion of women has to happen not just at the general level but in the core teams. Women-centric bodies that collect information about the particular issues of women in a disaster context should be constituted not just at the level of the execution of the plan but also at the policy formation level. Further, the facilities that are required for them such as special care for infant children, safe and secure sanitation facilities, and medical facilities for old and pregnant women and help in community cooking in the immediate aftermath of a disaster must be executed on priority basis. Also, women personnel must be included in the post-disaster relief tasks both at the administrative level and at the rescue forces level. It is important to ensure that these facilities once provided are also put to use by the women. Presence of other women from the administration and in forces would make the accessibility of these resources secure and feasible. Through these several ways, one can weave their way to a gender just approach to disaster mitigation and planning.

Bibliography

Brody, A., Demetriades, J., and Esplen, E. (2008). "Gender and Climate Change: Mapping the Linkages—A Scoping Study on Knowledge and Gaps". BRIDGE Institute of Development Studies (IDS), UK.

Enarson, E., and Morrow, B. H. (Eds.). (1998). *The Gendered Terrain of Disaster: Through Women's Eyes.* Miami, Florida: Laboratory for Social and Behavioural Research, Florida International University.

Ho, M. C., Shaw, D., and Lin, S. (2005). "Risk Perception of Flood and Landslide Victims in Taiwan". Unpublished Paper. Retrieved March 6, 2016 from http://webarchive.iiasa.ac.at/Research/RMS/dpri2005/Papers/MingchouHo.pdf.

Locke, R. A. (2010). *Rescued, Rehabilitated, and Returned: Institutional Approaches to the Rehabilitation of Survivor of Sex Trafficking in India and Nepal.* Denver: University of Denver.

Monzini, P. (2001). "Trafficking in Women and Girls and the Involvement of Organised Crime, With Reference to the Situation in Central and Eastern

Europe". *First Annual Conference of the European Society of Criminology*. Lausanne, Switzerland.

Narson, Elaine, and Morrow, Heam Betty. (1997). "A Gendered Perspective: The Voices of Women in Hurricane Andrew". In Walter Gilles Peacock, Betty Hearn Morrow, and Hugh Gladwin (Eds.), *Ethnicity, Gender and Sociology of Disaster*. London: Routledge.

Neumayer, E., and Plumper, T. (2007). "The Gendered Nature of Natural Disasters: The Impact of Catastrophic Event on the Gender Gaps in Life Expectancy, 1981–2002". *Annals of the Association of American Geographers* 97(3): 551–566.

Peek, L., and Fothergill, A. (2009). "Using Focus Groups: Lessons from Studying Daycare Centers, 9/11, and Hurricane Katrina". *Qualitative Research* 9(1): 31–59.

Wiest, R. E., Mocellin, J. S., and Motsisi, D. T. (1994). *The Needs of Women in Disasters and Emergencies*. Winnipeg and Manitoba: The University of Manitoba.

Structural Inequality and Disaster Risk Reduction: An Indian Experience

Binod Kumar

Introduction

Existence of a varied and stratified social structure has compelled the Indian society to experience challenges in accessing equality for all. Media reporting in contemporary era has unraveled the horrific stories on caste discrimination. However, not much attention has been paid to interlink disaster and caste discrimination. The reality of discrimination continues to haunt the lower caste at the time of crisis/disaster. It is a classic case of double jeopardy in social life, where the lower caste populace faces the wrath of natural disaster on one hand, and caste discrimination on the other.

Our progressive constitution gives fundamental rights against any form of discrimination; however, such discrimination persists in the society. Social prejudices and inequality, which exist in normal times, makes disaster even worst for Dalits, as it not only prevents their participation in disaster risk reduction but also extends to discrimination in post-disaster situations. In fact, studies by Human Rights Law Network, Dalit Watch and People's

---

B. Kumar (✉)
Assistant Professor, Dr. Bhim Rao Ambedkar College, University of Delhi, New Delhi, India

B. Zutshi et al. (eds.), *Disaster Risk Reduction*,
https://doi.org/10.1007/978-981-10-8845-2_13

Watch have shown that many people have been excluded in the process of relief and rehabilitation in context of Tsunami of 2004. State's failure to integrate people from different strata of the society in stages of disaster planning have unleashed untold miseries for the people residing in the lowest social order and have caused serious human right violations.

Hazards, Risks and Disasters in Society

Disaster reduction has been studied from two competing perspectives: Risk and Vulnerability. It has further been extended into Natural Hazard Paradigm and Social Vulnerability Paradigm. The natural hazard paradigm focuses on reducing vulnerability through building of better physical structure which can withhold the natural hazard (White 1974). This approach has been questioned by other schools of thought which became influential during 1980s, namely social vulnerability perspective which was inspired by the work of O'Keefe et al. (1976) and Hewitt (1983). Social vulnerability paradigm focuses on interventions which increase the coping capacity of social group to recover from disasters, acknowledging the fact that disaster has differential impact on different social groups (Akerkar and Devavaram 2015: 80). This school considers that people's susceptibility to risk is mediated through social, economic and political marginalization and has nothing to do with nature as such (Wisner et al. 2004: 6). Caste shows the prevalent social, economic, and cultural differentiation within societal structures and shapes people's relations with larger structure like state and their ability to access the resources for disaster risk reduction (Akerkar and Devavaram 2015: 81). Hence, it is social vulnerability and people's differential capacity to prepare, cope, and recover from disasters that make a hazardous event a disaster for some and not for others.

Discriminatory State Practices and Disaster

State and those monopolizing the power act mostly in disfavor of the marginalized community. It is evident at the time of emergency that relief materials are distributed to dominant and most powerful in the area (Gaillard 2012: 466). Prima facie, it may look that people's response to natural hazard and disaster is determined by caste but it is partially true. Role of the state is central to bringing resilience among marginalized people. However, sometimes development and disaster reduction policies left the vulnerability of marginalized group unattended and allow them to settle in hazard-prone areas without proper means of protection. The reasons may be related to livelihood or economic capacity to afford the land on higher price at some other places. By doing so, the state excludes them from decision-making, claiming compensation or insurance claim to mitigate economic impact of disaster, strategically.

Usually, people settled in these areas do not have legal entitlement over the land and houses; hence they are excluded from loss estimation and finally compensation. This has been a rampant problem in slum areas of many cities; like in the event of Mumbai flood in 2005, large numbers of people were not able to claim the governmental compensation despite severe damage to their life and property as they did not hold any legal entitlement over land and tenement.

Resources and strategy for natural hazard are often replaced by standardized Disaster Risk Reduction (DRR) Policies. These policies have often been drafted without consultation of local population, who have been facing disaster historically. These policies often ignored the indigenous knowledge on disaster risk reduction and lack cultural sensitivity.

DRR policies have been drafted around command and control top-down approach which gives little attention to marginalized groups. The Disaster Management Act (DMA) of 2005 does not have any provision, when it comes to relate the issue of caste and disaster risk reduction, except for Section 61 which states that "While providing the compensation and relief to the victims of disaster there shall be no discrimination on the ground of sex, caste, community, descent or religion".

Legally speaking, the said provision is of little use without having details on how to bridge the impact of caste during disaster planning and relief operation. Additionally, we also have provision of non-discrimination on the same grounds under Article 14 of the Indian Constitution. As constitutional jurisprudence has the capacity to regulate all other laws of the land, hence we do not need special laws related to discrimination. In place of a new law being placed there is need for incorporation of programs through which equal accessibility of resources could be fostered in the event of disaster. Marginalized caste groups are those who bear the brunt of disasters and the same group are often last and least consulted in drafting disaster-related policies. To make DRR more effective, sustainable, acceptable and culturally sensitive, it is important to incorporate indigenous knowledge of the people on DRR.

Navigating Disaster Law and Caste in India

Section 12 (ii) of DMA of 2005 mentions special provision for widow and orphan but caste has been ignored despite being a social reality which affects a huge section of the population. Section 13 reads as "The National Authority may, in cases of disasters of severe magnitude, recommend relief in repayment of loans or for grant of fresh loans to the persons affected by disaster on such concessional terms as may be appropriate." As provision goes, it has no preferential policy regarding disbursement of loan to weaker section to rebuild their houses and livelihood despite the fact that accessibility of loan is a big hurdle for lower caste, even in normal times.

Further State Executive Committee under section 24 (j) and district authority under Section 34 (l) may "Ensure that non-governmental organizations carry out their activities in an equitable and non-discriminatory manner." It is ironic that DMA 2005 has provision to ensure equity and non-discrimination but it is applicable for NGOs working in the field of disaster but it is silent in context of highhandedness of the state authority.

Section 51 of the DMA, 2005 posed a grey area in case of caste discrimination. According to this section—those who refuse to comply with the orders of the central government, state government or district authority or create obstruction in disaster work shall be punished with a maximum two-year term (this also includes unequal treatment of citizens and unfair distribution of benefits based on caste). The state on the other hand under the Schedule Caste and Tribe (Prevention of Atrocities) Act, 1989 provides harsher and stricter punishment for discrimination based on caste. The discretion is in the hands of the police to register the case of discrimination in event of a disaster under whichever law they wish.

Ironically, DMA, 2005 stipulates punishment of maximum two years term and fine for false claim under Section 52 which affects marginalized section negatively. In absence of sufficient documents and convincing power, they fear to claim compensation under the act.

Further, the act also specifies under Section 71 that action of national, state and district authorities as well as central and state government under DMA, 2005 cannot be challenged except in Supreme Court and High Court (Sarkar and Sarma 2006). It places restriction on poor and marginalized people, as they will not be able to access High Court and

Supreme Court in most of the cases. Good faith clause under Section 73 grants high level of immunity to the officials which may encourage them in the activities which goes against vision of the act.

Research Methodology

Present paper is based on secondary data which has been collected from various governmental and non-governmental agencies. The paper discusses cases related to caste-based discrimination in event of a disaster, which have been highlighted through the primary research by organizations like National Campaign for Dalits Human Rights, Dalit watch, Human Rights Law Network, International Dalit Solidarity Network, and People's Watch. The limitation of the paper lies in terms of the limited availability of data related to caste and disaster. With limited data, an attempt has been made to bring injustice to the forefront, faced by the Dalits in both pre and post-disaster situations.

Disaster and Caste Discrimination—Evidences from Case Study

People belonging to Dewar Caste (also known as Kaibarta) located in different coastal district of Odisha were discriminated by the upper caste during the Super Cyclone 1999 and also in the process of relief and rehabilitation. There have been instances of exclusion of lower caste people from the beneficiary list. In many occasions, lower caste people were not able to approach the officials who were preparing the list of beneficiary. Besides, they were also not able to convince the officials for want of necessary documents. It is a fact that people belonging to the lowest social order are most vulnerable in disaster and most of the time their entitlement documents are lost, which have furthered their exclusion for claim of compensation.

In case of Gujarat earthquake, the compensation was paid to the victims on the basis of owner-driven approach which, means that if you do not own any property then you will not get the compensation. This process of identification of victims left many people out of the purview of the relief and rehabilitation. Another flaw during Gujarat earthquake comes under relocation scheme where people were given option of relocation or compensation. People from upper caste got together and purchased their own land and decided to relocate, however it was not an option for the lower caste people. This resulted in the exclusion of lower castes to avail such schemes. In post rehabilitation period, where upper caste was able to attract the basic infrastructure due to their social networks; poor and marginalized were left without any basic amenities and infrastructure. Although, Gujarat Reconstruction and Rehabilitation Policy included some special initiatives for schedule caste but there was no specific help for them.

Another interesting instance on which the paper delves on is the case of Tsunami 2004 relief operation in Tamil Nadu. Being in lower social strata, Dalits in Tamil Nadu Tsunami operation of relief and rehabilitation were ignored not only by the upper caste but also by state authority. Whole cleanup effort in post Tsunami situation was done with the help of fisher folks. However, they were not paid for the same. In fact, they were not able to access potable drinking water put up by UNICEF, as the upper caste believed that they would pollute the water. In immediate aftermath of tsunami, Dalits were brought from other parts to remove dead bodies but they were not given safety gears like gloves, masks, and boots. Community kitchen established to distribute the food was divided into two parts: one for higher caste Indians and the other for Dalits. There were also instances of police not reporting the case of missing Dalit in Tsunami. It is unfortunate that even in the event of disaster, caste prejudices of the upper caste takes the upper hand. National Campaign for Dalit Human Rights also documented that in some parts of Tamil Nadu and Andhra Pradesh, Dalits were provided with less relief material and Dalit colonies were the last to get electricity and water supplies in rehabilitation efforts (Louis et al. 2007).

Disaster Cycle and Reproduction of Caste System

Caste is rooted in the Indian society in such a way that even in every cycle of disaster, it represents strong footprints. Where in the event of disaster, untouchables and Dalits have restricted access to relief and rehabilitation; they are also excluded through state's legal apparatus which have been discussed in legal analysis of the DMA of 2005. In post-disaster situations, claiming compensation and insurance for lower caste is an onerous task for want of entitlement records. The question of citizenship comes into the picture, as they rarely have ability and facility to keep the document safe. Furthermore, insurance claim is highly biased toward landholders in the case of crop insurance as it recognizes the landholder, despite knowing the fact that most of the lower caste and landless laborers are sharecroppers. Consequently, despite bearing the loss in the event of disaster, they are not able to claim compensation or insurance. In the rebuilding and risk mitigation phase of disaster cycle, lower caste especially Dalit and tribes are highly marginalized. Due to their illiteracy and backwardness, they hardly participate in the process of risk mitigation. Hence, they perpetually reside in hazard-prone zone and are severely affected by the disaster every time. In the rebuilding phase, due to domination of upper caste at local level, Dalits and tribes hardly gets their due share in regaining infrastructure and amenities in their locality and colonies. It is always Dalits' colonies which get last attention in the process of rebuilding (Fig. 1).

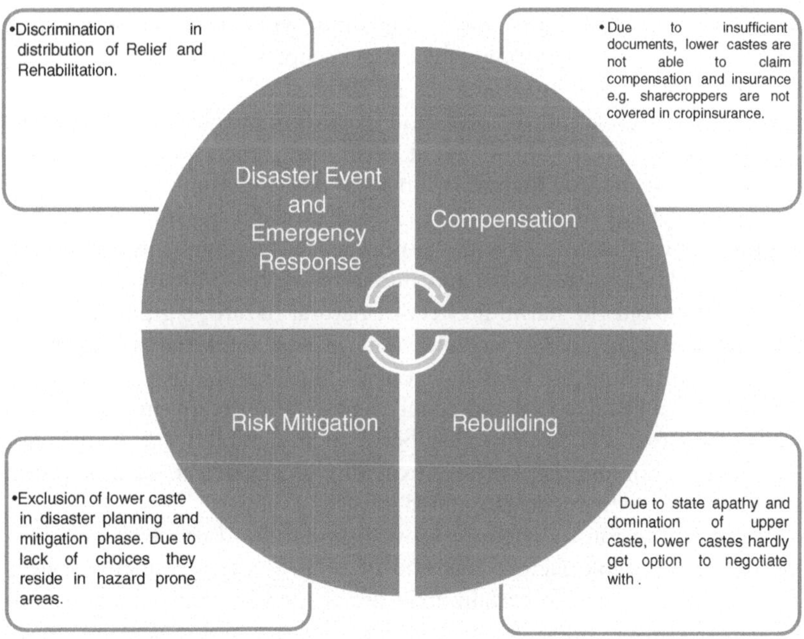

Fig. 1 Reproducing caste inequality in different phases of disaster cycle

Discussion

Largely, impacts of disasters mirror the everyday condition of marginalized people in India. Hence, ability of Dalit to face natural disaster reflects his positions within a larger society. Those who enjoy social, economic and political power in society are less in count among the people affected by natural disaster because they enjoy choices to live in safe areas. Consequently, disaster most often, detrimentally, affects the marginalized section of society. Differential access of resources by certain section of society is completely reflected in post-disaster relief program. Bosher (2007) observed that it is not exclusion from society which impacts poverty rather inclusion within the caste-based hierarchical system which excludes people in accessing resources and livelihood, this also gets reflected in the event of disaster. Organization of traditional Indian villages perpetuates the marginality of Dalits and untouchables as their habitation clusters most often situated at a place on the margin of the village. Caste affiliations bear vulnerability to floods and cyclone through segregation of home in rural India. In Ganges, river delta villages are often settled around central elevated grounds where more powerful caste resides. In contrast, less powerful caste occupies low-lying areas which are vulnerable to floods (Wisner et al. 2004: 239). Consequently, lower caste have lesser access to common resources within the village system and it is reflected in the event of disaster, Dalit and untouchables are not able to harbor themselves in safer places during the event of an upcoming natural disaster. Bosher (2007) through his study in Andhra Pradesh shows that lower caste, who often are the poorest in the society, have limited access to private and public resources. These resources are essential to avoid the harmful effect of natural hazard, e.g., land, house, livestock, savings, credit, farming, and fishing implements. Access to resources in multi-caste context is especially difficult where upper caste control the political, social and economic assets to the detriment of powerless lower castes. Thus, interplay between vulnerability and caste is not only reflected in identity and resources but also in power relation within a larger society which gets magnified in case of disaster.

People's decision to live in or move from a hazard-prone areas is constrained by multiple factors and one of them is caste. Sometimes people's decision to stay in certain areas are shaped by their attachment to a particular place. Such places have long history and it is especially in case of tribals in Indian context. Such locations bear a strong sentimental value apart from livelihood which largely overcomes the potential threat of natural hazard. Places like Jharia in Jharkhand face potential threat, as the risk of underground coal fire is common. However, the threat of such natural disaster is ignored by the indigenous community which has been residing in these areas.

Village organization in India is strongly based on the caste system. From occupation of an individual to his residence in the village, all is based on his caste. This generates an atmosphere where restrictions are imposed on access to resources and freedom of making choices. This discrimination enables the upper caste to exploit them even in a situation like disaster which has a long-lasting effect ranging from pre-disaster to post-disaster operations. Traditionally settlement patterns, housing structures, social networks, fishing and forestry activities have been decided on the basis of caste which make people more or less vulnerable while facing natural hazard. Hence, caste system places crucial role in people's vulnerability in Indian context. Besides, different caste groups also resort to different resources and methods to cope with natural hazard. These are deeply embedded in historical, geographical and occupational values. For example, fisherfolk in Pondpeta village of Ganjam district of Odisha did not vacate their village during Hudhud; however, same villagers vacated their house in case of Phallin, just one year ago. Reason of their movement, or not to move, from native place in case of natural hazard is based on their accurate estimation. Historically, fisherfolk has been staying in coastal villages and by the passage of time they have developed scientific indigenous knowledge related to cyclone—on the basis of wave of the sea, color of sea water and sky. Thereby the marginalized people have developed strategies to face and combat natural hazards through local and indigenous knowledge accumulated through ages.

Conclusion and Recommendations

The paper unravels questions of "why" and "how" the Indian lower caste population continues to face limited or restricted access of basic resources even during natural hazard, despite country's significant progress in disaster risk reduction. Why lower caste groups are more vulnerable during disasters? Caste is often neglected by those collecting data on disasters, why so? Eliminating such pertinent questions in the Disaster Management scholarship has resulted in lacunae in documenting pertinent plight of victim, especially the marginalized. Few reports, that of National Campaign on Dalit Human Rights and International Dalit Solidarity Network, have provided a glimpse of such atrocities but no extensive research has been done in this area. Representative from National Commission for Schedule Caste, National Commission for Schedule Tribe and National Commission for Women and Children should be included in National Executive of Disaster Management to voice the concerns of their constituency in pre and post-disaster situations.

Bibliography

Akerkar, Supriya, and Devavaram, John. (2015). "Understanding Rights Based Approach in Disasters: A Case of Affirming Human Dignity". In Andrew E. Collins, Samantha Jones, Bernard Manyena, and Janaka Jayawickrama (Eds.), *Hazards, Risks and Disasters in Society*. Amsterdam: Elsevier, p. 80.

Bosher, Lee. (2007). *Social and Institutional Elements of Disaster Vulnerbility: The Case of South India*. Bethesda: Academica Press.

Gaillard, J. C. (2012). "Caste, Ethnicity and Religious Affiliation and Disaster". In Ben Wisner, J. C. Gaillard, and Ilan Kelman (Eds.), *The Routledge Handbook of Hazard and Disaster Risk Reduction*. Oxon: Routledge, p. 466.

Hewitt, K. (1983). "The Idea of Calamity in a Technocratic Age". In K. Hewitt, (Ed.), *Interpretations of Calamity: From the Viewpoints of Human Ecology*. Boston, MA: Allen and Unwin, pp. 3–32.

Louis, M., Mathew, Koshy, and Tiphange, Henry. (2007). *Compounding Disaster: Conformability of Post Natural Disaster Relief and Rehabilitation Process with Human Rights Standards*. Madurai: People's Watch.

O'Keefe, P., Westgate, K., and Wisner, B. (1976). "Taking the Naturalness Out of Natural Disasters". *Nature*, 260.

Roy, Tirthankar. (2012). *Natural Disasters and Indian History: Oxford India Short Introductions*. New Delhi: Oxford University Press.

Sarkar, Subhradipta, and Sarma, Archana. (2006). "Disaster Management Act, 2005: A Disaster in Waiting?" *Economic and Political Weekly*, 3763.

White, G. (1974). *Natural Hazards: Local, National and Global*. New York: Oxford University Press.

Wisner, Ben, Blaikie, Piers, Cannon, Terry, and Davis, Ian. (2004). *At Risk: Natural Hazards, People's Vulnerability and Disaster*. London: Routledge.

Linking Gender Vulnerability and Disasters

Rahila Sikandar and Asif Khan

Introduction

According to the Disaster Management Act 2005, disasters means, "a catastrophe, mishap, calamity or grave occurrence in any area, arising from natural or man-made causes, or by accident or negligence which results in substantial loss of life or human suffering or damage to, and destruction of property. Or damage to, or degradation of, environment, is of such a nature or magnitude as to beyond the coping capacity of the community of the affected area" (GOI 2005). People were killed and many got injured and affected during the occurrence of numerous disasters occurred in the past, however, it affected men and women in various different ways. "Disasters affect men and women in different ways, with greater impacts on women and children" (Dasgupta et al. 2010). The ferocity of the disasters occurred in any or every form leads to the total devastation of the community as a whole. They interrupts

R. Sikandar (✉)
Special Centre for Disaster Research, Jawaharlal Nehru University, New Delhi, India

A. Khan
Indian Institute of Management, Kashipur, Uttarakhand, India

B. Zutshi et al. (eds.), *Disaster Risk Reduction*,
https://doi.org/10.1007/978-981-10-8845-2_14

progress and destroy the outcome of developmental efforts over several years, often pushing nations in expedition for progress back by several decades. India has seen the both natural and man-made disasters during its last 10 years. Thus, effective reduction of disaster risks, rather than a mere response to their occurrence, has in recent times, received increased attention both within India and abroad. Today's complexity in this interconnected world is to understand the proximity and the result of the disasters and to reciprocate it in the same manner, however, it is sometimes very difficult to inform people who live in far-flung areas and they easily become the victim of high-density disasters. Women and men belonging to different age and socioeconomic strata have distinct vulnerabilities and face disasters and also recover from them being in their respective strata (UNDP 2010).

The gender of an individual influenced the role they have in the time of a disaster and its effects on their lives. Gender must, therefore, be an essential consideration in disaster reduction and response. The World Health Organization was of the view that gender is a determinant of health, and others denote it as "relational" because it modifies how all other determinants of health, such as education or income, are experienced by men and women. In the aftermath of a disaster, gender takes on even greater significance, as social structures and processes are laid bare, revealing a "gendered disaster vulnerability", the term coined by noted scholars namely Enarson and Morrow in 1998. Applying a gendered analysis to disaster planning and recovery can highlight how gender can be either mitigated or aggravated by class, ethnicity, disability, and other factors. Across the globe, women are at greater risk than men in disasters (Parkinson et al. 2011; Alston 2009). A considerable amount of literature of past decade reveals the extent to which gender inequalities, expressed in women's social and economic marginalization vis-à-vis men, often result in women bearing a disproportionate burden of the costs of disasters and therefore, gender remains a peripheral concern in the field of disaster preparedness, relief, and management; and left this work upon only a few non-profit agencies and civil society groups (Mehta 2007). In this paper, the focus will be on the natural disasters that occurred in India and how does it make its impact on women.

Across the globe, women are at greater risk than men in case of disasters, not only in developing countries, but in developed countries also the risk of mortality for women is much higher than men. During disasters, women and children are the most vulnerable groups among the victims. It is expected that women should take the sole responsibility for the family and property because socially determined roles push them to be separated from their male relatives, as male members are more likely to be called upon to provide manpower in communities that have experienced a natural disaster. The care of children and elderly people is a gendered phenomenon too, and generally falls upon women, in the times of crisis as in daily life (Rivers 1982; Neelormi and Ahmed, n.d.). Moreover, "Gender based prejudices and divisions in many societies mainly affect girls and women as these are based on views of them physically and emotionally weak, inferior in comparison to men and boys" (Ariyanbandu 2009). The situation become worse when they are more like treated as stranded or helpless. Gender inequalities evolving from the disaster are apparent in caring responsibilities both during and after the disaster. Consequently, economic inequalities too, emerge in the aftermath, with recovery post-disaster being predominantly directed to projects involving male labor, while women in disaster-prone areas are often employed in low-status jobs. Economic insecurity contributes to increased vulnerability to violence for women (Parkinson et al. 2011).

While catastrophic natural events obviously pose a threat to and affect everyone caught in their proximity, both the risks to and the impacts of disasters are disproportionately carried by those who are already socioeconomically and physically disadvantaged by virtue of age (the elderly and children), condition (the disabled, ailing, women who are expecting), and gender. Though not invariably, this last category refers to women and girls. Context and culture also play a large role in determining outcomes, and there are evidences that says, it is men who suffer higher mortality and morbidity. Nonetheless, a substantial evidence highlights the extent to which gender relations place women in socially, economically, and politically marginalized positions vis-à-vis men that serve as a limiting "pre-condition" of people's ability to anticipate, prepare for, survive, cope with, and recover from disasters (Mehta 2009). Consequently, women in developing countries are more vulnerable to pay greater prize whenever a disaster strikes. This unwelcome situation of vulnerability is partly the result of poverty. In result to it, women are more likely to live in areas that are more susceptible to disaster and habitations that are poorly constructed, are less likely to have the resources to escape if a disaster threatens (Dasgupta et al. 2010; Henrici et al. 2010). India, being a developing nation, also is one of the most vulnerable countries to natural disasters. It has faced a number of natural disasters in the last decade that have taken lives of thousands of people and incurred heavy economic losses as well. Moreover, it is observed that more than half of the victims of the past disasters were women, because of having no technical knowledge about disaster occurrence and preparedness of the same in general (Gokhale 2008). The vulnerable situation of women becomes further worse as in certain cultural contexts, women such as widows, female heads of households, the disabled, orphan girls, and others who does not fit in conventional ideas of women, also find themselves without sufficient social supports. In crisis circumstances, this can translate into social exclusion, destitution, and even death (Philips et al. 2010; Mehta 2009).

It was reported that 42% more females died in the 1991 Cyclone in Bangladesh because their husbands had the decision-making power and they did not allow to leave their habitations without their husbands' permission (Fothergill 1996). In the recent disaster of Tsunami in the Indian Ocean, 80% of the 300,000 deaths were women and children from 13 nations (Philips and Morrow 2008). But local cultural practices alone are not the only responsible reason of women's vulnerabilities. Due to flawed assumptions, biases and prejudices and ignorance also result in women facing a host of discriminatory and exploitative situations in post-disaster situations. Lessons from the field have long illustrated how typically women's specific health and sanitary needs are not addressed in disaster response efforts, their rights as individuals are overlooked, their leadership roles and knowledge are not adequately recognized, and often their pressing needs for personal security is ignored (Mehta 2009). They are being ignored, even in public policies for disaster management at large which leads to wastage of their valuable contribution in disaster management. Thus, this paper focusses on some of the critical issues and concerns relating to the vulnerability of women at the time of disasters in a decade starting from 2006 to 2016 in India. Table 1 shows disasters that occurred in the last 25 years in India indicating the place of occurrence and the amount of human loss (Chart 1).

Understanding Gender and Disaster

The term gender generally reflects women. It is used as a synonym for women and girls. It also states the relationship between men and women and their roles in the social construction. Gender does not mean women. However, given that women are often in a disadvantaged position in many developing and developed countries, the promotion of gender equality

Table 1 Indicates the disasters occurred in the last 25 years (1990–2015)

S No.	*Name of the event*	*Year*	*State and area*	*Fatalities*
1	Floods	October 2014	Jammu and Kashmir	
2	Cyclone Hud Hud	September 2014	Andhra Pradesh	
3	Odisha Floods	Odisha 2013	Odisha	21
4	Andhra Floods	October 2013	Andhra Pradesh	53
5	Cyclone Phailin	October 2013	Odisha and Andhra Pradesh	23
6	Floods/Landslides	June 2013	Uttarakhand and Himachal Pradesh	4094
7	Cyclone Mahasen	May 2013	Tamil Nadu	08
8	Cyclone Nilam	October 2012	Tamil Nadu	65
9	Uttarakhand Floods	Aug–Sep 2012	Uttarkashi, Rudraprayag and Bageshwar	52
10	Assam Floods	July–Aug 2012	Assam	–
11	Cyclone Thane	December 2011	Tamil Nadu, Pudducherry	47
12	Sikkim Earthquake	September 2011	Sikkim, West Bengal, Bihar	60
13	Odisha Floods	September 2011	19 districts of Odisha	45
14	Sikkim Earthquake	September 2011	North East India with epicentre near Nepal Border and Sikkim	97 people died (75 in Sikkim)
15	Cloudburst	2010	Leh, Laddakh in J&K	257 people died
16	Drought	2009	252 districts in 10 States	–
17	Krishna Floods	2009	Andhra Pradesh, Karnataka	300 People died
18	Kosi Floods	2008	North Bihar	527 deaths, 19,323 livestock perished, 2,23,000 houses damaged. 3.3 Million persons affected
19	Cyclone Nisha	2008	Tamil Nadu	204 deaths
20	Maharashtra Floods	July 2005	Maharashtra State	1094 deaths, 167 injuries and 54 missing

(continued)

Table 1 (continued)

S No.	*Name of the event*	*Year*	*State and area*	*Fatalities*
21	Kashmir	2005	Mostly Pakistan, partially Kashmir	1400 deaths in kashmir (86000 deaths in total)
22	Tsunami	2004	Coastline of Tamil Nadu, Kerala, Andhra Pradesh, Pondicherry and Andaman and Nicobar Islands of India	5640 persons missing. 2.97 million people affected. 11,827 hectare of crops damaged. 3,00,000 fisher folk lost their livelihood
23	Gujarat Earthquake	2001	Rapar, Bhuj, Bhachau, Anjar, Ahmedabad and Surat in Gujarat state	13,805 deaths 6.3 million people affected
24	Orissa Super Cyclone	1999	Orissa	Over 10,000 deaths
25	Cyclone	1990	Andhra Pradesh	967 people died, 4,35,000 acres of land affected

Source https://ndma.gov.in/en/disaster-data-statistics.html

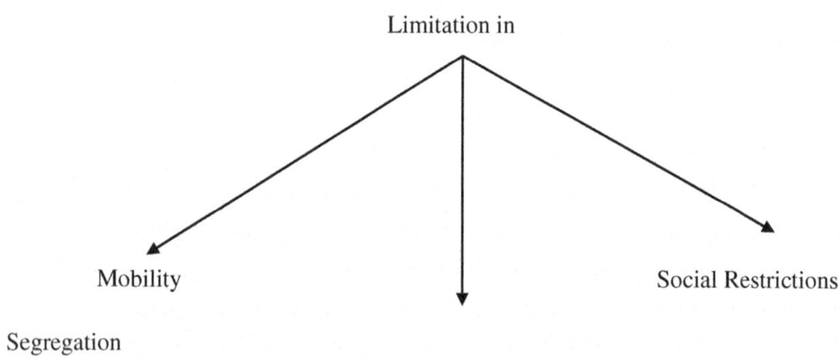

Chart 1 Represents limitations (*Source* Ariyanbandu and Wickramasinghe 2003)

implies an explicit attention to women's empowerment (UNDP 2010). Gender, the state of being male or female is typically used with reference to social and cultural differences rather than biological ones (Ashraf and Azad 2015). Disasters are the "a serious disruption of the functioning of a community or a society causing widespread human, material, economic, or environmental losses which exceed the ability of the affected community or society to cope using its own resources" (UN/ISDR 2014). These differences have put women and girls into chaos during disasters making their situation more vulnerable. Following is the chart representing the factors which are responsible for locating them into more risk.

By limiting women's access and by putting social restrictions on women, they are pushing them toward a higher degree of risks. By these sorts of activities, women are confined into a particular societal norm from which they cannot come over and can easily become the victim of the disaster (Chart 2).

Due to restricted mobility, women are confined into a particular society, thus they have low access or no access to the warning information, and however, if they get it by any chance they do not have the ability to respond to the warning information and can easily become the victim. During disasters, women became a very easily available commodity for sexual fouls. They merely became the victim of sexual abuse and domestic violence. During disasters, the gender-based discrimination is different from male to female,

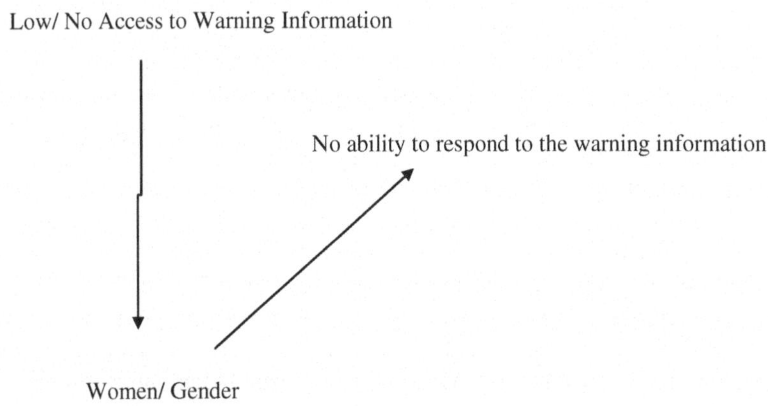

Chart 2 Represents low access and response (*Source* Ariyanbandu and Wickramasinghe 2003)

as female has to suffer not only domestic and sexual abuse, but also the societal pressure to not to move alone under any circumstances. Moreover, *pregnancy* and related *factors of childbirth* also play a major role in putting women at greater risk (Ariyanbandu and Wickramasinghe 2003). Other factors, such as a very ***high level of illiteracy and no access to information*** for any disaster or related activities lead them near to the high-risk zone. Due to illiteracy, they do not get access to any source of knowledge like easily available newspapers, information provided by internet, or any other source providing early signs of warning. In result of it, their resilience power to the losses occurred of disasters are extremely low and therefore, women pay a heavy price for higher disaster risks. It is not only the above factors which moves gender toward risk, there are other factors also such as biological and physical which plays there part in moving gender towards greater risk.

Disaster and Family Violence

Although, family violence may be unrecognized and unrecorded during and post-disaster. It is reported many times that women have experienced violent behavior from their partners post disasters, mainly due to changed habitation and employability, financial constraints, depression, and alcohol abuse aftermath. This increased male aggression may be operating alongside women's intensified vulnerability as a result of the same post-disaster pressures affecting men. Additionally, the demands of traditional female roles are intensified as women care for partners, children, parents, and others who are suffering after the disaster strike (Parkinson et al. 2011). In countries, where gender discrimination is tolerated, women and girls are particularly vulnerable to natural hazards. Not only is the percentage of women and girls who die higher in these countries, but the incidence of gender-based violence including rape, human trafficking, and domestic abuse is also known to increase exponentially during and after a disaster. Most disasters place an undue burden on women and girls who are responsible for unpaid work, such as providing care, water, and food for households (UNDP 2010) (Table 2).

The difference between women and men within the same household and within the same culture that are socially and culturally constructed changes over time. Moser also described that these differences are reflected in: roles, responsibilities, needs, perceptions, and views held by both women and men (Moser 1993). He further argued that, gender is not a synonym for women, but consider both women and men and

Table 2 Gender relations in society: Reflected in

Identities	A combination of physical and behavioral characteristics which set apart boys from girls, men from women;
Perception	Views as to how they are differentiated in their roles as men and women;
Attitude	Actions guided by the perceptions and;
Status	The place occupied by the men and women in family, community, and society

Source Ariyanbandu (2009)

their interdependent relationships. As gender is never irrelevant and must always be examined and reflected in practice, for men and boys as much as women and girls (Enarson and Meyreles 2004).

Gender issue comes into all spheres of life like personal, social, economic, and political, they are not equitable and prejudiced toward girls and women (Ariyanbandu 2009). This makes the condition of women and girls more vulnerable and leads to another complex interaction between poverty and gender. Different sexes behave differently during disasters, it depends upon them how they relate them to the disaster, their behavior and how do they respond during disasters. The culture of protection of women, young girls, and children in the pre-disaster situation by the family and social networks breaks down during disasters and both male and the females are not prepared for the new situations. It is not only the girls and women who can be the victim of sexual violence and trafficking, but also young boys who became the victim of sexual assault, rape, and trafficking during these conditions.

De Silva and Jayathilaka (2014) in their research paper entitled "*Gender in the context of disaster Risk Reduction: A case study of a Flood Risk Reduction Project in the Gampaha District in Sri Lanka*" was of the view that losses incurred due to disasters can be reduced if disaster management intervention consider gender perspective in its policies and actions.. The study was conducted in two *Grama Niladari* divisions: *Pamunuwila* and *Galedanda* in the *Gampaha* district where the project namely "Cleaning Natha Ela" was executed when flood hit the region. The objective of that intervention was to address issues related to floods in the area with special focus upon gender perspectives. It was revealed that even though the community experienced an important step when the project got success in reducing damages occurred due to floods. However, the project could have achieved more success, if it also covered gender perspective too, and the results could have been much greater if the project had considered the gender aspects related to floods. The study thus emphasized that gender should also be considered while framing developmental policies. In the case of cleaning the Natha Ela project, gender has not been considered important from project's starting till the end of the same. Thus, the study explicitly mentioned that though, disasters affect both the genders, however, the impact of the same could be entirely different. Therefore, in order to mitigate the losses resulted due to frequent disasters; gender must be taken into priority, so that men and women can be made resilient to disasters equally.

Ginige et al. (2014) conducted a study entitled "Tackling women's vulnerabilities through integrating a gender perspective into disaster risk reduction in the built environment". In this study, they have focused on the indispensable need to have "built environment" in order to tackle the fatal disasters. Based on their study, it was recommended that the gender perspective should also be incorporated in policies and plans related to disaster risk reduction. Under the process of constructing a built environment, the awareness level and knowledge of women should be assessed forthwith and then, in the second step, the gaps should be filled up by timely intervention.

However, Neumeyer and Plumper (2007, cited in Ariyanbandu 2009) states that, natural disasters do not affect people equally. It is extremely difficult to predict that how the increased disaster impacts will affect women in particular. There are incidences which reveal women's experiences of disasters are different from men and boys. They further explain that the biological and the psychological differences between the sexes are unlikely to explain large-scale gender differences in disaster situation. Observation speaks that there are people whose behavior changes and they became more aggressive and stressed during disasters and any mishap. Basically, these aggressive people have counted more males than females. And women, girls, and boys became the victim of aggression very easily resulting in rape, assault, and violence. Consequently, it is women who bear the aggression in the form of unwanted pregnancy, the psychological trauma of rape and assault, while both men and women run the risk of Sexually Transmitted Diseases like HIV/AIDS (Ariyanbandu 2009: 09). The following table states the causes for gender difference in disaster.

However, Table 3 reflects that the higher vulnerability of women is socially constructed and it is due to the differences into the socioeconomic differences between men and women and not because of biological and psychological reasons at large.

Table 3 Causes for gender difference in disaster

First	Biological and Psychological differences between men and women may at times disadvantage women in their immediate response to the disaster
Second	Social norms and the role behavior may lead to a behavior of women that increases their vulnerability in the immediate course of the disaster
Third	Disasters may lead to shortage of resources of basic need as well as a temporary breakdown of social order, in which case the competition between individuals becomes fiercer and existing forms of discrimination can emerge

Source Neumeyer and Plumper (2007)

In Indian perspectives, gender has been neglected in almost all the interventions of disaster mitigation. When the Gujarat earthquake came at Bhuj district, it affected women severely. However, many non-governmental organizations like the Self Employed Women Association (SEWA) and Disaster Mitigation Institute (DMI) played a key role in making women resilient and empowered to disasters. The mentioned agencies trained women for seismic-resilient techniques and provide them skills so that they can become economically stable too. All such efforts proved to be fruitful in the long run. It was observed that heavy damage occurred due to the disaster, started to be recovered by trained women of the region. In disaster management, gender should not be neglected, otherwise, women will continuously face off the brunt of neglect and nature (Mehta 2007) (Fig. 1).

Gendered vulnerability does not derive from a single factor, such as household headship or poverty, but reflects historically and culturally specific pattern of relations in social institutions, culture, and personal lives. Further, Canon predicted if there is no serious progress in reducing poverty, then it can be assumed that women will become increasingly affected by the impact of intensified hazards, in terms of their ability to resist and recover from them (Ashraf and Azad 2015: 139). For example, in Post Tsunami disaster, no special care was provided to pregnant women. There were cases of miscarriages and premature births, and others continuing their pregnancies under conditions of severe deprivation (Ariyanbandu 2009: 11). It was reported that no treatment was provided to those women who suffered from breast milk clotting (after losing their babies) at the camps. However, Byrne and Baden viewed that it is necessary to ask a gender perspective in disaster into relief intervention (Ashraf and Azad 2015: 137) as it highlights women or gender perspective highlights their capacity to deal with it.

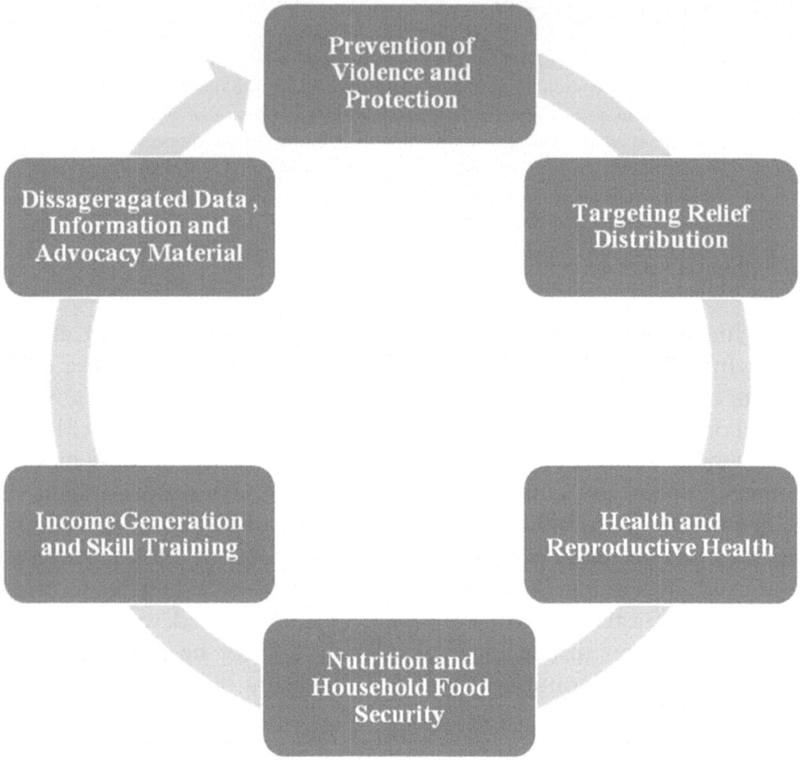

Fig. 1 Represents the understanding *about the gendered vulnerabilities and the areas of concern* which might be addressed accordingly (*Source* Ashraf and Azad 2015)

What Can Be Done?

First of all, participation of women in disaster planning, designing, executing, and monitoring, emergency programs and rehabilitation projects plans should be encouraged. With a vision to build a safe and disaster resilient India, the Government has adopted a holistic, proactive; multi-hazard oriented and technology driven strategy by promoting a culture of prevention, mitigation, preparedness, and response. In the above context, the Ministry of Home Affairs has undertaken several initiatives to achieve sustainable reductions in disaster risk. These encompass multi-hazard planning, use of technology for disaster mitigation, and empowerment of various stakeholders by enhancing their capacities to prepare, respond, and mitigate hazards (GOI-UNDP 2011). If programs and policies would be executed in their letter and spirit, vulnerability of women in disasters can be minimized effectively. As suggested by UNDP (2010) Gender Strategy Programs, the same destructive forces of disasters also create opportunities for women as agents of change. Disasters can also provide an opportunity to redress gender disparities. For example, during the recovery period following a disaster, long-standing biases against women can be challenged by programs that are sensitive to their needs and that involve them as equal partners in recovery work. However, if women and girls are left out of planning for disaster response or risk reduction measures, the special talents, skills and knowledge of 50% of the population are not capitalized upon and the needs of the most affected are unlikely to be met. Creating awareness about disaster risks and vulnerabilities, as well as advocating for comprehensive risk reduction among all stakeholders would be effective steps in making women able to survive in catastrophic situations as well. In order to mitigate adverse impacts of disasters upon women, gender issues must be urgently and effectively integrated into disaster research, planning, and organizational practice. Women's participation in emergency relief, resettlement, and reconstruction efforts following a disaster acquires significant knowledge and expertise that can greatly benefit communities that subsequently experience similar crises.

Conclusion

Thus this paper states about the vulnerable conditions of women and girls during disasters, as inaccurate gender assumptions by policymakers and practitioners and insensitivity to women issues and gender perspective in development planning and disaster risk reduction means that intervention falls short of reaching both the development and the disaster situations that render women of all age groups more vulnerable to disaster (Ariyanbandu 2009: 16). Therefore more gender inclusive policies are needed with appropriate strategy, plans, and program in order to bring empowerment to the nation and communities to successfully build the resilience against disasters and to bring down the level of vulnerability among gender especially women. A main concern in public policymaking would be the need to be prepared. This could be very well responded if every event that could happen studies carefully and to prepare a response accordingly for the future disaster. Those eliminated during the earlier one must be a part of the future public policy in case of disasters i.e., mainstreaming of those who were eliminated. The focus should be on the ability to respond as soon as the disaster occurs. Lack of knowledge, lack of resources, lack of time, and lack of preparedness makes disaster bigger than it was. Thus, the conclusion is that resilience during the disaster can be managed only by understanding the fact that how the unexpected can be managed and how to make communities resilience possible accordingly. As far as women safety and violence is concerned proper measure must be introduced by the state and there must be training centers to make women from a disaster-prone area to learn and make them empower to in order to overcome with the circumstances.

Bibliography

Alston, M. (2009). "Drought Policy in Australia: Gender Mainstreaming or Gender Blindness?" *Journal of Gender, Place and Culture* 16(2): 13–154.

Ariyanbandu, M. M. (2009). "Sex, Gender and Gender Relations in Disaster". In E. Enarson and P. G. D. Chakrabarti (Eds.), *Women, Gender and Disaster: Global Issues and Initiatives*. New Delhi: Sage.

Ariyanbandu, M. M., and Wickramasinghe, M. (2003). *Gender Dimensions in Disaster Management: A Guide for South Asia*. Colombo: ITDG South Asia.

Ashraf, M. A., and Azad, Md. A. K. (2015). "Gender Issues in Disaster: Understanding the Relationships of Vulnerability, Preparedness and Capacity". *Environment and Ecology Research* 3: 136–142. https://doi.org/10.13189/eer.2015.030504.

Dasgupta, S., Siriner, I., and Partha, S. D. (Eds.). (2010). *Women's Encounter with Disaster*. London: Frontpage Publications.

De Silva, K., and Jayathilaka, R. (2014). "Gender in the Context of Disaster Risk Reduction: A Case Study of a Flood Risk Reduction Project in the Gampaha District in Sri Lanka". *ScienceDirect* 18: 873–881. Elsevier.

Enarson, E., and Meyreles, L. (2004). "International Perspectives on Gender and Disaster: Differences and Possibilities". *International Journal of Sociology and Social Policy* 24(10–11): 49–93.

Fothergill, A. (1996). "The Neglect of Gender in Disaster Work: An Overview of the Literature". *International Journal of Mass Emergencies and Disasters*, 11–25. Retrieved May 31, 2016, from www.reserchgate.net/publication/238347161.

Ginige, K., Amaratunga, D., and Haigh, R. (2014). Tackling Women's Vulnerabilities Through Integrating a Gender Perspective into Disaster Risk Reduction in the Built Environment. *ScienceDirect* 328–335. Elsevier.

Gokhale, V. (2008). "Role of Women in Disaster Management: An Analytical Study with Reference to Indian Society". *The 14th World Conference on Earthquake Engineering*. Retrieved June 1, 2016, from http://www.iitk.ac.in/nicee/wcee/article/14_10-0049.PDF.

Government of India and UNDP Disaster Risk Reduction Programme. (2011). *State Level Programme for Strengthening Disaster Management in India*. New Delhi: Ministry of Home Affairs, GOI.

Henrici, J. M., Helmuth, A. S., and Braun, J. (2010). *Women, Disasters and Hurricane Katrina*. Institute for Women's Policy Research Fact Sheet (Vol. IWPR#D492). Washington: Institute for Women's Policy Research.

Mehta, M. (2007). *Gender Matters: Lessons for Disaster Risk Reduction in South Asia*. Kathmandu: The International Centre for Integrated Mountain Development.

Mehta, M. (2009). Reducing Disaster Risk Through Community Resilience in the Himalayas. In E. Enarson and P. G. D. Chakraborti (Eds.), *Women, Gender and Disaster: Global Issues and Initiatives*. New Delhi: Sage, pp. 57–74.

Moser, C. O. N. (1993). *Gender Planning and Development: Theory, Practice and Training*. London: Routledge.

Mulilis, J. P. (1999). "Gender and Earthquake Preparedness: A Research Study of Gender Issues in Disaster Management: Differences in Earthquake Preparedness due to Traditional Stereotyping or Cognitive Appraisal of Threat?" *Australian Journal of Emergency Management* 14(1): 41–50. Retrieved May 20, 2016, from http://ajem.infoservices.com.au/items/AJEM-14-01-10.

Neelormi, S., and Ahmed, A. U. (2007). *Climate Change in Bangladesh: Concerns Regarding Women and Special Vulnerable Groups*. Dhaka: Climate Change Cell UNDP, Centre for Global Change.

Parkinson, D., Lancaster, C., and Stewart, A. (2011). "A Number Game: Lack of Gendered Data Impedes Prevention of Disaster-Related Family Violence". *Health Promotion Journal of Australia* 22: S42–S45. Retrieved May 31, 2016 from http://www.genderanddisaster.com.au/wp-content/uploads/2015/06/Doc-016-A-Numbers_Game.pdf.

Philips, B., Jenkins, P., and Enarson, E. (2010). Violence and Disaster Vulnerability. In D. S. K. Thomas, B. D. Philips, W. E. Lovekamp, and A. Fothergill (Eds.), *Social Vulnerability to Disasters* (2nd ed.). Boca Raton: CRC Press.

Philips, B. D., and Morrow, B. H. (Eds.). (2008). *Women and Disasters: From Theory and Practice*. USA: International Research Committee on Disasters.

Rivers, J. (1982). "Women and Children Last: An Essay on Sex Decimation in Disasters". *Disasters* 6(4): 256–267.

United Nations Development Programme. (2010). *Gender and Disasters*. New York, USA: Bureau for Crisis Prevention and Recovery, UNDP. Retrieved May 30, 2016 from http://ndmindia.nic.inDM-Booklet-080211.pdf.

UN/ISDR. (2014). *Living with Risk: A Global Review of Disaster Reduction Initiatives.* Geneva: United Nations Inter-Agency Secretariat of the International Strategy for Disaster Reduction (UN/ISDR). Available at http://www.unisdr.org/eng/about_isdr/basic_docs/LwR2004/ch1_Section1.pdf.

Web Sources

http://www.undp.org/content/undp/en/home/ourwork/crisispreventionandrecovery/focus_areas/climate_disaster_risk_reduction_and_recovery.html.

http://ndma.gov.in/en/get-involved/community-based-disaster-management/17-on-going-programmes/37-goi-undp-drr-project.html.

Women and the Gaps in Framework of Disaster Risk Reduction

Priyanka Jha

Introduction

Disasters either man-made or natural goes on to have huge impact on the lives of all species in a manner of similitude. There is loss and destruction of life and property, more than that the kind of emotional toll that it takes on the lives is also of very high magnitude. Over a period of time there has been documentation of all these issues and factors, one can say that the material aspects of the loss has been well accounted and documented for and as a result there have been compensations that the state has provided for. There have been international relief, food aid, and supplies of medicines. There have been international and national policies that have been designed specifically for Disaster preparedness and mitigation. So on the surface of it, it looks all fine and perfect. But an essential question that looms large is whether policies and program, are they sensitive to the differential needs of the women.

One cannot escape from the truth that the discourse of disasters, like many other discourses are constructed with the male perspective. The inherent idea that one gets in situations of disasters has been male

P. Jha (✉)
School of Undergraduate Studies, Ambedkar University, Delhi, India

B. Zutshi et al. (eds.), *Disaster Risk Reduction*,
https://doi.org/10.1007/978-981-10-8845-2_15

centric or structured around the notion that beings are men. This raises a serious problem as it tends to marginalize and peripheralize the other species. This paper attempts to locate the gaps in the discourse of Disaster from the perspective of women. The paper is divided into two broad sections. The first section attempts to look at some of the major gaps that the Disaster discourse suffers from, as it brings forward the physical, infrastructural, ideational and legal/institutional shortcomings in the approach to disaster mitigation and management. The second section looks at some of the solutions that could be incorporated in national and international policies, making Gender a key category as the way forward.

Gaps in the Discourse of Disaster

First, Women and the needs of the women are not attended well as they are not taken into consideration. This occurs because of many reasons. Women needs are not understood as it is always Men who are in charge of all key responsibilities and they fail to realize that the needs of women being different from that of men. The most important need being that of the body. As a result, the bodily issues are not only neglected but many a times given a miss.

In many disasters, Women have spoken about the need for sanitary products and undergarments which are key requirements. One of the reasons that can be traced to the missing out of these key items from the aid and relief packages emanate from the fact that it is men who are in positions of responsibility. The reason why this is missed out is precisely because they don't have the kind of bodily requirements that women have, so it does not even strike to them that these are essential and key needs and requirement. body is treated as an entity in a very private space that many women are uncomfortable talking about and to expect them to ask for undergarments and sanitary napkins can be extremely humiliating. Body has always been treated as a private site, maybe through the discourse of Disaster, one needs to bring it out in the open and talk about it, as it is the body that becomes a very important site in terms of the damages which are caused by the disaster.

If one takes a look at the positions of responsibilities at all levels from preparedness to mitigation, one witnesses the absence of women. It is men who are in charge of the topmost position to the local site of work, from the Army officials to local community leaders to the emergency workers. This is a huge gap that needs to be addressed as there are problems of comprehending the issues which are related to women.

Second, when one argues that in the situations of Disaster there is absence of women in the key responsibilities and capacities, this can be understood from the fact that there is no participation of women at any level of disaster. Women are absent in decision making, prevention or recovery. This raises some severe and serious questions on the nature of equality that nations claim to uphold. One of the features of good governance embarks upon gender parity and is foundational to it but one witnesses that women are not given equal representation and space within this very overtly male handling of disasters. The question then can be asked that if women are brought in on board will the situation change. It can be answered in a very simple way, Yes it will because women understand their needs and body, and that of other women in better manner. This idea if located within the prism of experiential epistemology, knowledge of the experience and through the experiences becomes the biggest legitimizing factor in providing representation to women.

Thirdly, in situations of disasters the vulnerabilities that women face increases. One needs to take into consideration the kind of vulnerabilities that women face consists of rape, trafficking and violence. Risks cannot be understood as being homogeneous, they are of different kinds and one needs to understand what are the kinds of risks and dangers that increases and can have a direct bearing on the lives of people. One has witnesses that in the case of floods in Bangladesh, large number of women are trafficked in India and are forced into Flesh trade which was also the case of the earthquake in Nepal. In any case of Disasters, it is the Women are the worst victim. This brings to us the truth that there is differential impact of disasters. Vulnerabilities are of different kinds and magnitude, women feel more vulnerable than men, the reason can be traced to the fact that women feel extremely unsafe and unprotected. In different parts of the world women living in camps have complained about feeling unsafe in the camp areas. They feel that the very reasons which displaced them and uprooted from their houses have not really disappeared and feel that they still live in unsafe conditions. After the immediate post-disaster period when adequate relief did not pour in, women were bartered for food. In the late post-disaster period young girls were married off at a younger age to older men, women during and after disasters become increasingly prone to human trafficking, and the survivors of the recent quakes that hit Nepal led the human traffickers to increase their efforts to enslave young women for forced labor and sexual exploitation.

Fourthly, the central role of women facing the aftermath of disaster is totally neglected. India is one of the most vulnerable countries and has faced a number of natural disasters; it has been observed that in India more than half of the victims of these disasters were women; during the last major natural disasters it has been observed that in India women lack the technical knowledge about the occurrences of disaster and have had no participation in planning, designing and monitoring emergency and rehabilitation systems. Since they also have no permanent place in the decision-making mechanisms and are left out of planning for disaster response the skills and the knowledge of nearly 50% of the population has never been capitalized upon. As a result of this women are not regarded as active and resourceful disaster responders but as helpless victims.

Fifthly, Women are faced with a wide variety of situations both during and after the disaster their role in ensuring food security and their dependence on natural resources to do so, reinforces the impact of disasters, in the search of supplementary incomes women find themselves in situations where they are left to fend for themselves and they also assume greater responsibility for caring for their families. In the post-disaster situations, the responsibilities of women increase manifold in comparison to men as their care giving roles expand. Despite their access to the sources of recovery is constrained, even in cases where women have had access to shelters they have had to work harder than usual and in defiance of their own safety imperative to gather resources, their special health needs specially those of pregnant and lactating women. After a disaster it has been seen that women are more prone to depression and emotional disturbances most of these women are compelled to set aside their own losses and assume the roles of the care taker and bread earner for which they might not be prepared psychologically and physically thus with the trauma and stress added burden of responsibility make women prone to physical, mental and emotional distress.

Legal and Institutional Gaps: There are limited provisions in the national and international legal and governance frameworks which deals with the Disaster-related vulnerabilities and risks or is geared toward to removal of these structural barriers. There remain key gaps in mainstreaming gender in designing, planning and resourcing disaster management. Despite variety of international conventions invoking Gender equality and parity like Convention on Ending all Forms of Discrimination (COEFD), the Beijing Declaration and Platform for Action as well as agreements which focus on disaster risk reduction (DRR), including the Hyogo Framework for Action (HFA 2005) and the more recent Sendai Framework for Disaster Risk Reduction (SFDRR 2015), nations have failed to accomplish the objectives of gender parity and their progress remains fairly limited and unsatisfactory. The need is to identify the gender gaps and formulate specific strategies to fill them, especially in the Sendai Framework.

Consistent efforts led to the recognition of the constructive role of women, the SFDRR[1] tries to focus on identifying progress achieved in engendering DRR and gaps in mainstreaming gender into the DRR design, planning, resourcing, and management. The framework calls for an all-inclusive engagement of society, including "gender, age, disability and cultural perspective in all policies and practices; and the promotion of women and youth leadership."

Furthermore, the framework states that "Women and their participation are critical to effectively managing disaster risk and designing risk reduction policies, plans, and programmes; and adequate capacity building measures need to be taken to empower women for preparedness as well as build their capacity for alternate livelihood means in the post-disaster situations."[2]

However, the critical question being raised by some women activists and groups across the globe is implementation of these stated policies so that women are actually empowered to be active at all stages of the disaster cycle, from risk reduction, prevention, and preparedness to

[1] The Sendai Framework for Disaster Risk Reduction (SFDRR) and the Sendai Declaration were agreed upon on March 18, 2015 after negotiations spanning years and culminating in the Third World Conference on Disaster Risk Reduction (WCDRR3) held in Sendai, Japan, from March 14, 2015.

The SFDRR is built to last for the next 15 years and was one of the first of three major deals to be struck on climate change and sustainable development. The SFDRR replaces the Hyogo Framework for Action (2005) which provided a blueprint for disaster risk reduction including mention of the importance of the gender perspective for building resilience, calling for gender perspectives to be "integrated into all disaster risk management policies, plans and decision making processes, including those related to risk assessment early warning and information management and education, training."

[2] Sendai Declaration.

response and recovery. Women's leadership is included as the declaration impinges on gender-sensitive national policies and refers to sexual and reproductive health care services but unfortunately violence against women is not included anywhere in the framework despite the realization that disasters also leads to various kinds of violence and vulnerabilities for women.

The other problem with the Framework is that, women are included within the frame of marginalized groups, reinforcing the "victim" paradigm. The term gender equality does not appear in the text, nor is there a reference to women's human rights. There is a significant lack of stronger language around the gender-responsive policies.

Apart from the problems that women found with the declaration, it had several other shortcomings which a wide variety of critics feel that the targets adopted by the framework were "watered down" as they were made non-binding compared to the earlier proposals as percentage goals were rejected. The move therefore lacked clear commitments to finance prevention and resilience programmes in poor and vulnerable countries.

The framework also does not take into consideration the forces of globalization in society which tends to increase disaster risk. One major drawback of the framework is its vaguely defined targets; it includes seven targets intended to drive forward progress on protecting people and assets from natural hazards and to "substantially reduce" the number of people killed and affected by disasters as well as damage to infrastructure and services like health and education and to decrease economic losses all by 2030, which does not seem practically feasible. It is important to note that the framework does not define as to what exactly does "affected" mean and what constitutes "substantial". The framework also does not explicitly mention the most vulnerable sections. Another major roadblock toward the achievement of the goals of the Framework is the smooth transfer of technologies from the rich to the poor countries, as some countries held that such a policy could undermine intellectual property regimes. Ultimately, how the declaration is implemented over the next 15 years is more important than the text itself. The challenge is to maintain the political momentum generated at Sendai. Immense efforts have gone into the creation of a global framework for reducing disaster. Time will tell if at all these initiatives were productive or not, although it will be difficult to separate the results from what could have happened in the absence of such an instrument.

The Way Forward

Disaster risk reduction is a fundamental pillar of sustainable development, a gender perspective to DRR helps to focus on distinct specific capacities that could be used and not be wasted. Consistent advocacy over the last 10 years resulted in an increasing acknowledgement of constructive role of women in disaster reduction. The capacity and the knowledge of women enable the development of individual and community resilience. There has been a significant departure from the traditional view of women and girls being inherently vulnerable and passive recipients of disaster, however there are still many structural barriers to the long-cherished goal of gender parity and sensitivity to the differential needs of women.

Women as individuals and formally or informally through groups have participated and contributed immensely to emergency relief and reconstruction efforts. For example: The Gujarat quakes of 2001 saw remarkable contribution of women and women's groups which underwent training to take on their role to motivate households, build technical capacity, demonstrate arrangements and provide feedback as well as monitor reconstruction. This process led to the emergence of women as strong and confident community developers, monitoring basic services and voicing women's priorities initiating local projects and facilitating dialog. As a result, over 250,000 households were involved in relief work and over 4000 women took loans and over 1200 started their own businesses which led to the stabilization of livelihood and assets. Today more than 800 groups are working in these regions on health, education, sanitation etc.There is a need to do away with structural barriers to women's participation and leadership in decision making and strengthening their capacities to engage actively therefore the following aspects must be taken into consideration.

Insritutionalization

The need is to give priority to gender equality and leadership in the areas of governance at national as well as at local levels of planning. There is a need to ensure that the financing for Disaster Risk Reduction should be underpinned by gender-responsive budgeting and input based expenditure which should be based on sex disaggregating data and gender analysis. The need for sex-disaggregated data was also argued for by many women organizations in the preparatory conferences to Sendai. There is also a strong need to establish a gender-responsive accountability and monitoring mechanism to ensure that resources are reaching the right source. In the Nepal earthquake, many women activists argued that the relief money which was given to the male head of the family was used for alcohol consumption, instead if given to the women of the house would have been used for the right reasons. This links us to the second issue that in situations of disasters, the relief money should be given to the women than the men as the entire burden of running the errands lies at the shoulder of women and for this there is need for proper institutions.

Representation

Women as individuals and in organizational groups should be treated as leaders and agents of collaborative change. Women's representation in DRR can be promoted only through establishing a participatory decision-making mechanism which means that women should be taken in decision making positions. This can also be done by creating opportunities for collaboration between women and grass root organizations as well as strengthening women organizations and networks.

Participation and Capacity Building

There is a need to position women as active stakeholders. Their leadership can be facilitated only when they are systematically informed and included at the policy-making levels. Enhancing women's access to information, including early warning, training and education, capacity building will strengthen self-reliance.

Ensuring women's legal entitlements and practical access to assistance and services including sexual health, provisions for insurance, credit-based employment will facilitate the change. However, it is most significant to mediate the constraints of time, mobility and social security which impede their participation. Indigenous knowledge systems also need to be factored in or made foundational in the institutionalization of mitigation which on many occasions are with women, they being storehouse of local knowledge and wisdom.

Disasters by their nature rip apart the fabric of the state, decimating social safety nets diminishing the rule of law in the havoc that it ensues.

Women therefore need to rise to the forefront in confronting these disasters. A significant push needs to be given at all levels especially at the level of policy formulation with the most important step in this direction being to recognize women as active stakeholders in the process and not mere passive victims and recipients.

Bibliography

Dasgupta, S., Siriner, I., and Partha, S. D. (Eds.). (2010). *Women's Encounter with Disaster*. London: Frontpage Publications.

Enarson, E., and Chakraborti, P. G. D. (Eds.). (2010). *Women, Gender and Disaster: Global Issues and Initiatives*. New Delhi: Sage.

The Hyogo Framework for Action 2005–2015. www.unisdr.org/2005/wcdr/.../Hyogo-framework-for-action-english.pdf, www.unisdr.org/files/8024_8961BeijingAgendafinal22Apr091.

UNISDR. (2009, June). "Making Disaster Risk Reduction Gender-Sensitive: Policy and Practical Guidelines". Geneva, Switzerland: UNISDR, UNDP and IUCN. http://www.preventionweb.net/files/9922_MakingDisasterRiskReductionGenderSe.pdf.

United Nations Development Programme. (2010). *Gender and Disasters*. New York, USA: UNDP, Bureau for Crisis Prevention and Recovery. Retrieved May 30, 2016, from http://ndmindia.nic.inDM-Booklet-080211.pdf.

Post-disaster Economic Vulnerability of Female-Headed Households

Isaias S. Sealza

Introduction

Disaster strikes some sectors more severely than others. Few bounce back quickly, but many lag behind. Often the vulnerable ones are hardest hit, and the slowest to recover, if at all. Vulnerability, commonly the lot of the women sector all over the world, is worse for female household heads (Chambers 1985). Studies show that the standard of living of female-headed households (FHHs) is lower than male-headed households (Rajaram 2009; Chant 2007; Klasen et al. 2011).

The studies above examine the experiences of FHHs under common everyday conditions. This paper interrogates the circumstances of FHHs in a post-disaster context. The cumulative effect of being in a post-disaster environment and being in a FHH can be remarkably harsh. This paper is an attempt to pin down a point (or points) of entry for possible interventions from the outside to help the FHHs. The approach is to examine the relationship of the probable points of entry seen in

I. S. Sealza (✉)
Graduate School, Xavier University, Ateneo de Cagayan,
Cagayan de Oro City, Philippines

B. Zutshi et al. (eds.), *Disaster Risk Reduction*,
https://doi.org/10.1007/978-981-10-8845-2_16

the literature with a factor that can serve as a measure of an intervention's impact. Hence, this study attempts to understand the relationship between income (household earnings) that can be a measure of impact, and "structural factors" or factors that are particularly difficult to change; and, some "programmatic factors" or factors that are challenging but can be addressed within a reasonable period of time by development efforts from the outside. It argues that, while disaster affects people in the neighborhood in much the same way, the FHHs are likely to remain disadvantaged relative to the MHHs.

Conceptual Framework

The difficulties of women may not be necessarily because of their being biologically female but because of the social structure that circumscribes their chances to access better economic opportunities. The structural factors being referred to here are land ownership and level of education. In general, land for crops, for business, or housing is in short supply, and therefore very difficult to acquire. The social structure itself makes acquisition prohibitive for the poor (Elvinia 2013).

A similar state of affairs obtains in the case of education. Although basic education in the country is free, poverty prohibits school participation and increases drop-out rate such that about one-quarter of school-age population are out of school (UNESCO 2015; Uy and Pua 2015).

Since the two major means toward the attainment of better incomes, namely, land ownership and education are not readily accessible to the poor, perhaps the factors introduced from the outside can constitute other options. These programmatic factors refer to creation of work opportunities, chances to possess farm animals, and occasions for engagement in activities that build social capital.

Engaging in gainful work logically relates to income; and so does the possession of farm animals. Many people rely on farm and off-farm labor for subsistence. And farm animals can take the place of a bank account.

Social capital, for its part, facilitates improvement in income. It enables one to build alliances for business or labor network, or obtain succor in times of need.

The hypothesized relationships are shown in Fig. 1. The paper hypothesizes that: (1) FHHs have lower household earnings than the MHHs; (2) land ownership and education are associated with type of household; (3) Female household heads are likelier than their male

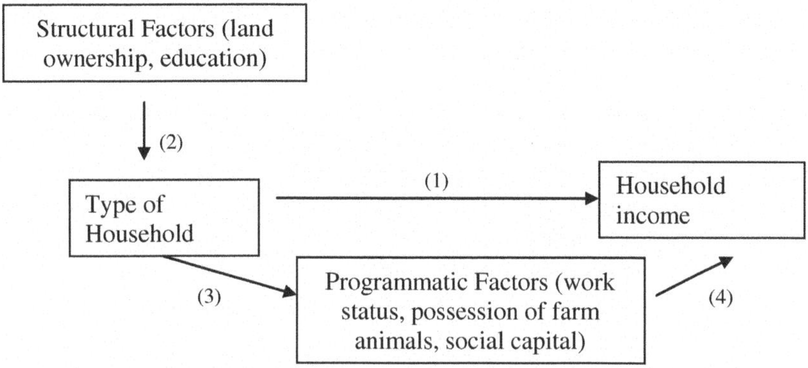

Fig. 1 Hypothesized relationships among the variables

counterpart to be unemployed, not to own farm animals, and not to participate in activities that build social capital; (4) households whose heads are working, have farm animals, or join organizations have higher incomes than household whose heads have no work, have no farm animals, or do not join organizations; and (5) the relationship, if any, between type of household and income remains the same even with the effects of variations in other important factors are held constant.

Figure 1 is a heuristic device. Structural factors are treated initially as antecedent variables, and the programmatic factors are taken as intervening variables, but their actual role in the model can be determined only after examination of the data.

Scope

This study covers resettlement sites for survivors of Tropical Storm *Washi*. Thousands of survivor households were resettled. Some three to four years after it happened, a complete count was conducted by the Archdiocese (an administrative district of the Catholic Church) of Cagayan de Oro. There were 18 resettlement sites, with 2550 survivor households comprising 12,242 persons. Data for this paper were taken from a simple random sample of that complete count.

The paper is a secondary analysis of an existing data set. Many of the variables that could have shed more light on the question of FFHs and household earnings were not part of the data set.

This paper contributes to the literature on feminization of poverty and is significant in at least two ways. First, it shows evidence on how the FHHs fare in comparison with the MHHs. And second, it directs attention to the possible entry point for development work among the internally displaced households. For example, if structural factors do not explain differences in earnings but programmatic factors do, then a clue is provided on initiatives to pursue to improve conditions. If structural factors explain the differences, then upgrading earnings would be predictably more strenuous. But in both instances, the Archdiocese and other stakeholders will have empirical basis moving forward.

The Following Are Important Terms

Education refers to the highest grade level completed in the formal school system by the household head.

FHHs have female as main bread-winner and decision-maker. Female headship should be due to widowhood, separation, single-motherhood resulting from pregnancy outside marriage, the husband's physical/mental incapacity, or headship in a non-nuclear family setup.

Income refers to total cash that gets into the household from all sources in an average month during the last three months. It is transformed into the ordinal scale by ranking households and getting the terciles.

Land ownership refers to possession or operation of a piece of farmland.

MHHs have male as main bread-winner and decision-maker.

Ownership of farm animals is indicated by possession of poultry (chicken, duck, turkey) or livestock (cow, horse, goat).

Social capital means membership in church-related organizations; or, membership in other types of social/occupational organizations.

Type of household is categorized into FHHs and MHHs, where headship is de jure and not due to temporary, work-related absence.

Work status is the attribute of working for cash or having regular source of livelihood during the last three months.

Review of the Literature

Natural disasters may result in the destitution of vulnerable groups, those whose resources are at the margin, such that in the event of misfortune (e.g., uncontrolled debts, political disorder, death of the only draft animal, lingering sickness or death of the chief earner, devastating flood, burning of the grain storage facility), recovery is extremely difficult. Women are among these groups, and if perchance they become household heads, moving out of poverty becomes almost impossible as cultural and other impediments begin to operate (Krishna et al. 2005; Reyes 2002).

The number of FHHs is increasing over time even as government tries to address inequalities by affirmative action.

At the individual level female disadvantages are apparent. In times of natural disasters more women than men die especially in countries where women have low socioeconomic rights (Neumayer and Plumper 2015). In ordinary times, being female often goes with disadvantages in the labor market, in pay and opportunities. For the same work, like providing the security, women earn less than men. Employers generally prefer men for physical strength, men do not go on maternity leave, and do not have to do childcare (Palan 2013).

Good quality education could have been an avenue for the upward social mobility of women. However, it is difficult to access. Basic education is free in the Philippines, but it is riddled with issues (big number of dropouts and repeaters, low grades, poor language skills, classroom shortage, and ill-equipped teachers). These problems in turn have resulted in a considerable number of illiterate Filipinos, out-of-school youth and graduates not ready for work (UNESCO 2015). According to Uy and Pua (2015), "elementary schooling is compulsory, but 24% of Filipinos of the relevant age group do not attend" mainly due to economic difficulties.

At the household level, disaster aftermath remains tragic for the surviving woman household head. Any household can dip into conditions of deprivation but FHHs are more prone than MHHs, *ceteris paribus*, and therefore, poverty tends to be "feminized". The literature attributes the feminization of poverty to various factors including the disadvantages of being female in the labor market and in other income generating opportunities as discussed above, in the multiple household tasks that women have to attend to, in the higher dependency burden of FHHs, in the limited access of women to economic resources like land and employment, and in the cultural norms that inhibit the social capital formation and therefore the productivity of women (Moghadam 2005). Hence, falling into and remaining in poverty appear to be the lot of FHHs.

The multiple tasks that a woman has to attend to in the household have some cultural roots. In the Philippines in general the wife is expected to care for the husband and children, and do household chores (e.g., doing the laundry and ironing, cleaning the house). These tasks have opportunity costs, as they compete with time for economically productive activities. Being a female household head in this context becomes a "double day burden" (Klasen et al. 2011; Moghadam 2005; Sealza 2013).

The higher dependency burden of FHHs may be understood as an offshoot of the fact that in the usual household system, the husband and the wife can be joint earners, hence burden is shared. In an FHH, there could be a single earner. Therefore, even for the same number of dependents, the burden would be greater for an FHH (Dungumaro 2008).

In many African countries where women are hard to put with land rights after divorce or death of the husband, the women definitely have difficulty with economic resources. In the Philippines, some money lending institutions do not allow women clients without husbands (Kimani 2008; Sealza 2008).

Sometimes cultural norms also inhibit women productivity. In rigidly patriarchal societies, women are powerless; decisions are made by men (Kimani 2008). Such are experienced also by poor women in India and the Philippines (Bhat 2002; Sealza 2013).

The number of households headed by women in the Philippines and worldwide is increasing (Indexmundi 2016; Bongaarts 2001). This trend is being attributed to factors like male long-distance migration, death of males due to armed civil conflicts and wars, un-partnered teenage pregnancy, and dissolution of marital bond or family disruption. Not all categories of FHHs are poor of course. De facto FHHs for instance where male partners are temporarily absent due to employment may not be poor compared to de jure FHHs that ensued for example from the husband's death or desertion (IFAD 2013).

International donor agencies and aid groups contend that "removing inequalities give society a better chance to develop. When men and women have relative equality, economies grow faster…" (AusAID 2010; World Bank 2010). For one, there would be a wider range of talents and added raw and polished labor as economic resource, and for another, there would be increased productivity across the gender divide.

Affirmative actions, or the policy preference of extending favor to members of a disadvantaged group, have been undertaken by many governments across the globe in order to arrest inequalities. These actions have bases in conference resolutions that found their way to legal pronouncements. The United Nations commission on women had organized conferences over the years: Mexico in 1975, Copenhagen in 1980, Beijing, 1995. In these conferences, member nations identified the gaps that prohibited the full implementation of agreements to help women. In the Beijing accord of 2005, the Commission adopted a platform of action which member states reaffirmed. The Philippines was a signatory to the Beijing Declaration (United Nations 2013).

Affirmative action in the Philippines comes in many forms. For example, there are laws that protect women, literacy training programs aimed at empowering women, and financing micro-entrepreneurial activities for women's income generation. The 1987 Philippine Constitution (Article II, Section 14) says that: "The State recognizes... the fundamental equality before the law of women and men", and that "The state shall protect working women by providing safe and healthful working conditions... and opportunities that will enhance their welfare...". These Constitutional provisions paved the way for the Magna Carta for Women (RA 9710 of 2009) that provides for the formulation of "plans, policies, programs, measures, and mechanisms to address discrimination and inequality in the economic, political, social, and cultural life of women and men"; for the Philippine Commission on Women; and, for the Women in Development and Nation Building Act (RA 7192 of 1992) that provided for self-reliance of women in improving income.

Programs to pursue affirmative action, however, move at a snail's pace because of impediments embedded in the social structure. For example, the Agrarian Reform program of the Philippines (e.g., Republic Act 9700), aimed at redistributing land resources to poor tenant-farmers has been carried out by different administrative regimes over the years but it has never been completely successful. Elvinia (2013) said, "reform laws have been tainted with vested interest of the landed elite in enacting the law, making the reform implementation difficult and derailed".

Since efforts to address structural impediments have not been successful, going by what the literature is saying, hope may come from external programmatic interventions. Much of the disadvantages of women have been about access to economic resources and work opportunities. If these are provided to them, things might change (Sealza 2003, 2007, 2013, 2014; Bhat 2002). In addition to access to economic resources, social capital was seen as facilitative of improvement in income. Social capital facilitates many things from alliance in building business network, to aid in times of crisis. It is viewed more as intentional assistance and mutual aid than as simple give-and-take or symbiosis (e.g., Van Vugt and Snyder 2001). It is built mainly upon membership in organizations.

Women welfare is among the UN's Sustainable Development Goals. In the Philippines, the well-being of women hitherto, remains wanting (Sealza 2013). If women under ordinary everyday circumstances are needy, one could expect that they should be more so after a disaster.

It is apparent in the literature that gainful employment is important, along with physical and financial capital, education, and social capital to help the FHHs.

Research Methodology

Data were taken from a sample drawn at random from the complete count of 2204 households, involving some 12,242 persons in 18 resettlements for *Washi* survivors. Data collection was commissioned by the Cagayan de Oro Archdiocese in order to: (1) get the profile of the households; and (2) determine how many Roman Catholics were there and how many have already received the Sacraments. A sample size of 337 was drawn by the simple random sampling procedure (Margin of Error = 0.05).

The study was approved by the Institutional Ethics Review Board of the Research Institute for Mindanao Culture (RIMCU), Xavier University, composed of research professionals with varied backgrounds (Public Administration, Medical Anthropology, Sociology, Ethnography) trained on ethics reviewing. The members are organic at RIMCU, except the Chair.

Signed informed consent was obtained from the respondents who were given names and organizational affiliation of the researchers along with contact numbers, informed of the purpose and uses of the study and their rights as respondents, assured of data confidentiality, and other consent requirements.

Face-to-face interview was conducted using an interview schedule (questionnaire). Questions covered social, economic, and demographic attributes of the households, family, organizational membership, amenities, and health and sanitation.

Principal Component Analysis (PCA) was used to construct a level of living index (LLI) to identify the main dependent variable. Indicators for

LLI were filtered from a wide array of ordinal and interval level variables with the linear correlation procedure. Cell phone ownership, considered material in distinguishing those with high LLI from those with low, was used as benchmark indicator in examining the level and direction of correlation coefficients. Indicators exhibiting high and significant correlation (0.05 level or better) with cell phone ownership were included in the PCA procedure. Then, factor analysis followed.

An acceptable set of 10 indicators surfaced after eight runs: socio-demographic (gender, education and age of the household head), sanitation (source of potable water and type of toilet being used), house ownership (house and lot ownership status), asset ownership (cell phone, refrigerator) and household monthly income (Henry et al. 2000).

Values of the communalities from the Component Matrix in this factor analysis ranged from 0.36 to 0.67 which satisfied the >0.30 value recommended by Burt-Banks. The Measure of Sampling Adequacy which was 0.60 also met the recommended value of 0.60 or more by Kaiser-Meyer-Olkin. The Bartlett's Test of Sphericity was 613.101, significant at Chi-Square 0.001 or better, df = 45. The eigenvalue of the LLI component was 1.964 (recommended value >1.0) (Henry et al. 2000).

The FHHs exhibited lower LLI than the MHHs. From composite index LLI, income was selected as focal dependent variable because it is the component of the index with a highest factor loading, in which household categories were differentiated, and which would likely be sensitive to initiatives introduced from the outside (Rosenberg 1968; Aneshensel 2013).

Income distributions are usually skewed and widely dispersed. This paper sidesteps the normality assumptions that often go with parametric models by using non-parametric (distribution-free) alternatives along with positional measures, median and the tercile. The households are ranked according to income. Then terciles are obtained dividing the distribution into three ordered categories [3rd tercile labeled "high income group", 2nd tercile the "middle income group", and 1st tercile the "low income group"].

The analysis begins first with the terciles of FHHs and the MHHs each juxtaposed with the tercile distribution of the entire sample. Second, type of household head is cross-tabulated with income. Third, income is cross-tabulated with other variables of interest, namely (1) programmatic factors: social capital (membership in church organizations, and membership in other social organizations); ownership of farm animals (livestock and poultry); and, work status during the last three months (with regular work; or, without regular work); and (2) structural factors (land ownership, and education of the household head). And finally a closer examination is made with strong and statistically significant relationships. The Chi-Square is used as significance test and Cramer's V as measure of the strength of relationship.

Findings

Tropical Storm *Washi* brought formidable flashflood to urban Cagayan de Oro, the Philippines, in December 2011. It washed away poor housing communities along the river banks. Official estimates pegged the casualty at 2000 people dead or missing (independent estimates gave a much bigger number), and thousands of internally displaced persons (IDPs). Property damage was estimated at PhP1.3B (US$29.5 M) (Tan 2011).

Relief agencies, both local and international, set up relief centers and housing structures to help the survivors. Eventually, IDPs were resettled and provided with skills training and livelihood opportunities. This was to help them sustain life after the disaster.

There were 50 (14.8%) FHHs and 287 (85.2%) MHHs in this study. The average income of the total sample was PhP5798 (US$126 at PhP46 to the dollar) (SD = 2913.77), and the median was 5900. Lower average than median income and the wide dispersion mean that the skewness of the distribution is high at 2.152 (Bulmer 1979). As usually is the case, it is skewed to the right because there are more poor than non-poor.

Some 93% of the household heads did not own or operate a piece of land; 7% did. About 26% had elementary education or lower; 56% had high school/some high school; 18% had college/some college education. Seventy-five percent had gainful work; and, 25% had none during the past three months. Thirteen percent raised livestock, 87% did not. Twenty-three percent raised poultry; 77% did not. Thirty-three percent were members of social organizations; 67% were not. Nineteen percent were members of church organizations; 81% were not.

The average income of FHHs was 5350, that of the MHHs was 5876 per month. Again, the intention of this paper is to analyze differences in the positions of the FHHs and MHHs in the income distribution, and posits that, given the "feminization of poverty" thesis, the FHHs would tend to be at the lower end of that distribution. The income data have been converted into tercile distribution: low equals 5000 and below; middle equals 5001–6000; high equals 6001 and above per month (Fig. 2).

Figure 2 shows the income terciles for the total sample. The terciles of course are of approximately the same height. Figures 3 and 4 are the individual terciles of FHHs and MHHs respectively which exhibit how they depart from the terciles in Fig. 2. Only 6% of the FHHs are in the high income tercile, while about 38% of the MHHs are in the high income tercile (Figs. 3 and 4).

Table 1 shows that indeed data support the argument that FHHs are at the lower end of the income distribution. Close to 38% of the MHHs belonged to the high income group; only 6% of the FHHs were in this income category. Ninety-four percent of the FHHs were either in the low or the middle income group. And the association between type of HHH and income categories is very strong (Cramer's V = 0.35, which is significant at the 0.01 level, Chi-Square test, df = 2) (Quinnipiac University Faculty 2015).

Test Factors

This section attempts to explain what "causes" income differences between FHHs and MHHs. Tables 2 and 3 show that of the test (programmatic) factors earlier identified to likely confound the relationship

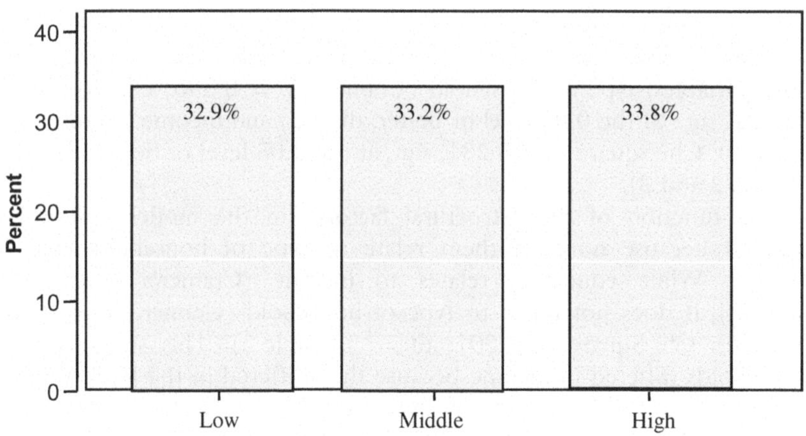

Fig. 2 Tercile distribution of the total sample

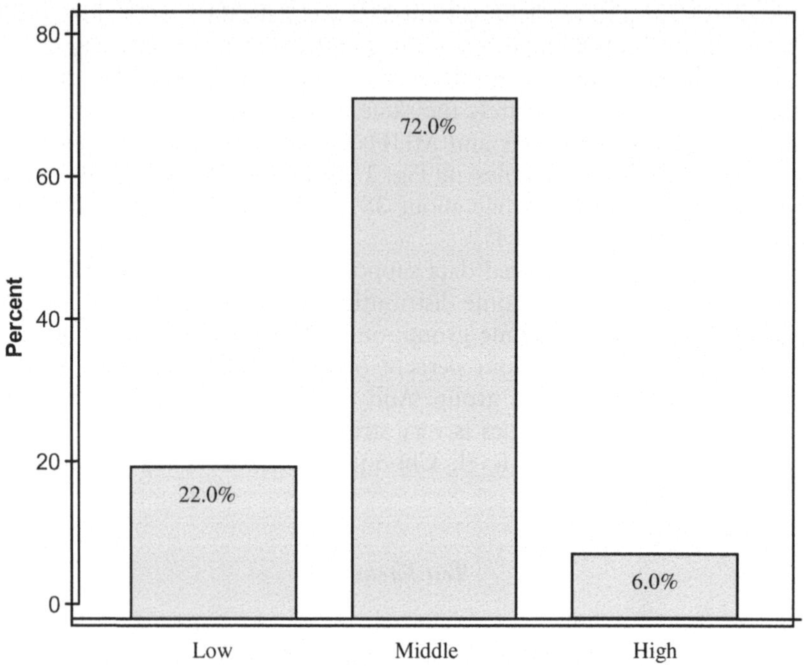

Fig. 3 Tercile distribution of FHHs

between type of household and income distribution, only work status surfaces as a likely candidate. Work status is the only factor that relates to both variables: type of household (Cramer's V = 0.396; Chi-Square = 52.932, sig., at the 0.05 level or better, df = 2) and income (Cramer's V = 0.660; Chi-Square = 145.281, sig., at the 0.05 level or better, df = 2) (Tables 2 and 3).

The function of the "structural factors" in the model cannot be assessed because none of them relate to type of household and to income. While education relates to income (Cramer's V = 0.14, Table 3), it does not relate to type of household (Cramer's V = 0.06, n.s. with Chi-Square = 1.301; df = 3, Table 2). This suggests that households differed in income because they differed in the level of education of the HHH, such that households whose heads had high education tended to belong to the relatively higher income group. However, FHHs and MHHs did not differ in level of education (Table 2).

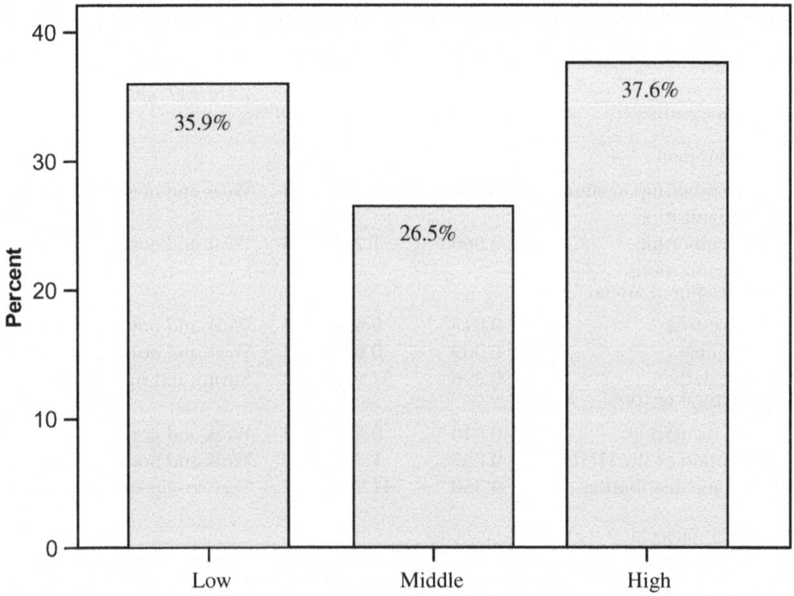

Fig. 4 Tercile distribution of MHHs

Table 1 Distribution of type of household by income group

Income group	*Type of HHH*		*Total*
	Female	*Male*	
High	3	108	111
	6.0	37.6	
Middle	36	76	112
	72.0	26.5	
Low	11	103	114
	22.0	35.9	
Total	50	287	337
	100.0	100.0	100.0

$\chi^2 = 41.910$, sig., at α 0.01 or better, df = 2; Cramer's V = 0.35

In other words, work status might explain why income distribution differs between the FHHs and the MHHs; that is, the FHHs are in the lower income category because they have no regular sources of income. This is examined next.

Table 2 Association between type of household and the test factors

Test factors	*Type of HH*			*Strength of association and statistical significance at α 0.05*
I. Programmatic	*Cramer's V*	*Chi-Square*	*Df*	
A. Social capital				
1. Membership in church organization	0.067	1.492	1	Weak and not significant
2. Membership organizations	0.060	1.2110	1	Weak and not significant
B. Ownership of assets				
1. Livestock	0.045	0.670	1	Weak and not significant
2. Poultry	0.009	0.027	1	Weak and not significant
C. Work status	0.396	52.932	1	Strong and significant*
II. Structural factors				
A. Land ownership	0.040	0.552	1	Weak and not significant
B. Education of the HHH	0.062	1.301	3	Weak and not significant
III. Income distribution	0.350	41.910	2	Very strong and significant*

*Statistically significant at the 0.01 level of better

Table 3 Association between the test factors and income

Test factors	*Income categories*			*Strength of association and statistical significance at α 0.05*
I. Programmatic	*Cramer's V*	*Chi-Square*	*Df*	
A. Social capital				
1. Membership in church organization	0.083	2.326	2	Weak and not significant
2. Membership in other organizations	0.016	0.083	2	Weak and not significant
B. Ownership of assets				
1. Livestock	0.027	0.252	2	Weak and not significant
2. Poultry	0.057	1.098	2	Weak and not significant
C. Work status	0.660	145.281	2	Very strong and significant*
II. Structural factors				
A. Land ownership	0.078	4.126	2	Weak and not significant
B. Education of the HHH	0.142	13.602	6	Moderately strong and significant*

*Statistically significant at the 0.01 level or better

Work Status

Some 25% of the household heads had no regular source of livelihood during the last three months. About 75% had work. Of those who did not have work, 33 (39.3%) were female, and 51 (63.7%) were male.

Table 4 presents the distribution of households by type and income category, with work status held constant. The original table was collapsed (low and middle income categories were put together) to minimize cells with zero values.

The Chi-Square may not be appropriate to test the significance of the association between type of household and income under the "no regular work" category of work status because a cell has an expected frequency of less than 5 (Table 4). One will notice however that the percentage difference between FHHs and MHHs in the "high" income category in Table 1, the original table (31.6%), has reduced to 11.8% in the "has no regular work" category, and 25.6% (43.2–17.6) in the "has regular work" category of work status, which implies that the original relationship did not hold. Under the "has regular work" category, the strength of the original relationship between type of household and income (Cramer's = 0.35; Chi-Square = 41.910, sig., at the 0.01 level or better, df = 2) (Table 2) has diminished (Cramer's V = 0.11; Chi-Square corrected for continuity = 3.28, not sig., at the 0.05 level, df = 1). These findings suggest that the income difference between the FHHs and MHHs is due to difference in work status.

Table 4 Type of household and income level under separate categories of work status

Income level	*Has no regular work/income source*[a]		*Has regular work/income source*[b]		*Total*
	FHH	*MHH*	*FHH*	*MHH*	
High	0	6	3	102	111
	0.0	11.8	17.6	43.2	
Low	33	45	14	134	226
	100.0	88.2	82.4	56.8	
Total	33	51	17	236	337
	100.0	100.0	100.0	100.0	

[a]One cell has zero value

[b]Chi-Square corrected for continuity = 3.28, not sig., at the 0.05 level, df = 1. Cramer's V = 0.11

Looking at Table 4 one finds that, at the household level, the cumulative burden of being poor plus having experienced disaster, plus living with the FHH is the lot of some 10% ($n = 33$) of the households in the resettlements.

What specific types of work were there? Work engaged in by MHHs have been mostly skilled occupations: massage therapy, masonry, *aircon* repair, baking, operating big machines, driving, carpentry, welding, and so on. Some engaged in casual work as waiter, pedicab driver, and car washer.

Of the 50 FHHs, 34% had regular work; 33 (66%) had none. Regular work includes: self-employment (vending: candle, vegetable, cigarettes/candies, food; and, skilled work: lending, village health work, massage therapy, beautician). Women who said that they had no work include those whose occupations were non-regular: doing the neighbor's laundry, house cleaning, and babysitting.

Conclusion and Recommendation

In conclusion, the hypothesis, "FHHs are likelier than MHHs to belong to the lower end of the income distribution", is supported by the data. The difference was mainly due to regular work engagements of MHHs in skilled occupations like masonry, aircon repair, baking, driving, carpentry, and the likes. Most of the FHHs had to rely on intermittent work opportunities in the neighborhood.

The economic vulnerability of FHHs, therefore, is associated with lack of sufficient work opportunities for women. The feminization of poverty thesis finds support in this study. Some 10% of the households are apparently affected by the cumulative effects of extremes like poverty, unemployment, and perhaps a host of other disadvantages that go with being female. More attention and better planning and execution should be made to address women concerns.

The findings provide a glimmer of hope especially since the differences between types of households in terms of income are not due to "structural" factors that are difficult to correct; but are rather due to differences in work status that can be addressed, albeit with some struggle still, by development programs from the outside. This provides an indication to the entry point for development work. It should be recognized by government planners and policymakers, business groups, aid agencies, civil society organizations and the Church that the women remain at the disadvantageous side of the male-female divide.

In particular, the Archdiocese of Cagayan de Oro is influential in the local population of mostly Catholics (about 83%). It has been into numerous activities and programs, mostly on youth empowerment, interfaith dialogue, natural family planning, and other faith-related endeavors (Sealza et al. 2011). It is time for the Church to look more seriously alongside religious concerns the work and livelihood concerns especially of the FHHs in the disaster resettlement areas.

There is of course the necessity for training needs assessment among the FHHs, and for product marketing where serious research work is necessary.

Theoretically, work status functions as "intervening" variable that explains income difference between FHHs and MHHs.

The obvious entry point for development work is the creation of opportunities for the sources of livelihood for women. And livelihood opportunities may demand the requisite training for skills on entrepreneurship, startup capital and market study, and training for skills in technical-vocational work.

Bibliography

Aneshensel, C. S. (2013). *Theory-Based Data Analysis for the Social Sciences* (2nd ed.). Singapore: Sage.

AusAID. (2010). "Gender Equality and Development". Retrieved December 4, 2012, from http://wwww.ausaid.gov.au/keyaid/gender.cfm.

Bhat, R. (2002). "Feminisation of Poverty and Empowerment of Women—An Indian Perspective & Experience". Paper presented at the Women's Conference, Townsville, Australia, July 3–7, 2002. Retrieved October 15, 2015, from http://www.adfvc.unsw.edu.au/Conference%20papers/TIWC/BhatRashmi.pdf.

Bongaarts, J. (2001). "Household Size and Composition in the Developing World". Working Paper No. 144. New York: The Population Council.

Bulmer, M. G. (1979). *Principles of Statistics*. London: Dover.

Chambers, R. (1985). *Rural Development*. New York: Longman Inc.

Chant, S. (2007). "Children in Female-Headed Households: Interrogating the Concept of an 'Inter-Generational Transmission of Disadvantage' with Particular Reference to the Gambia, Philippines and Costa Rica". New Working Paper Series, London School of Economics, Gender Institute ISSN No: 1470-8515. Retrieved May 15, 2015, from http://www.lse.ac.uk/genderInstitute/pdf/children.pdf.

Dungumaro, E. (2008). *Journal of Comparative Family Studies* 39(4): 429–451.

Elvinia, J. (2013). "Is Land Reform a Failure in the Philippines? An Assessment on CARP". Retrieved October 14, 2015, from http://ir.nul.nagoya-u.ac.jp/jspui/bitstream/2237/15881/1/10_Jose%20Elvinia_2.pdf.

Henry, C., Sharma, M., Lapenu, C., and Zeller, M. (2000). *Assessing the Relative Poverty of Microfinance Clients.* Washington, D.C.: International Food Policy Research Institute.

IFAD. (2013). "Enabling Poor Rural People to Overcome Poverty". Retrieved May 12, 2015, from http://www.ifad.org/gender/learning/challenges/women/60.htm.

Indexmundi. (2016). "Philippines—Female Headed Households". Retrieved May 27, 2016, from http://www.indexmundi.com/facts/philippines/female-headed-households.

Kimani, M. (2008). "Women in North Africa Secure More Rights". *Africa Renewal.* Retrieved May 14, 2015 from https://www.un.org/africarenewal/magazine/july-2008/women-north-africa-secure-more-rights.

Klasen, S., Lechtenfeld, T., and Povel, F. (2011). "What About the Women? Female Headship, Poverty and Vulnerability in Thailand and Vietnam". Retrieved May 14, 2015, from www.oecd.org/dev/pgd/46982877.pdf.

Krishna, A., Gibson-Davis, C., Clasen, L., Markiewicks, M., and Prez, N. (2005). "Escaping Poverty and Becoming Poor in Thirteen Communities in Rural North Carolina". Retrieved from http://www.mdcinc.org/docs/poverty.pdf.

Moghadam, V. M. (2005). "The 'Feminization of Poverty' and Women's Human Rights". SHS Papers in Women's Studies/Gender Research. No. 2. UNESCO.

Neumayer, E., and Plumper, T. (2015). "The Gendered Nature of Natural Disasters: The Impact of Catastrophic Events on the Gender Gap in Life Expectancy, 1981–2002". Retrieved May 15, 2015, from http://www.lse.ac.uk/geographyAndEnvironment/whosWho/profiles/neumayer/pdf/Disastersarticle.pdf.

Palan, S. (2013). "Survey Finds Many Australian Employers Value Men Over Women, Prefer Workers Without Children". ABC News. http://www.abc.net.au/news/2013-07-21/survey-businesses-prefer-male-workers/4833586.

Quinnipiac University Faculty. (2015). "Guidelines for Interpreting Cramer's V". http://faculty.quinnipiac.edu/libarts/polsci/Statistics.html.

Rajaram, R. (2009). "Female-Headed Households and Poverty: Evidence from the National Family Health Survey; Evidence from the National Health Survey". Retrieved May 15, 2015, from www.frbatlanta.org/-media/Documents/news/conferences/2009/.

Reyes, C. M. (2002). "The Poverty Fight: Have We Made and Impact?" PIDS Discussion Paper Series No. 2002-20. Makati: Philippine Institute for Development Studies.

Rosenberg, M. (1968). *Logic of Survey Analysis.* New York: Basic Books.

Sealza, I. S. (2003). "Tibod sa Barangay Project of MILAMDEC". Research report prepared for MILAMDEC. Cagayan de Oro City: Research Institute for Mindanao Culture, Xavier University.

Sealza, I. S. (2007). "Women Organization: The Case of MAKAKABUS". Research report prepared for Philippine Development Assistance Programme. Cagayan de Oro City: Research Institute for Mindanao Culture, Xavier University.

Sealza, I. S. (2008). "Poverty Roots and Causes". Paper for Loyola Chair Award (Unpublished Manuscript).

Sealza, I. S. (2013). "Women Through Learning and Livelihood Project in Southern Philippines". In A. Singh, E. T. Gonzalez, and S. B. Thomson (Eds.), *Millennium Development Goals and Community Initiatives in the Asia Pacific*. New Delhi: Springer.

Sealza, I. S. (2014). "Recovering from the Effects of Natural Disaster". *European Journal of Sustainable Development* 3(3). https://doi.org/10.14207/ejsd.

Sealza, I. S., Sealza, L. S, Manalo, M. L., and Almaden, C. R. (2011). "An Assessment Study of Diocesan Pastoral Plans in the Light of the Second Vatican Council of the Philippines" (Unpublished Manuscript).

Tan, K. J. (2011). "'Sendong' Deadliest Cyclone in 12 Years with Over 1000 Dead". GMA News. Retrieved March 20, 2012, from http://www.gmanetwork.com/news/story/242343/news/nation/sendong-deadliest-cyclone-in-12-years-with-over-1000-dead.

UNESCO. (2015). "Philippine Education for All 2015: Implementation and Challenges". Retrieved October 14, 2015, from http://planipolis.iiep.unesco.org/upload/Philippines/Philippines_EFA_MDA.pdf.

United Nations. (2013). United Nations Entity for Gender Equality and the Empowerment of Women. Retrieved April 20, 2013, from http://www.unwomen.org/en.

USAID. (2010). "Muslim Religious Leaders Issue a National Fatwah in Support of Family Planning". Retrieved from http://philippines.usaid.gov/he_success_stories4.html.

Uy, J., and Pua, J. (2015). "Poverty: Causes, Effects and How We Can Solve It". Retrieved October 14, 2015, from http://sites.my.xs.edu.ph/joseph-pua-14/social-science/poverty-causes-effects-and-how-we-can-solve-it.

Van Vugt, V., and Snyder, M. (2001). "Cooperation in Society: Fostering Community Action and Civic Participation". *American Behavioral Scientist* 45: 765–768. Retrieved June 24, 2014, from http://www.professormarkvanvugt.com/files/CooperationinSociety-AmericanBehavioralScientist-2001.pdf.

World Bank. (2010). "Gender and Sustainable Development". Retrieved March 3, 2015, from http://web.world.org/WBSITE/EXTERNAL/TOPICS/EXTSOCIALDEVELOPMENT/0,contentMDK:22511004~pagePK:148956~piPK:216618~theSitePK:244363,00.html.

Child Centered Disaster Risk Reduction (CC DRR)/Resilience

Ray Kancharla

Definition of Resilience

"Resilience is the ability of an individual, a household, a community, a country or a region to withstand, to adapt, and to quickly recover from stresses and shocks. The concept of resilience has two dimensions: the inherent strength of an entity—an individual, a household, a community or a larger structure—to better resist stress and shock and the capacity of this entity to bounce back rapidly from the impact. Increasing resilience (and reducing vulnerability) can therefore be achieved either by enhancing the entity's strength, or by reducing the intensity of the impact, or both. It requires **a multifaceted strategy** and a broad systems perspective aimed at both reducing the multiple risks of a crisis and at the same time improving rapid coping and adaptation mechanisms at local, national and regional level. **Strengthening**

Reducing Risk, Saving Lives—Save the Children's approach to Disaster Risk Reduction and Climate Change Adaptation—2014.

R. Kancharla (✉)
Save the Children India, Gurgaon, India

B. Zutshi et al. (eds.), *Disaster Risk Reduction*,
https://doi.org/10.1007/978-981-10-8845-2_17

resilience lies at the interface of humanitarian and development assistance"—(European Union 2012).[1]

John Twigg in '**Characteristics of a Disaster Resilient Community** (2009)'[2] outlines that community resilience can be understood as the capacity to:

- anticipate, minimize and absorb potential stresses or destructive forces through adaptation or resistance
- manage or maintain certain basic functions and structures during disastrous events
- recover or "bounce back" after an event (see Footnote 2)—***rapidly through Build Back Better methodologies.***

Children and Youth have not always emerged as a visible stakeholders in defining resilience. Most recent effort of John Graham in an attempt to define Child Centred has proposed the following definition:

> Ability of a system, community, household and child to prepare for, manage, and recover and improve from recurring and protracted shocks.

Why Children?

Children have always been there and they will be part of the large mass of people. Is there any relevance to dedicate specific focus on children in disaster situations? Children will always be part of the families, hence, why make much ado about this? Adults will look after them anyway. While this may be true at a surface level, this is often a naïve position. Due to these children have been ignored, neglected and at times even severely abused. Children and women are more vulnerable to disasters. Disasters exacerbate their impacts on children and women.

[1] The Eu Approach to Resilience: Learning from Food Security Crises, 2012—page 5.

[2] John Twigg, Characteristics of a Disaster Resilient Community, 2009, v2, page 8. http://www.actionaid.org/sites/files/actionaid/characteristics_of_a_disaster_resilient_community.pdf.

Systemically, in the way the needs of children are addressed, one tends to assume that "children are treated as miniature adults". It is about time that both governments and society at large recognize children as a specific stakeholder in the scheme of things, more so in disaster situations.

> Disasters such as the Asian tsunami of 26 December 2004, can affect even more children, causing them to lose their homes, their families, their schools, their access to adequate food, water and sanitation and even their lives in a matter of minutes. Despite these statistics, however, the protection of children remains a second concern for the international community in all phases of emergency response. The failure to protect children from these escalating threats not only results in personal tragedy but carries a long-term social cost as well as, including the spread of HIV/AIDS, an elevated maternal and infant mortality rate, a loss of education and a generation of marginalized youth. (**BOND DRR GROUP**)[3]

Indian sub-continent promulgated a Disaster Management Act in 2005. This is a landmark not only in India, but in the region. However, it has not made a mention of children and what are the protocols that need to be adhered to in the case of children.

Who is a Child? One is amazed at the perspectives the families and society harbor as an understanding about who is a child. According to UNCRC, a child is someone under 18 years of age. Further, it is important to disaggregate children according to their formative stages and address specific needs of children in disasters through pre-, post and during disaster phases.

-09 to 24 months – 1,000 days	**03 to 05 years -ECCD**	**06 to 14 years – School going**	**15 to 18 years –Adolescence**

At each of the stages mentioned above, children have specific needs and protocols to be developed and put in place. The real mantra to be used as a talisman when addressing children in humanitarian crises contexts is—**'Children Can't Wait' (CCW)**.

[3]UNCRC—Article 12.

Child Centred Disaster Risk Reduction (CC DRR)—*A Niche Approach*

Save the Children in India initiated a new position in the larger Disaster Risk Reduction Community in terms of promoting a niche approach "Child Centred Disaster Risk Reduction (CC DRR). The evidence for such an approach emerged in post-Kosi Floods scenario where in more than 100 villages across 4 districts of Bihar, a district each in West Bengal and Odisha experienced this unique approach. The CC DRR approach had envisioned a 5-pillar action:

PREMA—Preparedness, Resilience, Early Warning, Mitigation, and Adaptation.

In the recent years, there has been a growing debate on the concept and approach to resilience building. The Inter Agency Working Group of BOND (British Organization of NGOs in Development) drafted a working paper on Characteristics of Resilience, in which it is highlighted that the "*principles of resilience ought to be Socio-ecological systems-based, Conflict Sensitive, Inclusive, Enhance the agency of most vulnerable people, Accountable and politically engaged, Conducted in partnership and Long-term*"[4] Key factors remain context-specific, diversified and people-driven.

The envisaged outcomes of such a process would ensure that there is an overall Preparedness Plan which is clearly owned by the children, communities and local governance. This is accentuated by the demonstrable evidence of a functioning Early Warning System. The communities as part of their overall risk reduction integrated into development agenda have also undertaken mitigation and adaptation measures, based on the risk mapping and climate vulnerabilities identified. More often than not, there is a strong limitation when it comes to allocation of resources toward adaptation and mitigation, while preparedness planning and early warning systems often do not need huge financial investments, but a strong political will. Resilience is the innate capacity that has been there and invariably matured itself through the actions in preparedness, early warning, mitigation, and adaptation. Often, this is what becomes a strong driver for emergence of "culture of preparedness".

Five Pillars of Action in CC DRR

- Children and Youth Participation and Leadership (CYPL)
- Community Level Actions (CLA)

[4] http://www.preventionweb.net/english/professional/trainings-events/events/v.php?id=4255.

- Institutional Strengthening (IS)
- Policy Level Engagement (PLE)
- Mainstreaming DRR into Development.

Children Participation as an essential ingredient of risk mapping, planning and active engagement in the task forces as well as overall governance at the local. This is founded on Children's Right to Participate as enshrined in United Nations Convention on the Rights of the Child (UNCRC) (see Footnote 4) Article 12.1 emphasized that "*States Parties shall assure to the child who is capable of forming his or her own views the right to express those views freely in all matters affecting the child, the views of the child being given due weight in accordance with the age and maturity of the child.*"

The second India Disaster Management Congress organized by Ministry of Home Affairs and National Institute of Disaster Management held by Government of India (see Footnote 5) (in New Delhi from 4 to 6 November 2009 in Vigyan Bhavan, New Delhi) has included this agenda and acknowledged that Children in the age group of 13–18 years can effectively be represented on local disaster management committees.

This is evidence in the states of Bihar, West Bengal, Odisha, Delhi, Andhra Pradesh, Tamil Nadu as well as in Andaman Nicobar Islands in the context of village level disaster risk reduction initiatives as well as school safety initiatives.

Formation of Children Group (CG)—an identified group of children (approximately 12–15) including girls and boys based on their understanding and capability are designated to come together and function as a Children Group on behalf of the village/institution. Their role is to help, support and guide other children in matters of protection, education (school attendance) and manage CFS/TLC. Children Group has regular meetings to understand the situation of children in the village and the problems/issues that confront them in their day to day lives such as neglect, abuse, violence, absenteeism from school, child labor, child marriages, child trafficking, etc. The leadership group among children will try to address it at their level. Where they fail, they will represent it to the ***Children Resilience and Protection Committee (CRPC)*** which is a composite body of governance representatives, duty bearers and other opinion leaders in the village. Two children from the Children Group, namely, a girl and a boy will also be designated as representatives to the ***Children Resilience and Protection Committee (CRPC)***.

Community Level Actions (CLA)

This is a process of empowering the vulnerable households to map their risks, latent capacities and emerge with systems and mechanisms to cope with disaster risks.

- *Risk and Vulnerability Mapping* (RVM): This is conducted through participatory vulnerability capacity analysis (PVCA) where in the children and communities are guided to undertake their social, economic, political, cultural mappings as well as disaster history of the village.
- *Identification of Capacities* (IC): Further, local capacities both institutional, human and material resources are mapped toward enhancing local resilience.
- *Formation of Village Disaster Management Plan* (VDMP): RVM and identified capacities are analyzed and converted into a village disaster management plan which would be ratified at the village council meeting. Based on this planning, communities undertake local risk reduction initiatives, small-scale mitigation and adaptation actions.
- *Children Resilience and Protection Committee* (CRPC): The CRPC comprises of elected representatives of the community, teachers, early child care unit worker, health workers as well as self help group/farmers group representatives who have grasp of local conditions and have the ability to contribute to resilience. Two children with leadership capacity from the Children Group (a boy and a girl) are co-opted into CRPC, so that the children's concerns and issues are taken into planning and addressed.
- *Task Force Teams* (TFT): Five types of local task forces are formed and trained to be always alert and ready to respond in case of any hazard or disaster. The types of task forces are: (1) Search and Rescue Group; (2) Early Warning Group; (3) First Aid Group; (4) Psycho-Social Group; and (5) Relief Management Group. These task forces function as the first line of defense in the event of any disaster or risk.

- *Grain Bank* (GB) to ensure food security: Mothers form a grain bank committee in order to save food grains during non-disaster days to ensure that their children do not need to starve when disaster strikes the area. Grain Bank Committee develop their rules of governance and sustainability of the grain bank. Chatarghat village in Bihar became a model for local area in the flood risk communities as early as 2007. The local self-governance leader has strongly testified to the fact that in their village nobody will starve for first days after a disaster, even if no external help arrives (see Footnote 5).

Institutional Strengthening (Panchayat Raj Institution—School—ICDS Centre, Primary Health Centre, etc.).

Child Centred DRR process and actions would ensure that the local self-governance and other life-line/key institutions are capacitated to strengthen their resilience. These comprise of the following:

- Affirmative Support from Panchayat Raj Institutions (Local Self Governance in India) and adoption of plans.
- School Risk Map and Safety Planning including capacity building.
- ICDS Risk Maps and Safety Planning including capacity building.
- Primary Health Centre Risk Map and Safety Planning.
- Local Implementing Partners' Capacity Building.
- Engagement with Governmental Departments and Duty Bearers.
- Establishment of Panchayat Disaster Management Resource Centre (P-DMRC)—*a technology-driven innovation that would emerge as one-stop shop for disasters and development agenda.*
- Establishment of School Disaster Management Resource Centre (S-DMRC).

Policy/Planner Level Engagement (PLE)

The aim of CC DRR is to ensure that the actions undertaken at local level are incorporated into the local district level planning as well as validated at all levels of disaster management.

The following nodal institutions are vital for the ownership of CC DRR outcomes:

- District Disaster Management Authority to incorporate the results and outcomes of CC DRR

- State Disaster Management Authority to validate the process and enable their sustainability
- Coordination with Block level authorities and line departments so that local resilience plans are supported and budget allocated annually
- Strengthening coordination with ICPS, Education Authorities in order to ensure that safety and protection of children is mainstreamed into integrated child protection system as well as educational system.

Mainstreaming DRR into Line Departments and Development (M3D)

The process and outcomes of CC DRR can only be sustained if they are incorporated into the rural development and other social welfare schemes meant for the vulnerable and marginalized children and communities. Hence, it is crucial to

- Link with social safety net schemes.
- Promote convergence inter-departmentally (WASH, NREGS, Education, ICDS, SSA, etc.).

Some pioneering Child Centred Disaster Risk Reduction and innovations
From 2008 onwards, Save the Children in India partnered with Children, communities and local governance institutions in developing innovative models in Child Centred DRR.

- Building Models of Child-Friendly Multi-Hazard Disaster Resilient structures (CFMDRS)—11 CFMDRS have been built in 3 states (5 in Bihar, 3 in West Bengal and 3 in Odisha) with a capacity to accommodate at least 1000 persons/approximately 2–3 villages in its radius. They have designated spaces for children (class rooms, play areas) as well as child-friendly toilets. These local structures are equipped with standard kits needed for search and rescue, early warning, first aid as well as for managing relief at community level.
- Innovative Panchayat Level Disaster Management Resource Centres (P-DMRC) have been conceptualized and piloted in three states of India. They are emerging to be "one-stop-shop" for local governance, children and communities in ensuring that they are

disaster prepared, there is an early warning task force active who are aided by a system, and that there is a tracking of safety net that is being delivered to the marginal and poor households.
- Children and Youth at the heart of Resilience: 550 Villages have trained and active task forces with participation and leadership of Children and Youth.

School Safety, DRR, and Education

Save the Children in India had been facilitating school safety planning with the overarching approach of "five pillar action" outlined earlier in this. Children, teachers, school management committees in Odisha, West Bengal, Bihar, and Assam have been convinced in undertaking pioneering initiatives in this area. The evidence of such a dynamic process and outcomes were demonstrated to National Disaster Management Authority when a flagship project titled National School Safety Project was initiated.

- September 1–2, 2011—*National Disaster Management Authority launched its flagship programme titled National School Safety Project* (NSSP). Children from Odisha, Bihar and West Bengal projects of Save the Children participated and shared their knowledge and practice of School Safety. This has been highly appreciated by central and state disaster management authorities. As a result of the children's genius of experience sharing which outlined the processes of risk mapping, preparedness planning as well as active participation in task forces, enabled a most profound statement placed before all—"ANY SCHOOL SAFETY INITIATIVE MUST BEGIN WITH CHILDREN FIRST AND MUST BE SUSTAINED BY CHILDREN AND SCHOOL COMMUNITIES".
- Further, Safe School Constructions in Leh and Kargil after cloudburst of August 2010 were also models of how a safe school with disaster resilient features needs to be undertaken.

  The technology was very much the people's own—Passive Solar Technology with new dimensions of Disaster Resilience and Child-Friendly Features. School Safety Planning is an integral part of it all.
- Mock Drills have been a regular phenomenon.
- On October 29th, National DRR day Rallies by School Children have been a regular phenomenon.

Comprehensive School Safety (CSS)[5]

The Goal of CSS is to:

1. Protect learners and education workers from death, Injury and harm in schools
2. To plan for educational continuity in the face of all expected hazards and threats—including everyday risks
3. To safeguard education sector investments
4. To strengthen risk reduction and resilience through education.

There are three key pillars to CSS:

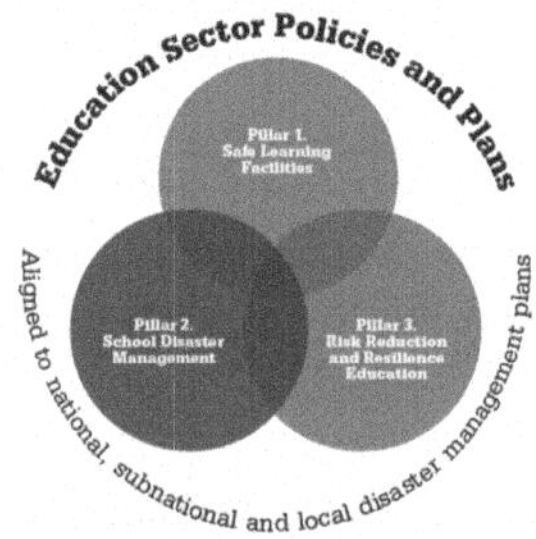

Pillar 1: Safe Learning Facilities:

- Building maintenance
- Non-structural mitigation
- Fire safety

Pillar 2: School Disaster Management:

- Multi-hazard risk assessment
- Education sector analysis
- Child-centred assessment and planning

[5] Comprehensive School Safety—A global framework in support of The Global Alliance for Disaster Risk Reduction and Resilience in the Education **Sector** and **The Worldwide Initiative for Safe Schools,** in preparation for the 3rd U.N. World Conference on Disaster Risk Reduction, 2015.

Pillar 3: Risk Reduction and Resilience Education:

- Structural safety education
- Construction as educational opportunity
- Household disaster plan
- Family reunification plan

Case Studies from the Heart of Humanitarian Crises

6 Districts of Uttarakhand were devastated by unprecedented flooding and landslides in June 2013 (14–16) and displaced hundreds and thousands of hill population. Children were at severe risk in post-disaster situation with homes and schools destroyed and life line of the villages completely broken.

In the midst of such a chaotic situation, Relief teams went about identifying most vulnerable children and responded to their life-saving needs. Santosh was at severe risk due to a surgery he underwent. Only when he received timely medical attention, his life could be saved. Risk reduction has to be timely.

Child Friendly Spaces (CFS)/Temporary Learning Centres (TLC) were created by Save the Children to provide a space where children can learn, laugh and be protected while their parents attempt to restart their lives in the aftermath of the disaster.

- Children related assessments have always exposed and reiterated the fact in the post-disaster situations, Children's vulnerabilities are heightened. As their protective layer is shattered, often with government's departments becoming disabled, children's exposure to neglect, abuse, trafficking, etc. heighten.

- With schools damaged or destroyed, schools are not functional for long periods of time; hence, children become school drop-outs. Often, parents resort to sending children into labor, as supplementary earners.
- CFS/TLC are niche approaches to ensure children are protected and their education is restored immediately.
- Children and communities have recognized the value of CFS/TLC and have voluntarily come forward to manage them effectively.

- CFS/TLC have addressed a coordinated approach in terms of education, protection, life skills, health and nutrition which were the critical needs of the hour.
- The educated youth including girls emerged as potential resource for initiating programmes for children.
- Uttarakhand children have demonstrated their leadership in reducing their own risks and pro-actively engage in Disaster Risk Reduction as well as Climate Change concerns and issues.

Srilatha (name changed for protecting identity), 16, *lived in a beautiful village in Urgam Valley, Joshimath. She came to the relief distribution site supported by her friend. She introduced herself confidently in fluent English. The Relief programme officer asked her to name the white snow covered mountain seen at a distance. She did not reply. On repeated instance, she maintained utmost silence. Then, the Child Friendly Space facilitator informed "ye dekh nahi sakti" (she is visually challenged).*
Srilatha narrated, "I used to go to school in class III. I have lost my sight due to illness in 2012. I want to study again. I need your support"—with sparkles in the eyes.
Her admission to a Government Institution for the visually impaired has been realized, where she is pursuing her dream.

In the midst of numerous challenges, Children of Child Friendly Spaces in Ukhimath block of Rudraprayag district experienced the great excitement when huge quantities of colorful pencils arrived for them from the Government of Japan. Children jumped in joy and happiness, as they got to know that each one of them received a pack to give vent to their creativity. For days and weeks, Children practiced coloring, writing with myriad pencils they received. This enabled them to experience a huge relief of their psyche (a psycho-social reliever).

Neha says, I love this pencil, these are colorful and good, I draw very well with these pencils.

In addition, the parents of these children were supported with Disaster Resilient Livelihoods, Safe Water Systems, Supplementary

Nutrition for malnourished children along with a Cash Transfer to attend to diverse needs. Such Integrated Approach is the best to ensure the journey from Risk to Resilience.

What Is the Way Forward to Enable Children and Youth as Future Leaders?

In March 2015, World Conference on Disaster Risk Reduction in Sendai has outlined a new framework for DRR replacing the erstwhile Hyogo Framework for Action. The current framework has made mention of children and other vulnerable sections of the society as key stakeholders. In the light of this, it is important to undertake the following:

Child Centred DRR to Be incorporated in Government's planning and implementation:

- Participation of children in resilience building activities needs to be promoted through activities of Education Ministry and other lifeline ministries (Women and Child Development, Water, Sanitation, Agriculture, et al.).
- A vulnerability-based approach to risk, resilience and adaptation, that is child-centred, needs to be adopted.
- DRR action plans for "make my city resilient" and climate action plans need to incorporate this approach and make it work for children in urban locations.
- Deprivation, disruption, disasters are the manifestations that compound poverty and risk. Not addressing these with an assured delivery of finances to the most vulnerable will hamper the progress and in turn, increase the risks. Any future investment must be made after considering this as a "*conditio sine qua non*" for ensuring the safety and progress of our future leaders.

- Government's smart cities plan needs to embrace this approach that has children at its center. Further, innovative family tracing and re-unification platforms and mechanisms for separated, unaccompanied and orphaned children need to have a place in the urban agenda.
- Comprehensive school safety, early child care centers as well as "safety of out-of-school children" are critical aspects of school disaster management and hence, need to be prioritized.
- Children and youth as tomorrow's leaders of change is the first step toward delivering safety and progress for the next generation.

Accountability to Global Framework for Disaster Risk Reduction

Sendai Framework for Disaster Risk Reduction which has promulgated the 15-year framework for action has clearly articulated the space and role for the future citizens and leaders in our world.

"Children and youth are agents of change and should be given the space and modalities to contribute to disaster risk reduction, in accordance with legislation, national practice and educational curricula" SFA, Role of Stakeholders—Page 20 # (ii) (More on WCDRR: http://www.wcdrr.org).

Bibliography

Child-Led Disaster Risk Reduction—A Practical Guide—Save the Children Publication—ISBN number 978-979-16424-0-8—www.savethechildren.net.

Communication from the Commission to the European Parliament and the Council—The EU Approach to Resilience: Learning from Food Security Crises—Brussels, 3.10.2012—COM(2012) 586 final.

Comprehensive School Safety Framework (CSS)—A framework finalized by UNESCO, Save the Children, IFRC, et al.

Feeling the Heat—Save the Children UK, 2012.

Good Practices on Disaster Risk Reduction—A compilation of Case Studies by ECHO Partners from 2007 Flood Response projects—March 2009.

Reducing Risk as Equals—Regional Review of Child Centered Disaster Risk Reduction in South Asia—Save the Children and RedR—2013.

SAARC Framework for Children and Disasters—10 point Agenda for Action—SAARC Disaster Management Centre, May 2011. http://saarcsdmc.nic.in/pdf/Publications/SAARC%20Framework.PDF.

Sendai Framework for Disaster Risk Reduction. (2015–2030). http://www.wcdrr.org.

The Future Framework for Disaster Risk Reduction—A guide for decision-makers—ODI working paper—June 2014.

The Resilient Future We Want: Children and Disaster Risk Reduction in Post 2015. www.childreninachangingclimate.org.

Transforming Childhood in Vulnerable Situations: A "Braided Approach to Resilience" (BAR)—Save the Children UK, 2013.

UNCRC—United Convention on the Rights of the Child—1989—Ratified by all countries except two Reducing Risk, Saving Lives—Save the Children's Approach to Disaster Risk Reduction and Climate Change Adaptation—2014.

Reflections on Resilience

Béné, C., Godfrey Wood, R., Newsham, A., and Davies, M. (2012). *Resilience: New Utopia or New Tyranny? Reflection About the Potentials and Limits of the Concept of Resilience in Relation to Vulnerability Reduction Programmes.* Brighton: IDS.

Twigg, J. (2009). "Characteristics of a Disaster Resilience Community: A Guidance Note, Version 2". In *DFID Disaster Risk Reduction Interagency Coordination Group*. London: DFID.

Government Policy and Position Papers

DFID. (2011). "Defining Disaster Resilience: A DFID Approach Paper". http://www.fsnnetwork.org/sites/default/files/dfid_defining_disaster_resilience.pdf.

EC. (2012). "The EU Approach to Resilience: Learning From the Food Security Crisis". http://ec.europa.eu/europeaid/what/food-security/documents/20121003-comm_en.pdf.

USAID. (2012). "Building Resilience to Recurrent Crisis: USAID Policy and Program Guidance". http://transition.usaid.gov/resilience/USAID-ResiliencePolicyGuidanceDocument.pdf.

Useful Websites

http://www.preventionweb.net/english/?logotext.

www.eldis.org.

http://www.resalliance.org/.

For a more detailed discussion of different definitions and implications, see the 2010 DFID funded report: The Resilience Renaissance community.eldis.org/.59e0d267/resilience-renaissance.pdf.

PART IV

Disaster: Socio-economic Impact and Post Disaster Recovery

Disaster Risk Reduction and Resilience Building Among the Coastal Zone Communities in Lakshwadeep and Kerala

Mondira Dutta

Context

Over the recent years' disaster risk reduction (DRR) is being increasingly realized as a major phenomenon in handling all kinds of crisis management. According to the United Nations over 700,000 people have lost their lives, more than 1.4 million have been injured, an estimated 23 million have been left homeless bringing the overall affected people to the tune of 1.5 billion. In terms of economic loss, the total value was estimated to more than $1.3 trillion. During the period between 2008 and 2012, around 144 million people were displaced by disasters. This does not include another estimated 21 million people who have been trafficked

The author likes to acknowledge the University Grants Commission and JNU for extending research grant under UPOE that enabled the study for undertaking a field visit and conduct a primary survey in Lakshadweep and Kerala in October 2017.

M. Dutta (✉)
Jawaharlal Nehru University, New Delhi, India

B. Zutshi et al. (eds.), *Disaster Risk Reduction*,
https://doi.org/10.1007/978-981-10-8845-2_18

during disasters (IFRC 2014). However, field-based studies (Dutta 2010, 2011, 2015) and firsthand information from the several thousands of people rehabilitated reveal that this is only the tip of the iceberg!

The World Conference on disaster started gaining focus and subsequently has been held thrice in Japan. The first world conference was convened in Yokohama in 1994. The second conference was convened in Kobe 2005 and finally the third one was held in Sendai in 2015. The third conference referred to as the Sendai Framework for DRR (2015–2030) is a follow up to the Hyugo Framework of Action (2005–2015). The UN Office for Disaster Risk Reduction (UNISDR) served as the coordinating body for the second and third UN World Conference on Disaster Reduction. These conferences have made an effort toward synergizing the strengths of the government officials and other stakeholders, such as NGOs, Civil Society Organizations, institutions, local government, and various stakeholders from around the world in order to build community resilience and achieve DRR. The Sustainable Development Goals (SDG)s also recognized the importance of DRR as an integral part of sustainability and Conference of Parties (COP21) Paris Agreement on Climate change. The UN conferences on Disaster and Climate Risk Management in the context of sustainable Development thus came under focus.

It was in concurrence with the World Conference convened in Kobe, that the Disaster Management Act (DMA) was passed in 2005. This laid down policies, plans, and guidelines for disaster management in India. Subsequently, the National Disaster Management Authority (NDMA) was established followed by the State Disaster Management Authority (SDMA) and the District Disaster Management Authority (DDMA). The National Disaster Response Force (NDRF) got created for the purpose of specialist response to the very severe and threatening disaster situation. Additionally, the National Institute of Disaster Management (NIDM) was established to mandate funds for Disaster Mitigation at various levels. The Sendai Framework for DRR was adopted on 18 March 2015. This was a call for action to make the world a safer place in the next 15 years. The target was to reduce the manmade and natural hazards to a minimum by adopting early warning systems and focusing upon the four priority areas namely understanding disaster risk, strengthening risk governance, investing in DRR for resilience and creating disaster preparedness. The Sendai framework is a critical part of the post 2015 development agenda. It supports the 17 goals of SDG and the climate change agreement. Although each state has the primary responsibility, but all citizens need to be engaged and risk informed. The UNISDR has been tasked to support the implementation, follow-up, and review of the Sendai Framework. In alignment with the Sendai Framework for DRR, SDGs, and the COP21 Paris Agreement, recently in 2016, the National Disaster Management Plan (NDMP) was formed. The NDMP has short, medium, and long-term measures that can be completed within a time frame of five, ten, and fifteen years. Some of these are implemented concurrently rather than consecutively.

Objectives

The present research attempts to explore the disaster preparedness among the lives of coastal zone communities when disasters hit the regions of Lakshadweep and Kerala coasts in particular. It attempts to study how far community resilience in the region has helped in DRR among the fishing community. The main objectives of this paper include:

- Identification of the most affected zones
- Training program and capacity building measures adopted by the community
- Whether the training program was particularly oriented toward the fishing community
- Community resilience Measures adopted for DRR
- Understanding of the traditional knowledge system adopted to mitigate disasters.

Area of Study

The region of South Asia in general is known to be among the most affected area when disaster and climate change occurs. As per the fourth assessment report of 2007 by the International Panel on Climate Change (IPCC) a range of impacts has been listed that has hit the region from time to time. These include enhanced frequency of droughts and floods, negatively affecting the overall local production; a rise in the sea levels subjecting coasts to increasing risks like coastal erosion and growing human-induced pressures on coastal areas; melting of glaciers in the Himalayas resulting in more flooding and rock avalanches; etc. There has been a prediction of a decrease in crop yields by 30% in Central and

South Asia by the middle of the twenty-first century. Within South Asia, Bangladesh has been stated as the most vulnerable country because of its regional connectivity through geophysical and hydrological features and its livelihood reliance on trade (ELIAMEP 2008).

India's unique geo-climatic and socio-economic conditions have made it vulnerable to a large number of natural as well as man-made disasters. The most frequent disasters include floods, droughts, cyclones, earthquakes, landslides, avalanches, and forest fires. As a result, out of 36 states and union territories in the country, 27 of them are disaster prone (NIDM 2014). On an average, about five or six tropical cyclones form in the Bay of Bengal and Arabian sea and hit the coastal areas every year. Out of this at least two or three turn out to be quite severe.

The coasts of Kerala situated between the Arabian seato the west and the western Ghats to the east, runs for some 580 km in length, while the state itself varies between 35 and 120 km in width. Most of the state (notwithstanding the isolated regions) is subject to comparatively little seismic or volcanic activity. Kerala is prone to several natural hazards, the most common of them being flooding, landslides, lightning, drought, coastal erosion, earthquakes, Tsunami, windfalls, and epidemics (Yesodhara et al. 2007).

The 2004 Indian Ocean Tsunami significantly affected the Kerala coast. The coast located in the shadow zone with respect to the direction of propagation of the tsunami encountered unexpected devastation. The extremely fragile Western Ghats region is highly prone to natural calamities. The tsunami affected parts of Kerala coast, maximum devastation was reported in the low coastal land of Kollam, Alappuzha and Ernakulam districts, particularly a strip of 10 km in Azhikkal, of Kollam district. This left around 176 people dead and 1600 injured in the coastal belt. Further, the tsunami affected 187 villages affecting nearly 250,000 persons in Kerala. Several thousands of dwelling units were completely destroyed and 142 relief camps accommodated the distraught people after tsunami. Kerala in particular have been encroached upon for construction of hotels and in the name of development activities all along the western coast especially in the districts of Alappuzha, Thiruvananthapuram, Kochi, and Munnar. These areas had been identified earlier based on scholarly works such as that of Kasturirangan (2013) and Gadgil (2015) under the Ministry of Environment and Forests (MoEF). The Kasturirangan panel had been constituted by the MoEF basically to look into the recommendations of an earlier report on the Western Ghats submitted by the Ecology Expert Panel which was headed by the ecologist Madhav Gadgil. The Gadgil panel was constituted in 2010 by MoEF. Their main task was to study the impact of population pressure, climate change and development activities on the Western Ghats. The panel recommended declaring almost the entire Western Ghats as ecologically sensitive area (ESA). This report was opposed by the states and by politicians and farmers' organization stating it was detrimental to the development process of the region. Hence MoEF constituted the Kasturirangan panel in August 2012 to revisit the recommendations based on the ground realities. The Kasturirangan panel stated that almost 41% of the Western Ghats is "natural landscape" with a low population impact and rich biodiversity. The remaining 59% was "cultural landscape" mainly dominated by human settlements and agricultural fields. As a result, the panel recommended that 90% of the region demarcated as "natural landscape" should be protected.

Some of the major natural hazards in Kerala are depicted in Table 1 (Sanjith 2017).

The disaster affected areas show that already new constructions had replaced old structures such as the schools, buildings, and hotels that were earlier close to the sea and had been completely drowned. The cyclone Ockhi that took place during the period 30 Nov–6 Dec 2017, caught the community members completely unaware and were hardly disaster prepared. The fisherfolk were not advised with any early warning system resulting in some casualties. The gradual impact spread over to the neighboring Lakshadweep islands creating havoc. People were in anger as 178 fishermen went missing.

Lakshadweep is a group of coral islands in the Indian Ocean. It has several islands and islets together with submerged banks covering a geographical area of 32 sq. km (Prakash et al. 2015). According to the 2011 census, Lakshadweep has a population of 64,429. The islands are comprised of 11 inhabited and 16 uninhabited islands. There are some newly formed/submerged sand banks and reefs. The Inhabited islands are Kavaratti, Agatti, Bangaram, Amini, Kadmat, Kiltan, Chetlat, Bitra, Andrott, Kalpeni, and Minicoy. Malayalam is spoken in all the islands except Minicoy where people speak Mahi which is spoken in the Maldives as well. Bitra is the smallest island with a population of only 271 persons (Census of India 2011). The uninhabited island Bangaram has been enumerated during 2011 census operation and has a population of 61 persons (MSME 2016). Bangaram is the only island where

Table 1 Recent major natural disaster in Kerala

S.No	*Type of disaster*	*Areas vulnerable*	*Deaths*	*Damages*	*Remarks*
1	Landslides, Nov 2001	10 Taluks—Landslide at Amboori, Thiruvananthapuram	38	–	Major event in the last 20 years
2	Tsunami, 2004	36.6% of Kerala's 580 km coastline is susceptible to coastal erosion with Thiruvananthapuram being the most vulnerable	171	–	Major event in the last 20 years
3	Floods, 2013	27 Taluks	145	Rs 899 crores	Major event in the last 20 years
4	Drought, 2016	50% of Kerala's land area is moderately to severely susceptible to drought	–	–	It was declared that the State was facing its worst drought in 115 years and the govt even considered artificial rain
5	Cyclone Ockhi, Nov 30, 2017	Coastal areas	22 (Lakshadweep) 218 (Kerala and Tamilnadu)	178 fishermen went missing	No warnings were issued

consumption of liquor is permitted by the tourists. Agatti is the long air strip for flights to land and take off with a long runway. Kavaratti is the largest island in Lakshadweep consisting of around 17,000 people and is the capital headquarters of the island. These islands are open to storms, cyclones, and heavy rains since they are surrounded by the vast oceans. In spite of its smallness in terms of geographical structure, according to the Lakshadweep official website, Lakshadweep has a total lagoon area of 4200 km, large territorial waters (about 20,000 sq. kms) and an exclusive economic zone (about 400,000 sq. kms). The largest ecozone is from Andaman to Lakshadweep.

Unfortunately, most of the major infrastructure and economic activities are located nearby the coasts which are highly vulnerable to cyclones and sea level fluctuations. History of disasters in Lakshadweep shows that the islands have been frequently hit by cyclones causing considerable damage. The major flood prone areas of the islands are Karavatti, Kiltan, and Amini which are among the most low-lying areas.

Field Visit Survey

A field visit was undertaken in the month of October 2017 to the islands of Agatti, Kavaratti islands in Lakshadweep and a few other areas along the Western Ghats and other vulnerable areas of Kerala. Interviews and focused group discussions were held among the shopkeepers, fisherfolk, women and local authorities. Besides the focused group discussion with fisherfolk, interviews were held with Mr. Farooq Khan, the Chief Administrator of Lakshadweep. A visit to the desalination center was also undertaken. This Center has been highly affected due to Ockhi in Dec 2017. The Disaster Management Institute was also visited in Thiruvananthapuram and a focused group discussion was held among the fisherfolk in Kochi, Kerala. A research group working on DRR were interviewed at the Centre for Development Studies, Thiruvananthapuram.

The primary survey shows that nothing is cultivated in Lakshadweep. Everything is imported from mainly Kochi. The population mainly belong to the Islamic belief and hail from the Malabar region. The roads in the villages are well made, cemented and firm so that they can withstand the heavy rains. They are narrow but enough for conveyance to ply on.

Goat breeding is prevalent with hardly any cows due to a lack of vegetation. Lifestyles are tough with few offices. All inhabited islands have a medical dispensary. Sports is practiced in some parts. Permission for entry into the islands is mandatory for the tourists visiting Lakshadweep. A survey had been conducted to capture the high-end tourists. The carrying capacity ranges from 200 to 350 tourists. But the authorities are careful in restricting the number of tourists to the lower range. Their waste disposal also is planned accordingly.

During the Tsunami of 2004, when Lakshadweep was badly affected, the coral reef was damaged to an extent of 40%. These have been revived to back by 20% with strict and active district administration policies being firmly in place. The Chief Administrator stated that fishing nets are not allowed within forty kilometers of the shore. Closer to the shore only fishing rods are allowed so as to not disturb the corals which have almost revived after the Tsunami hit the region in 2004 unlike the rapid destruction that takes place in the Great Barrier Reef of Australia. In fact, this was a blessing in disguise as the local government realized the need to have sustained monitoring of the reefs and help in the rapid recovery. The government of India also understood the importance of involving the local community toward this endeavor. During the field visit, it was observed that even little boys who were splashing about in the sea were observant and sensitive to pick up any polythene bags floating in the sea and put it in the garbage bins which were along the shores. The islands were kept as clean as possible. The youngest member of parliament in the Lok Sabha, Hamdullah Sayeed, is thirty years of age and hails from Lakshadweep.

Every household in the island has one member in government job. It has 100% literacy, with minimum birth and death rate, infant mortality and maternal mortality are almost nil. Education is free. All children above class VI get bicycles. There are fifty-six schools in Lakshadweep. Every island has a degree college. Four students are in IIT and their expenses are fully paid. Meritorious students are sent out for higher education free of cost. Emergency medical facility is available round the clock. There are choppers available in three islands to transport the patient even in the middle of the night.

Disaster History

Surrounded by vast oceans, Lakshadweep islands are prone to severe natural disasters throughout the year. Among the major hazards that have hit Lakshadweep one of the earliest natural calamities as recorded was in April 1847, when the great storm struck the islands. It commenced in Kalpeni about 8 pm. On 15th April, passed on to Andrott and finally reached Kiltan after devastating these two islands. All the houses in Kalpeni were damaged or washed away (http://lakshadweep.nic.in/depts/revenue/index.htm accessed on 7 Jan 2018). It took the lives of 246 people out of a population of 1682, who were drowned or washed away in Kalpeni. In 1891, another violent storm burst upon Kavaratti island causing considerable damage to the coconut trees. The storm caused damage in Agatti and Amindivi group of Islands also. On December 1, 1922, Kalpeni island was again hit by a severe cyclone. The waves completely washed over the narrow northern end and the sea poured across the island into the lagoon. Although there was no loss of life but the impact of the cyclone was felt in other islands also. Subsequently yet another major storm struck the islands in 1941.

According to the Disaster management authority, severe calamity struck Lakshadweep in 1963 and 1965 causing considerable loss of properties in Andrott, Kalpeni, Agatti, and Kiltan. However, loss of lives was not reported. The cyclone of 1977 in Kalpeni and Andrott also was severe but without causing any loss of life. The major brunt of the loss was on a large number of coconut trees which were uprooted and some houses that got damaged. In 1981, again a violent storm struck upon Kavaratti bringing in considerable damage to Agatti and Aminidivi groups of Islands. The last severe cyclone to have hit the islands was in May 2004, which affected Kavaratti, Amini, Kiltan and Agatti Islands. The storms and cyclones have thus hit some of the islands over the years in 1847, 1891, 1922, 1963, 1977 and during May 2004. The main problems are due to the hitting of waves during such occasions which result in sea erosion and uprooting of the coconut trees and other vegetation besides damaging of properties (http://lakshadweep.nic.in/depts/revenue/index.htm accessed on 7 Jan 2018).

Lack of Disaster Preparedness

The goal of disaster preparedness is to achieve a satisfactory level of readiness to respond to any emergency situation through programs that strengthen the technical and managerial capacity of the Governments, organizations, and communities. During preparedness phase, governments, organization, and individuals develop plans to save lives, minimize disaster damage, and enhance the disaster response operations. These measures include preparedness plans; emergency/exercise training; warning systems; emergency communication systems; evacuation plans and training; resource inventories; emergency personnel/contact list; mutual aid agreements; and public information/education.

It was as late as Dec 2017, when the severe cyclonic storm called Ockhi touched down on Lakshadweep around the 29 November. This was the third and strongest storm of the year 2017 which originated in the Gulf of Thailand on November 21. Not only did the cyclone uproot coconut trees but also caused extensive damages to houses, power lines and other infrastructure in the islands. The disaster preparedness was hardly existing.

More than 245 people died out of which 218 were from India alone. Lakshadweep islands had 22 casualties. It left at least 551 people, mainly belonging to the fisherfolk missing. The three major islands of Lakshadweep that were worst hit due to Ockhi were Minicoy, Kalpeni, and Kavaratti islands that suffered major damages. The Lakshadweep islands met with massive coastal erosion, power disruption, and extensive damage to property with life and also suffered from drinking water shortage. An estimated loss to the tune of Rs 200 crores was reported following the collapse of the "breakwater"—a barrier built out into the sea. This was basically to protect the coast from the force of waves at Kalpeni and save the corals from damages. A desalination plant at Kavaratti also got damaged due to cyclone Ockhi. According to a member of the Parliament, Mohammad Faizal, the Lakshadweep islands suffered losses over Rs 500 crores (*Financial Express*, December 3, 2017b).

Disaster relief materials to the tune of 4 tons were distributed that included necessities such as rice, pulses, salt, potatoes, water, blankets, raincoats, disposable clothes, mosquito nets, etc. Local administration was in-charge of distributing these materials. Some other dry provisions and ready to eat meals were dispatched to Bitra Island by chopper from the Kavaratti. This relief material was to last for a week for 2000 people, according to the government (NDTV 2017).

The huge losses incurred from the cyclone forced the government of Kerala to start maintaining a registry of all its fishermen and install GPS and other tracking systems as well as life-saving gadgets on their vessels so as to limit the damages during times of natural disasters. Huge protests from the residents of the coastal regions were found as a result of the slow response and disaster unpreparedness by the government (*Hindustan Times* 2017a). The Indian Navy provided humanitarian assistance and disaster relief to people affected by cyclone Ockhi in Lakshadweep (*Hindustan Times* 2017b). According to the news report of December 12, 2017, the fisherman community of Kerala estimates that the number of missing person is much higher than the government figure of 70. After thorough and repeated search operation in Kerala the death toll increases to 48 with massive economic losses (*The News Minute* 2017).

The aftermath of cyclone did flow in donations from the Uttar Pradesh Chief Minister Shri Yogi Adityanath, Chief Minister of Bihar, Shri Nitish Kumar have come forward donating funds to the cyclone-affected people in Lakshadweep and other areas. The Prime Minister, Shri

Narendra Modi announced a relief package of Rs 325 crore for Kerala, Tamil Nadu, and Lakshadweep islands (*The Statesman* 2017). The Center also assured to help reconstruct 1400 damaged houses under the Pradhan Mantri Awas Yojana (Scroll.in 2017).

The southernmost tip of the country in Kerala and Tamil Nadu were severely affected by the cyclone. In Kerala, nearly 220 families were rehabilitated from the coastal areas of Kochi to escape the rough sea following the way of cyclone Ockhi. Chellanam in Kerala is highly affected by the tidal wave attack. Nearly 100 houses, mainly in Puthenthodu and Bazaar areas, were flooded. Schools were turned into rehabilitation camps by the authority. According to a recent news of December 29, 2017, the Kerala government expressed happiness over immediate release of funds from the central government. The central governments assessment team visited all the cyclone affected regions of the state including Thiruvananthapuram, Kollam, Ernakulam, Thrissur, Malappuram, and Kozhikode (*Firstpost* 2017).

A total of 180 families in Chellanam, 17 in Kannamaly and 18 in Edavanakadu were shifted to rehabilitation centers. Essential items including medicine and food were being made available at the camps (*Deccan Chronicle* 2017). NDRF teams were deployed in vulnerable places. Warnings to fishermen not to go out to sea were issued and those out fishing were called back. Schools and college remained closed.

Economy

Lakshadweep Islands have no mineral deposits. The community members stated that petrol and diesel had to be purchased at high rates. Petrol was bought at 150/per liter privately while govt rates were only Rs 90/per liter. There is no forest area in the Lakshadweep Islands. Lakshadweep is a no-industry area. It has neither large-scale nor medium scale industries. There are no public-sector units either. The main reason being non-proximity to raw materials required, technical institutions and lack of skilled manpower. The union territory (UT) is abundant in fish catching and Coconut based activity. There are potential for development of the items in these sectors to fetch overseas market. In the islands of Karavatti, Agatti, Kiltan, Amini, and other areas, livelihood and living becomes a threat and comes to a standstill every year due to storms, floods, and natural disasters. As most of the Lakshadweep islands are low-lying and due to heavy rain people residing in such part of the island have to face extreme consequences. Due to the constant hitting of waves during a cyclone/Tsunami, sea erosion takes place along with uprooting of coconut trees which is their major source of livelihood as well as damaging the little vegetation and properties. Venturing out into the sea is even more risky forcing the community to remain without any livelihood options for days on end. Lakshadweep islands import practically everything. Fishing and coconut plantation being their only means of production.

The major sectors of its economy thus consist of Fisheries (Tuna), Horticulture (Coconut), and Tourism (Island/Coral Reefs). The Islanders depend heavily on mainland for everything except for fish and coconut. There is ample scope for export of the processed fish from the UT of Lakshadweep. Fishing is one of the mainstays of the people of Minicoy Island. It has become the main source of income of the people, besides coconut cultivation. The sea around the island is highly productive. The island stands first in the country in per capita availability of fish. As per data available in the field, there are 872 fishermen who are actively involved in fishing activities such as fishing, transportation of fish, marketing, processing, fish curing, manufacturing tuna products, and its value-added products (http://lakshadweep.nic.in/depts/revenue/index.htm accessed on 7 Jan 2018). Fisheries are the second most important economic activity of the UT. The estimated potential resources of the sea around Lakshadweep are about 1 lakh tons of tuna and other fish like sharks, etc. Tuna production is worth 150,000 million dollars for business in Lakshadweep, however only 4% are being harvested now which is around 4000$. Plenty of potential left for being exploited.

All belong to the nuclear family system. Agriculture is the mainstay of the economy of Lakshadweep. More than 90% of the population depends upon agriculture. It is a Coconut Mono Crop territory. The entire cultivable land is used for coconut cultivation. It is observed that whenever a cyclone hits the islands, the first causalities are the coconut trees followed by the disruption of fishing activities effecting the livelihood of local people. Undoubtedly, these events have also lead to the losses of physical and social infrastructure in the longer run creating a set back to the pace of development on these islands.

Role of Women

Women are the most vulnerable during a disaster. They have an important social role in maintaining the family. As per the World Health Organization, women and children are particularly affected by disasters, accounting for more than 75% of displaced persons. Women's vulnerability is further increased by the loss of men. In Lakshadweep, most of the women wear a Hijab or a burkha. However, it has the largest number of households headed by women which is almost 44%. The islands are scattered and remote, each fringed by white coral sand and shallow lagoons protected by reef walls. They are the only coral islands in India. The island group's Gross Domestic Product from coconut, tuna, tourism, and sea-faring touches almost Rs 100 million, with the tuna canning factories concentrated on the Minicoy Island. Women often constitute a majority of the work force in these sectors. The unique feature of this region is that it is a Muslim society with 100% literacy where all the women are literate and most of them work in the natural products post-harvest industry. The Shariat law (Islamic law) exists there but this is in combination with matrilineal traditions. The local knowledge of the women about the general bio-diversity and where the concentration of coral, giant clam, bait-fish, and tuna is tremendous. The field visit undertaken made an effort in understanding the traditional knowledge systems and community wisdom among the fishing communities particularly, the women who responded to disaster mitigation. It has been seen that women are the worst hit whenever a disaster hits the region. They face the brunt of the severe weather events such as Tsunami and the super cyclones. Many of the coastal communities have weak undiversified and limited livelihoods that rely heavily on unsustainable utilization of natural resources. The women seemed not only well aware about conservation of marine life but also capable of building awareness among the community and mitigating disasters through traditional methods ushering in sustainability and conservation effectively.

In Orissa (India) and other parts of India, some interesting practices are observed for weather forecasting and disaster warning. For example, prior to an earthquake different behavior of a particular fish, like the *Singhi* fish, which comes to the top of the water level can be observed. Globally there have been numerous cases of local tradition and practices where people predict earthquake or storms by observing natural signs such as changes in animal behavior, abnormal weather phenomenon, and many others in DRR. It has been found that the diverse communities in different hazard-prone areas of this region have learned to live cope up with the risks from various natural disasters through the use of their traditional knowledge gained from their ancestors. For example, in South East Asia, such as in Indonesia, there are numerous early warning signs of cyclones that are displayed much in advance. Some of these include the nesting behavior of bees and hornets under the tree, changing currents and tides in the ocean and the appearance of particular birds all play a crucial role in DRR. In the villages of Fiji and Tonga noticeable changes in the shape of rainbows called half rainbows prior to bad weather or cyclone, behavior change of some kinds of birds prior to cyclones, and spiders making more webs before a cyclone comes. These traditions are alike and similar in many of the islands (Johnston 2016).

Likewise, in the Islands of Lakshadweep, mainly the women chose traditional signs as their first and most reliable ways of knowing that a cyclone was coming. The warning concerns the behavior of flora and fauna as well as winds and ocean. They include nesting behavior of bees and hornets where they build their nests closer to the ground one or two months preceding the cyclone (ndma.gov.in).

In fact, both in India and Bangladesh many believe that crying of dogs in a continuous motion is an early warning of flood in the region. Movement of ants to higher places also predict an oncoming flood. Cattle become restless and stop eating grass is also considered an early warning sign of cyclone (Howell 2003). The field visit survey revealed that women believe and practice many such aspects in Lakshadweep. Women need to be recognized as critical partners in both driving and delivering solutions. This becomes even more precarious in disaster-prone areas where daily activities and business induced male migration is a common feature. It is important to impart training and capacity building measures to the coastal communities especially the women for disaster risk mitigation. The need for a well-trained pool of manpower and organization including voluntary organizations dealing with disaster is a must.

In Kerala too, some communities follow the matriarchal system and have almost 23% homes headed by women. The land of Kerala lying along the coastline stretches north–south along a coastline of 580 km with a varying width of 35–120 kms. It is the land of monsoons and also one of the wettest places in the world. Although it has one of the highest literacy rate and a low female feticide resulting in healthy sex ratio in the country, the state is not devoid of Crime against women (Fig. 1).

According to the National Crime Record Bureau (NCRB) of 2015, a total of 461 cases were registered under the Domestic Violence Act, 2005. This showed an increase of 8.2% (from 426 cases in 2014 to 461 cases in 2015) where Kerala alone reported around 132 cases. However most of the cases go unreported in many states. In fact, the rate of cognizable crimes in Kerala is the highest in the country at 723.2 (cases per 1 lakh population), more than thrice the national average of 234, as per the 2015 data released by the NCRB.

Unfortunately, discrimination and exclusion continue to deprive the women of their social, economic, and environmental rights and opportunities. Their low status in the society, biased structures and economic vulnerabilities plays a greater strain on their adaptive capacity to disaster preparedness and ability to deal with shocks, stresses and change while mitigating a disaster. The semi structure interviews with women revealed that in spite of their contribution toward the society they are seldom involved in decision making and particularly in the planning process to

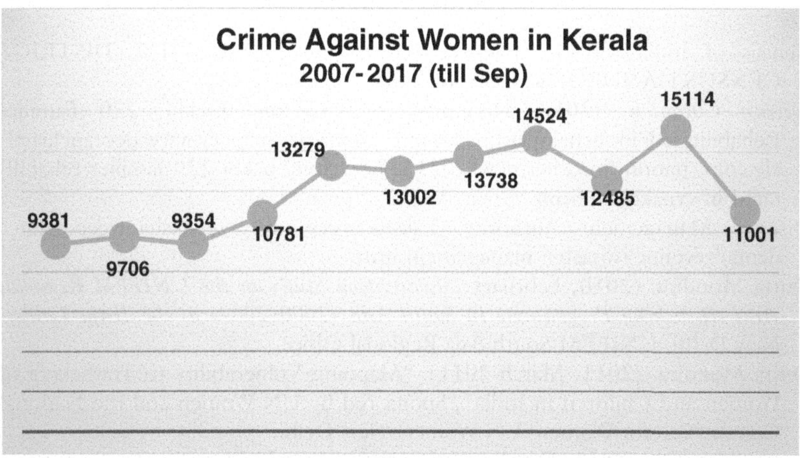

Fig. 1 Kerala, Crime Against Women, (2007–2017). *Data Source* http://www.keralapolice.gov.in/public-information/crime-statistics/crime-against-women

mitigate natural disasters. The prevalence of gender bias and discrimination is found to be integrated in the social structures of the Indian society, as a result of which a woman is not involved as an equal partner in the consultative process.

Disaster Management Policy

There exists hardly any DMA for this Territory. This UT is without any legislature and the rules of the Central Government are not implemented at all. However, in 2004, the administration has formulated a framework of Disaster Management Plan for Lakshadweep. This skeletal framework is the only basis which serves as the basic guidelines document toward cyclone preparedness and response mechanism. The institutional/administrative framework in the UT of Lakshadweep comes directly under the administration of Ministry of Home Affairs. The Disaster Management framework formulated by the administration thus serves as a guideline toward disaster preparedness. The authorities are presently working toward evolving a more dynamic disaster management policy in consultation with similarly placed states and UTs in terms of disaster risk mitigation and building community resilience.

Bibliography

Census of India. (2011). Lakshwadeep, Series-32, Part XII-B, DISTRICT CENSUS HANDBOOK, Govt. of India.

Deccan Chronicle. (2017, December 2). "Cyclone Ockhi: 220 Families Rehabilitated in Ernakulam". *Deccan Chronicle.* http://www.deccanchronicle.com/nation/in-other-news/021217/cyclone-ockhi-220-families-rehabilitated-in-ernakulam.html.

Disaster Management Authority, Lakshadweep. http://lakshadweep.nic.in/depts/revenue/disaster_management.htm.

Dutta Mondira. (2010, February). *Evaluation Study on the UNIFEM Regional Anti-trafficking Programme in South Asia (2000–2009) as the Team Leader.* New Delhi: UNIFEM South Asia Regional Office.

Dutta Mondira. (2011, March 2011). "Mapping Vulnerability to Trafficking of Women and Children in India", Sponsored by UN Women and the National Commission for Women, UN Women, New Delhi.

Dutta, Mondira. (2015, October). "Understanding Maldives in the Context of Trafficking". In Henrik Berglund, Mondira Dutta, and Per Hilding (Eds.), *Development and Regional Cooperation in Central and South Asia: Euro-Asian Perspectives.* New Delhi: Pentagon Press.

ELIAMEP. (2008). http://www.eliamep.gr/en/european-integration/stavros-costopoulos-research-fellowship/aies-european-academy-2008/.

Financial Express. (2017a). "Cyclone Ockhi: Lakshadweep Islands Suffer Over Rs 500 Crore Loss, Says MP Mohammad Faizal".

Financial Express. (2017b, December 3). http://www.financialexpress.com/india-news/cyclone-ockhi-lakshadweep-islands-suffer-over-rs-500-crore-loss-says-mp-mohammad-faizal/958852/.

Firstpost. (2017, December 29). "Cyclone Ockhi: Kerala Government Happy with Central Team's Assessment of Damages". *Firstpost*. http://www.firstpost.com/india/cyclone-ockhi-kerala-government-happy-with-central-teams-assessment-of-damages-4280601.html.

Gadgil. (2015). "Report of the Western Ghats Ecology Expert Panel Part I". http://www.moef.nic.in/downloads/public-information/wg-23052012.pdf (Accessed on 4 December 2017).

Hindustan Times. (2017a, December 3). "In Pics: Cyclone Ockhi Leaves a Trail of Death and Destruction in South India".

Hindustan Times. (2017b, December 6). "Aftermath of Cyclone Ockhi: Kerala to Maintain Fishermen Registry, Install GPS in Vessels". Retrieved December 06, 2017, from http://www.hindustantimes.com/india-news/aftermath-of-cyclone-okchi-kerala-to-maintain-fishermen-registry-install-gps-in-vessels/story-PRh5zv8GzzAl7Dt3LmrYwO.html.

Howell, P. (2003). "Indigenous Early Warning Indicators of Cyclones: Potential Application in Coastal Bangladesh", Disaster Studies Working Paper No. 6, Benfield Hazard Research Centre, London.

IFRC. (2014). http://www.ifrc.org/en/news-and-media/opinions-and-positions/opinion-pieces/2014/statement-on-migration-by-francesco-rocca/.

Johnston, Ingrid. (2016, May 11). *Rebuilding Sustainable Communities After Disasters: Remote Islands*. Newcastle upon Tyne: Cambridge Scholars Publishing, p. 254.

Kasturirangan. (2013). "Report of the High Level Working Group on Western Ghats Volume I". http://www.moef.gov.in/sites/default/files/1%20HLWG-Report-Part-1_0.pdf (Accessed on 4 December 2017).

Kerala Police Crime Statistics. http://www.keralapolice.org/public-information/crime-statistics/crime-against-women.

Kurian, Vinson. (2017, November 30). "Weather Update: Cyclone 'Ockhi' Heads for Lakshadweep, to Intensify Further". *The Hindu Buisness Line*. http://www.thehindubusinessline.com/news/cyclone-ockhi-moving-towards-lakshadweep-to-intensify-as-severe-cyclone/article9977465.ece.

Lakshadweep Official Website. http://www.lakshadweep.nic.in/FiveYearPlan_1.htm.

Livemint. (2017, December 27). "Govt Team Visits Cyclone Ockhi-Hit Areas in Tamil Nadu, Kerala to Assess Damage". *Livemint*. http://www.livemint.com/Politics/znH6njncHofLJFhJckOFVK/Govt-team-visits-Cyclone-Ockhihit-areas-in-Tamil-Nadu-Kera.html.

MSME. (2016). *Brief Industrial Profile of Lakshadweep '2015–16'*. Thrissur: M S M E–Development Institute, Ministry of MSME, Govt. of India.
ndma.gov.in. http://www.ndma.gov.in/en/kerala-sdma-office.
NDTV. (2017, December 8). "Navy Dispatches Relief to Ockhi Cyclone-Hit Lakshadweep as Navy Ships Carried the Relief Material to the Island of Minicoy on Sunday and the Kavaratti and Kalpeni Islands Yesterday". https://www.ndtv.com/india-news/navy-dispatches-relief-to-ockhi-cyclone-hit-lakshadweep-1783759.
NIDM. (2014). "National Profile". http://nidm.gov.in/easindia2014/err/pdf/country_profile/India.pdf (Accessed on 16 December 2017).
Prakash et al. (2015). *Geomorphology and Physical Oceanography of the Lakshadweep Coral Islands in the Indian Ocean*. Springer.
Sanjith, Sidhardhani. (2017, September 24). "How Prepared Is Kerala to Face Natural Disasters?" *Times of India* (City News).
Scroll.in. (2017, December 19). "Narendra Modi Meets Families Affected by Cyclone Ockhi, Announces Rs 325 Crore in Government Aid". Scroll.in. https://scroll.in/latest/862057/narendra-modi-meets-families-affected-by-cyclone-ockhi-announces-rs-325-crore-in-government-aid.
The Hindu. (2017a, December 2), "Cyclone Ockhi Continues to Batter Coastal Kerala, More Than 2600 Evacuated from Ernakulam District". *The Hindu*. http://www.thehindu.com/news/national/kerala/cyclone-ockhi-continues-to-batter-coastal-kerala-more-than-2600-evacuated-from-ernakulam-district/article21247995.ece.
The Hindu. (2017b, December 19), "Cyclone Ockhi: TN Seeks Central Relief Package of Rs 9300 Cr". *The Hindu Business Line*. http://www.thehindubusinessline.com/news/national/cyclone-ockhi-tn-seeks-central-relief-package-of-rs-9300-cr/article9996342.ece.
The Indian Express. (2017a, December 3). "Lakshadweep Islands Suffer Over Rs 500 Cr Loss Due to Cyclone: MP Mohammad Faizal". *The Indian Express*. http://indianexpress.com/article/india/cyclone-ockhi-lakshadweep-islands-suffer-over-rs-500-cr-loss-mp-mohammad-faizal/.
The New Indian Express. (2017b, December 3). "Cyclone Ockhi: Nitish Announces Rs One Crore Assistance to Cyclone-Hit Lakshadweep". *The New Indian Express*. http://www.newindianexpress.com/nation/2017/dec/03/cyclone-ockhi-nitish-announces-rs-one-crore-assistance-to-cyclone-hit-lakshadweep-1717594.html.
The News Minute. (2017, December 12). "Cyclone Ockhi: Bodies of Eight More Fisherman Found in Kerala, Death Toll Rises to 48". *The News Minute*. http://www.thenewsminute.com/article/cyclone-ockhi-bodies-eight-more-fishermen-found-kerala-death-toll-rises-48-73051.
The Statesman. (2017, December 5). "Cyclone Ockhi: UP CM Yogi Donates Rs 5 Crore to PM Relief Fund". *The Statesman*. http://www.thestatesman.

com/india/cyclone-ockhi-cm-yogi-donates-rs-5-crore-pm-relief-fund-1502539809.html.

The Times of India. (2017, December 26). "Central Team Visits Ockhi Affected Areas in Kerala". *The Times of India*. https://timesofindia.indiatimes.com/city/kochi/central-team-visits-ockhi-affected-areas-in-kerala/articleshow/62258733.cms.

Yesodhara, E. P., Kokkal, K., and Harinarayan, P. (Ed.). (2007). *State of Environment Report of Kerala 2007-Volume II: Natural Hazards*. Thiruvananthapuram: Kerala State Council for Science, Technology and Environment, Government of Kerala.

Posttraumatic Stress Disorder and Depression Among the Survivors of Nepal Earthquake: A Need Assessment Study

Md Zahir Ahmed

Introduction

Tremors have been among a standout among the most dangerous cataclysmic events all through delayed mankind's history. On April 25, 2015 at 11:56 neighborhood time, a tremor of extent 7.8 struck Nepal, with an epicenter 77 kilometers northwest of Nepal's capital Kathmandu in the Gorkha region. As of May 15, 8316 individuals had been accounted for murdered and 17,866 individuals harmed. The moderately shallow profundity of the seismic tremor, at 2 kilometers, was related with especially solid powers scattered over a wide geographic region. Oft on May twelfth at 12:50 nearby time, another seismic tremor (greatness 7.3) hit with an epicenter 76 kilometers upper east of Kathmandu in the Dolakha/Sindhupalchowk locale. This seismic tremor influenced locale as of now emphatically affected by the April 25th quake and

M. Zahir Ahmed (✉)
Policy Research Centre, Dhaka, Bangladesh

B. Zutshi et al. (eds.), *Disaster Risk Reduction*,
https://doi.org/10.1007/978-981-10-8845-2_19

new regions. As of May 15, 117 passings and 1961 harmed have been accounted for because of this seismic tremor (Nepal Earthquakes 2015: Desk Review).

The Geographic and climatological data of Nepal are:

- Nepal has high geological decent variety, covering three natural zones from nearly ocean level to the most elevated point on earth.
- Nepal is inclined to normal risks, including seismic tremors, streak surges in the storm time frame, avalanches, blasting of frigid lakes, changes because of a dangerous atmospheric deviation, and ecological debasement.
- Geographical and climatic highlights will hamper quake reaction, with Nepal having exceptionally poor street scope.

Nepal is inclined to various common risks, including tremors. The Kathmandu valley is situated in a standout among the most seismically dynamic regions on the planet. Seismic tremors in 1934 and 1988 executed more than 8500 and 721 individuals, separately.

Close by, regular perils in Nepal incorporate blaze surges, icy lake upheaval surges, avalanches, flames, hailstorms, and dry spells. A give an account of worldwide fiasco hazard positions Nepal eleventh on the planet as far as powerlessness to seismic tremors and 30th as far as water-instigated catastrophes. This hazard is relied upon to increment with the effect of a dangerous atmospheric deviation. In the vicinity of 1975 and 2006, Nepal's temperature expanded by 1.8 °C and the normal temperature rise recorded was 0.06 °C every year. Environmental change is another main consideration in Nepal's weakness to fiascos.

Notwithstanding passing, physical wounds and financial misfortunes, solid tremors regularly result in genuine psychological wellness results. The capable 7.3 greatness tremor that struck Nepal as of late left an as of now damaged populace held by considerably more profound dread, underlining worries that the nation is badly arranged to adapt to the mental reactions.

As cataclysmic events as often as possible effect in poor countries, Nepal's tremor has uncovered the gross insufficiencies of its emotional well-being administrations exactly right when they are generally required (Reuters 2015).

Study demonstrates that before the calamity, Nepal had one of the world's sickly mental social insurance frameworks, with just not more than 100 specialists and around twelve clinical analysts to serve a country of 28 million individuals, as indicated by government information. Inside Nepal's just government-run mental healing facility a huge split keeps running down one of the dividers caused by shudder a month ago. The patients live in cramped, faintly lit wards that stink of pee and junk scattered around the floor. The 50-bed doctor's facility in Kathmandu is overloaded to the point that patients can just remain there for up to a month. The readiness was not satisfactory to handle debacles like such size of size.

A few deadly mental issues occur because of the experience of catastrophic events. The predominance rates of mental issue among the hard-hit survivors and the individuals who have lost their relatives or nearest one were far higher than those of daintily hit survivors. Second, the initial couple of months (i.e., 0–6 months a while later) is frequently a high beginning stage for mental clutters. Also, notwithstanding PTSD and gloom being distinguished as the most well-known mental issues. In this present investigations, we are concentrating on the PTSD and its seriousness alongside the wretchedness among the survivals of Nepal seismic tremor.

Posttraumatic Stress Disorder

Posttraumatic stress disorder (PTSD) is a nervousness issue that can occur after a man is presented to at least one horrendous mishaps, for example, real anxiety, rape, fighting, or different dangers on a man's life. Manifestations incorporate exasperating repeating flashbacks, shirking or desensitizing of recollections of the occasion, and hyper excitement, proceed for over a month after the event of a horrible mishap. The vast majority who have encountered a damaging occasion won't create PTSD. Individuals who encounter attack based injury will probably create PTSD, instead of individuals who encounter non-strike based injury, for example, seeing injury, mishaps, and fire occasions. Kids are more averse to encounter PTSD after injury than grown-ups, particularly on the off chance that they are under ten years old. War veterans are normally in danger for PTSD. Moreover, cataclysmic event can reason for seriousness of PTSD.

Depression

Depression is a finding recorded in the DSM-V as an inclination issue depicting individuals who feel pitiful and skeptical. The vast majority feel tragic, disheartened, or "down" once in a while, however that is not sadness. In clinical gloom, these emotions keep going for quite a while and are joined with different side effects. Somebody with gloom may have these side effects: solid dismal sentiments or no solid emotions by any stretch of the imagination, feeling regretful, feeling vulnerable or miserable, tension, fear, low confidence or lost enthusiasm forever. These are indications of sadness when they keep going for quite a while (generally for over two weeks). A great many people who have not had melancholy don't totally comprehend its belongings. Rather, they consider it to be basically being dismal.

Need Assessment Study

Need Assessment study is a precise procedure for deciding and tending to necessities, or "gaps" between current conditions and wanted conditions or "needs". The inconsistency between the present condition and needed condition must be measured to properly recognize the need. The need can be a want to enhance current execution or to rectify an inadequacy. A necessities evaluation is a piece of arranging forms, frequently utilized for development in people, instruction/preparing, associations, or groups. It can refine and enhance an item, for example, preparing or benefit a customer gets. It can be a viable device to clear up issues and recognize suitable mediations or arrangements. Necessities appraisals are just successful when they are closes centered and give solid confirmation that can be utilized to figure out which of the conceivable intends to-the-closes are best and effective for accomplishing the coveted outcomes.

There are three points of view on require in a necessities appraisal; saw require, communicated need and relative need:

- Perceived necessities are characterized by what individuals think about their needs, every standard change with every respondent.
- Expressed requirements are characterized by the quantity of individuals who have looked for help and concentrates on conditions where emotions are converted without hesitation. A noteworthy

shortcoming of communicated needs expect that all individuals with needs look for offer assistance.

- Relative requirements are worried about value and should consider contrasts in populace and social pathology.

Rational of the Study

Cataclysmic event can possibly influence mental well-being. Many impacts are normal to a wide range of catastrophe. In any case, we have seen the particular cataclysmic event i.e. seismic tremor has interesting results for the emotional well-being of survivors. All things considered, 17 noteworthy quakes (size 7.0–7.9) and 1 seismic tremor of greatness 8.0 or above happen the world over every year. Quakes, contrasted with other catastrophic events, convey the most elevated danger of extreme harm and damage. Steady or repeating interruptions from the quake considerably add to proceeded with psychological well-being issues. Some examination has discovered that general misery levels following a quake seem to come back to ordinary after around a year, yet posttraumatic stretch responses don't blur until year and a half after the seismic tremor. The commonness of PTSD differs generally in quake survivors. In grown-ups, rates as high as 92% have been seen, while in kids, upwards of 95% and as few as 4.5% have been found to have PTSD after a seismic tremor. Additionally, the pervasiveness rate of gloom is likewise high among the survivals of catastrophic events i.e. quake.

It is essential to enable survivors to perceive the commonality of most anxiety responses and misery to debacle. The discoveries of this present examination might be associated with the emotional well-being care arrangement of Nepal for a superior change and deal with the extent of PTSD and Depression in permissible range.

Objective

The purpose of this study was to:

- Prospectively assess the beginning, cover, and course of PTSD and prior depression following the horrible seismic tremor of Nepal.

- The degree and power of injury presentation, particularly, mourning, damage to self or family and dread as basic outcomes of wrecking calamities, are chance components for poor psychological well-being among survivors; and
- Social and Psychological (Professional) bolster is a vital defensive factor for emotional wellness after debacles.

METHOD

Sample

Members were examined from Three (03) Districts of Nepal; Nuwakot, Kavrepalanchok, and Lalitpur. These three regions are for the most part influenced by the current seismic tremor.

A sum of 100 inhabitants were examined (men = 50, women = 50). Member age ran from 20 to 71 years, and the mean age was 44.14 years (SD = 12.46).

The extents of Nuwako people groups were 40% (n=40) of the specimen, were 29% (n=29) for Kavrepalanchok and were 31% (n=31) for Lalitpur.

The financial status were Upper Middle Class 3%, Lower Middle Class 40% and Lower Class 29%.

Measures

We have used Three (03) measures/scales to conduct the present need assessment study. They are:

PTSD Checklist-Civilian Version (PCL-S)

The psychometric properties of the PTSD Checklist (PCL), another, brief, self-report instrument, were resolved on a populace of 40 engine vehicle mishap casualties and rape casualties utilizing conclusions and scores from the CAPS (Clinician-Administered PTSD Scale) as the criteria. For the PCL all in all, the connection with the CAPS was 0.929 and demonstrative proficiency was 0.900 versus CAPS. Examination of the individual things indicated colossal estimations of individual thing relationships extending from 0.386 to 0.788, and with demonstrative efficiencies of 0.700 or better for manifestations. We bolster the estimation of the PCL as a short screening instrument for PTSD.

The PCL (Weathers et al. 1993) is an effectively regulated self-report rating scale for surveying the 17 DSM-IV manifestations of PTSD. It has phenomenal test–retest unwavering quality over a 2–3 day time frame. Inner consistency is high for each of the three gatherings of things comparing to the DSM-IV indication bunches and in addition for the full 17-thing scale. The PCL connects firmly with different measures of PTSD, for example, the Mississippi Scale, the PK size of the MMPI-2, and the Impact of Events Scale, and furthermore associates reasonably with level of battle introduction.

An aggregate score is processed by including the 17 things, with the goal that conceivable scores go from 17 to 85. Utilized as a consistent measure, the PCL has great demonstrative utility. Three renditions of the PCL are accessible, in spite of the fact that the distinctions are little. The PCL-M is a military form and inquiries allude to "an upsetting military ordeal". The PCL-S is a non-military form that can be referenced to a particular horrendous mishap; the inquiries allude to "the upsetting background". The PCL-C is a general nonmilitary personnel form that is not connected to a particular occasion; the inquiries allude to "an unpleasant experience from the past". The scoring is the same for each of the three form. The PTSD Checklist is an ordinarily utilized measure, with military (PCL-M), regular citizen (PCL-C), and particular injury (PCL-S) forms. Since we are contemplating on the certain occasion on the seismic tremor of Nepal and among the common people so we have utilized the PCL-S (Table 1).

National Stressful Events Survey for PTSD Short Scale (NSESSS)

National Stressful Events Survey for PTSD Short Scale (NSESSS-PTSD), another self-report scale for PTSD that is brief (9 things), which were decreased from a bigger pool of things that were regulated to a subsample of people with plausible DSM-5 PTSD analyze from a vast national specimen. The resultant scale included things from every rule and exhibited high inner consistency (Table 2).

ZUNG Self-Rating Depression Scale

Dr. William W.K. Zung had developed and introduced the Self-Rating Depression Scale in 1965. Which is regularly known as Zung SDS. The Zung SDS was intended for surveying dejection in patients whose essential analyses were that of a depressive issue. The scale ought to be comprehensive concerning side effects of the sickness, it ought to be short and straightforward, it ought to quantitate as opposed to qualitate, and it ought to

Table 1 Sample items from Nepali translated PTSD check list—civilian version (PCL-S)

PTSD CheckList – Civilian Version (PCL-S)

Patient's Name(विमारीको नाम): ____________________

Instructions(निर्देशन): Below is a list of problems and complaints that people sometimes have in response to stressful life experiences. Please read each one carefully, put an "X" in the box to indicate how much you have been bothered by that problem *in the past month.* *(तल समस्या तथा गुनासाहरु दिइएको छ जुन जनताल कहिलेकाहि गरेका तनावपुर्ण जिवनका अनुभुतिहरु हुन । विगतमा तपाईलाई दिगदारी भएका प्रत्येक भनाइहरु सावधानी पुर्वक पढनुहोस र कतिहदसम्म दिगदारी भएको हो कस चिन्ह लगाउनुहोस् ।)*

The event you experienced was ____________________ on ____________(date)

तपाइले अनुभुत गरेको घटनाको मिति...।

		Not at all कत्तिपनि होइन ।	A little bit एक दम थोरै ।	Moderately औषत ।	Quite a bit केहि हदसम्म ।	Extremely ज्यादै ।
	Score(अंक)	1(१)	2(२)	3(३)	4(४)	5(५)
1	Repeated, disturbing memories, thoughts, or images of a stressful experience from the past?(विगतको तनावपुर्ण दोहरीएको डिष्टर्ब गर्ने सम्झना, विचार तथा तस्विर?)					
2	Repeated, disturbing dreams of a stressful experience from the past?(विगतको तनावपुर्ण दोहरीएको डिष्टर्ब गर्ने सपना ?)					

act naturally directed and show the patient's own particular reaction at the time the scale is taken. The 20 things of the scale address each of the four most usually discovered qualities of melancholy: the inescapable impact, the physiological reciprocals, different aggravations, and psychomotor exercises. Ten things are worded emphatically and ten things are worded adversely. Everything is scored on a size of 1–4 (a tad bit of the time—more often than not) with reverse scaling for the contrarily worded things. This yields a general score of 20–80 and a changed over SDS Index score of 0.25–1.00.

In an investigation of the prejudicial energy of the Zung SDS, scores/ for clinically discouraged patients were fundamentally higher than typical controls. The mean file score for a gathering of patients who had experienced a treatment program was fundamentally lower after the program (0.39) than earlier (0.74) (Table 3).

Translated Nepali Version of the Measures (Scales)

In this present examination, three scales; PTSD Checklist-Civilian Version (PCLS)—Nepali Version, National Stressful Events Survey PTSD Short Scale (NSESSS)—Nepali Version and ZUNG Self-Rating Depression Scale—Nepali Version were utilized.

Table 2 Sample items from Nepali translated national Stressful events survey PTSD Short Scale (NSESSS)

Severity of Posttraumatic Stress Symptoms—Adult*

Name(नाम):____________________________ Age(उमेर): ______ Sex(लिङ्ग): Male(पुरुष) ☐ Female(महिला)☐ Date(मिति): ________________

Please list the traumatic event that you experienced(तपाइले अनुभव गरेको दुखद घटना उल्लेख गर्नुहोस्):

Date of the traumatic event(दुखद घटनाको मिति): ________________________

Instructions(निर्देशन): People sometimes have problems after extremely stressful events or experiences. How much have you been bothered during the PAST SEVEN (7) DAYS by each of the following problems that occurred or became worse after an extremely stressful event/experience? **Please respond to each item by marking (☐or x) one box per row.(सामान्यतया अत्यन्त दुखद घटनापछि मानिसमा समस्याहुने गर्दछ। तपाई कति हदसम्म घटनाको सात दिन सम्म देहाय बमोजिमका समस्याले दिगदार हुनुभएको थियो। क्रिपया तलका प्रत्येक भनाइमा कस चिन्ह लगाउनुहोस् ।)**

						Clinical Use
		Not at all कत्तिपनि होइन।	A little bit एक दम थोरै ।	Moderately औषत।	Quite a bit केहि हदसम्म ।	Extremely ज्यादै।	Item score
	Score	0(०)	1(१)	2(२)	3(३)	4(४)	
1	Having "flashbacks," that is, you suddenly acted or felt as if a stressful experience from the past was happening all over again (for example, you re-experienced parts of a stressful experience by seeing, hearing, smelling, or physically feeling parts of the experience)?(विगतको घटनाले झझल्को दिइरहन्छ ?)						
2	Feeling very emotionally upset when something reminded you of a stressful experience? ?(जब केहीले विगतको तनावयुक्त अनुभव सम्झायो भने तपाईंको अनुभव ?)						

Table 3 Sample items from Nepali translated ZUNG self-rating depression scale

Self-Rating Depression Scale

Patient's Initial(विमारीको नाम): ________________________________

Date of Assessment(जाँचेको मिति): ______________________________

****Please read each statement and decide how much of the time the statement describes how you have been feeling during the past several days.**

	Make check mark (✓) in appropriate column.	A little of the time	Some of the time	Good part of the time	Most of the time
1	I feel down-hearted and blue -मैले मुटुको धडकन झरेको र निलो अनुभव गरे।				
2	Morning is when I feel the best बिहान सबैभन्दा ठिक ठान्छु।				

Procedures

The examination was directed secretly. Verbal educated assent was acquired from every member preceding meeting. Composed educated assent was not gathered separately in light of two contemplations. To begin with, as a piece of the mental alleviation program, this examination was required to limit however much as could be expected the aggravation to survivors who have quite recently experienced seismic tremor injury. Second, as per perceptions of agents, with respect to oral responsibilities, nearby members may translate composed educated assent as doubt. An announcement affirming this exploring system and verbal assent was marked among examiners and nearby group units. We have directed a need evaluation examine 4 months after the tremor. Given that most residences fallen or were severely harmed, almost a large portion of the inspected occupants lived in transitory lodging. People who experienced mental impediment or real psychoses (e.g., schizophrenia, real depressive issue, and natural mental issue) were rejected from the example. In spite of the way that most members are conversant in either Nepali (Since a large portion of the member was with low instructive capability).

- Nepali students were involved in each group to help clarify the survey when required;
- When the nearest birthday system couldn't pick a qualified member from a family, family selection was rather utilized as an option testing technique. A relative who could completely comprehend the examination content and the technique was then picked as the member from such family units.

Statistic information included sex, age, conjugal status were incorporated while meeting.

Result

We have applied three diagnostic measures to study, whether the sample group is suffering from PTSD and Depression and if there is the presence of PTSD; what is the magnitude of severity.

For diagnosing PTSD we have used the PTSD Check List—Civilian Version (PCL-S)—Nepali Version to all of the sample. The 17 item's Check List determined as follows.

Table 4 Frequency and percentages of respondents diagnosed as having PTSD or not ($n=100$)

	Frequency	*Percentages*
Having PTSD	18	18
No symptoms	82	82

Table 5 Percentages of respondents at severity level of PTSD ($n=18$)

Severity Level of PTSD	*Moderate (2)*	*Severe (3)*	*Extreme (4)*
Percentages	44.44	22.22	33.33

In the above result (Table 4) we have seen the 18 percentage of the sample were suffering from PTSD syndrome. Here we have found the regardless of the PTSD among both male and female.

Since we have found the significance percentage of PTSD among the sample so then administrated the *National Stressful Events Survey PTSD Short Scale (NSESSS)*—Nepali Version which has 09 items.

The result Table 5 illustrate that, the proportion of the sample suffering from PTSD is suffering from the severe level of PTSD as well. Using the *National Stressful Events Survey PTSD Short Scale (NSESSS)*, we have found that the severity level among the PTSD sufferer ($n=18$) 44.44% are in moderate level of PSTD. The most alarming is 22.22% of them are suffering from Severe level of PTSD and 33.33% of them are Extreme level of severity which needs immediate professional help for a long time. Since the traumatized event has passed more than Four (04) months.

Afterwards, we have administrated the ZUNG Self-Rating Depression Scale which has 20 Items with 10 reverse scoring. We have applied this measure to all of our 100 participants. The result as following.

With using and scoring the ZUNG Self-Rating Depression Scale, Table 6 illustrates that, among the participants 56% of them are in Mild depression, 19% Moderate and 25% of them were suffering from Severe level of Depression. Since there is no case of no depression so we can easily illustrate the representation of both male and female.

Thereafter, if we look at the demographic analysis then we have found.

Form the result Table 7, we have seen the 4 of the sample of total 29 from Kavrepalanchok was having PTSD and 25 of them with no symptom of PTSD. From the statistics 13.79% of the sample was suffering from PTSD.

Table 6 Frequency and percentages of respondents diagnosed as having different levels of depression ($n = 100$)

Level of depression	*Frequency*	*Percentage*
Mild	56	56
Moderate	19	19
Severe	25	25

Table 7 Distribution of respondents diagnosed as having PTSD or not by districts ($n = 100$)

Name of the district	*Having PTSD*	*No PTSD*
Kavrepalanchok	4	25
Lalitpur	6	25
Nuwakot	8	32

Table 8 Gender differences in PTSD and depression

Psychological disorders	*Male*			*Female*			*df*	*t-value*
	N	*M*	*SD*	*N*	*M*	*SD*		
PTSD	50	46.84	9.59	50	47.34	8.91	98	−0.270
Depression	50	45.68	15.32	50	46.40	14.38	98	−0.242

For the district Lalitpur, 6 of the total 31 participants having PTSD and 25 with no symptom of PTSD. From the statistics 19.35% of the sample was suffering from PTSD.

For the last district Nuwakot, we have found 8 of the sample of total 40 were having PTSD and 32 of them with no symptom of PTSD. From the statistics, 20.0% of the sample was suffering from PTSD.

We have illustrated the Gender differences in PTSD and Depression among the earthquake survival of Nepal. The Gender differences as follows (Table 8).

For the PTSD, for male we have found that $n = 50$ with median 46.84 and Standard Deviation (SD) 9.59 and for the female $n = 50$ with median 47.34 and Standard Deviation (SD) 8.91. The df is 98 and t-value is −0.270. So from the statistics we have seen there is no significant Gender Difference/s in the PTSD.

For the depression, for male we have found the $n=50$ with median 45.68 and Standard Deviation (SD) 15.32 and for the female $n=50$ with median 46.40 and Standard Deviation (SD) 14.38. The df is 98 and t-value is -0.242. So from the statistics we have seen there is no significant Gender Difference/s in the PTSD.

At the end of the analysis of the Gender differences in PTSD and depression, we have found no significant difference/s.

Discussion

Two consecutive earthquakes brought a devastating impersonation to Nepal. With physical injury; fatal mental disorders also developed among the survivals of Nepal earthquake. The present investigation uncovered huge levels and pervasiveness of mental issues among the survivors of the Nepal seismic tremor. Reliable with past examinations on mental issues among to a great degree influenced survivors in the underlying phases of a seismic tremor. The present investigation found that around 18% individuals experienced likely PTSD. The pervasiveness rates of depression (34%) were even rest of the others are in mild depression. With superficial investigation, we have also found the anxiety among the participants. But there may be other reasons to develop the syndrome of anxiety so we have systematically avoided the issue.

This finding implies that the side effects of PTSD, despondency and a couple of uneasiness are basic in the intensely influenced areas. The predominance rates of PTSD were generally equivalent to results of seismic tremor related examinations that were directed in initial couple of months after quake among the seriously influenced survivors. It ought to be called attention to, in any case, that the decent varieties in numerous perspectives in tremor related examinations, (for example, ruinous degree, influenced populace, appraisal approaches, instruments, and timing of examination) made it hard to direct enough correlation on bleakness and hazard components of mental issue crosswise over seismic tremor catastrophes. But according to the report of Reuters (May 13, 2015) Four (4) percentage of the total victims were suffering from PSTD but after Four months the prevalence rate goes higher to 18%. Such high pervasiveness rates are not shocking considering that this seismic tremor was a standout among the most dangerous around there in the close past. The examination of PTSD commonness found in most catastrophe contemplates, the present investigation likewise found that tension and discouragement are regular emotional wellness issues in the beginning times after the seismic tremor. From a mediation and well-being upgrade viewpoint, the present examination additionally gives valuable data to reconstructing and help work. Individuals at high hazard for mental clutters, particularly ladies and those with less social connections and less social help, ought to be particularly considered. Our discoveries propose that recouping and enhancing emotional wellness is once in a while pretty much mental administration; furthermore, psychological well-being ought to be identified with remaking homes, groups and social emotionally supportive networks. Given the outrageous absence of brain research look into on this social condition, building sufficient mental administration destinations and preparing nearby individuals for mental alleviation work is required instantly.

On April 25, 2015 at 11:56 neighborhood time, a tremor of size 7.8 struck Nepal, with an epicenter 77 kilometers northwest of Nepal's capital Kathmandu in the Gorkha area. As of May 15th, 8316 individuals had been accounted for murdered and 17,866 individuals harmed. The generally shallow profundity of the quake, at 2 kilometers, was related with especially solid powers scattered over a wide geographic range.

Oft on May twelfth at 12:50 nearby time, another quake (greatness 7.3) hit with an epicenter 76 kilometers upper east of Kathmandu in the Dolakha/Sindhupalchowk regions. This seismic tremor influenced locale as of now firmly affected by the April 25th quake and new regions. As of May 15th, 117 passings and 1961 harmed have been accounted for because of this seismic tremor (Nepal Earthquakes 2015: Desk Review).

For studying the mental disorders among the survivals of Nepal, we have chosen Kavrepalanchok, Lalitpur, and Nuwakot among the 75 administrative districts of Nepal. We have selected these as they are mostly vulnerable among others. The findings recognized a few related hazards for Two mental issue. Among the socioeconomics, sexual orientation contrasts in the predominance rates were the most also investigated. In the present study luckily we have found no differences in gender while exposed to earthquake trauma. Although several studies agree with the gender differences but this study does not support that.

Gendered methods for adapting have likewise been examined in Nepal. Men underwrite more access to social help than ladies. This has been believed to enhance psychological wellness results for men contrasted with. Be that as it may, men are more probable than ladies to take part in destructive liquor and medication utilize. One investigation presumed that ladies are educated to stifle outrage and are raised to be detached and leave, instead of battle, when looked with challenges in their lives. In spite of this, some major nongovernmental and common society associations drove by ladies are championing the headway of ladies in Nepal.

The aftereffects of the present investigation demonstrated that the power of starting apprehension anticipated positive side effects of PTSD and Depression. We have systematically avoided the anxiety since it may have other reason to develop. Steady with the consequences of late examinations, the underlying sentiment fear was a genuinely vigorous indicator of mental issue.

The force of this underlying apprehension speaks to one's close to home understanding to a calamity and is incorporated as PTSD Criterion A2 in the DSM-IV (1994).

In the first place, the force of starting apprehension is connected to injury introduction when all is said in done. Individuals with the most grounded fears have regularly endured the severest injuries. Second, the power of introductory dread may connote powerlessness of particular identity to catastrophe and calamity. People high in neuroticism are more responsive to unfavorable occasions; in this way, they might probably create mental disarranges.

This present study found, 18% with the positive syndrome of PTSD and among them 44.44% were moderately 22.22% severely and 33.33% were extremely suffering PTSD.

While demographic analysis, we have found that from Kavrepalanchok 13.79% of the sample was suffering from PTSD. For the district Lalitpur, 6 of the total 31 participants having PTSD and 25 with no symptom of PTSD. From the statistics, 19.35% of the sample was suffering from PTSD. For the last district Nuwakot, we have found 8 of the sample of total 40 were having PTSD and 32 of them with no symptom of PTSD. From the statistics, 20.0% of the sample was suffering from PTSD.

The prevalence rate of Nuwakot was a bit higher because the damage was higher here than other districts and being remote, the aid reached there in late. For the depression, we have found all of the participants were in different level of depression.

A number of recommendations have made in the recommendation section to build strong community support to cope with the mental disorders.

Recommendations

Since we have found 18% of the sample suffering and among them 33.33% are with extremely severe level of PTSD and high prevalence rate of depression among all of 100% participants; so some immediate response is needed to aide to lessen the commonness rate of PTSD and Depression and alongside other mental issue.

- Mental Healthcare camp: To provide clinical treatment to the victims Mental Healthcare Camp need to be established. Although only 100 professionals are working in Nepal it is necessary. In this regard, Nepal can take professional help from other countries. In the camp, the fundamental medicines for individuals with PTSD are psychotherapy ("talk" treatment), meds, or both.
- Besides the therapy; sometimes prescribing medicine might be needed or both of the way of treatment altogether.
- Since it is unknown the origin of psychological disorders like PTSD and Depression so future longitudinal study will be needed to reveal the pattern and prevalence of psychological disorders i.e. PTSD, Depression.
- The course of PTSD after catastrophes stays vague. Our survey demonstrates that keeping in mind the end goal to give dependable gauges obviously of post debacle psychopathology and to empower examinations between thinks about, generous advance is required in every aspect of estimation (e.g. instruments and strategy for organization), time periods for subsequent meet-ups and examining systems.
- Our present study all the more unequivocally and unmistakably recognizes populaces in danger (e.g. minorities, elderly, kids, coordinate casualties, people on call). Our audit recommends that while it is sheltered to expect a noteworthy decrease in psychopathology in the all-inclusive community, various gatherings remain profoundly powerless against PTSD for the time being and may well have an alternate long haul course of PTSD and endure the worst part of the social and monetary results related with this condition.
- Special arrangement for the most vulnerable population i.e. children, old person need to ensure as early as possible.
- Special care need to be ensured for those who have lost their close one. In this regard, "Child Care Unit" can be introduced for the orphans due to the earthquake.

- Besides the mental health; physical health needs to take care. The victims with severe injury need to be treated carefully. Artificial limb/s need to be adjugated.
- Since the mental illness is related to many other factors, so reconstructions may be required for those who have lost their property.
- Good communication may also help to make the scnerion normal at large scale.
- Since there is no mental healthcare system or hospital developed in private sector in Nepal, so this study highly recommend the involvement of private sector in mental healthcare network.
- Strengthening National Mental Health Care System.

Limitations of the Study

While studying and analyzing the collected data, we have faced a series of problems. Which we have identified as the limitations of this present study. They are as following:

- Although prepared specialists directed our overview sufficiently, this review was basically a self-report evaluation. There were no clinical judgments of practical debilitations or PTSD and sorrow side effects, and the PCL-S, ZUNG Self-Rating Depression Scale, and Stressful Events Survey PTSD Short Scale (NSESSS), don't really demonstrate the nearness of psychopathology.
- The need evaluation contemplate is neither approved nor repaid by outsider payers notwithstanding when it is demonstrated for moral clinical practice and sound hazard administration.
- The investigation inspected a destroyed area, subsequently, the conclusions ought not to be summed up to delicately influenced areas. Be that as it may, we have attempted to pick such regions are generally influenced by the seismic tremor.
- The Need evaluation contemplate is not totally effective to show to explain completely the customer's issues since social event data in detail of the member/s is not generally such simple.
- One significant issue in the approval procedure is that commentators regularly work from institutionalized approval conventions that endorse proper clinical criteria expected to approve mental appraisal, including institutionalized time or administration units inside which to finish the administration.

- The research design most commonly employed in the analyzed studies also deserves attention. Since this is a very scientific process; so we could not robustly change in the design. Like there was the reverse scoring in the ZUNG Self-Rating Depression Scale. Which needed time to make everyone understand related to the study.
- Although open-ended questions are best for the need assessment study; but due to socioeconomic conditions we had to use the yes/no (✓) type questions.
- We have faced the reluctance of the participants.
- Despite our way to-entryway technique, the example circulation did not totally compare to the populace dispersion in this area; particularly.

Bibliography

Basoglu, M., Kilic, C., Salcioglu, E., and Livanou, M. (2004). "Prevalence of Posttraumatic Stress Disorder and Comorbid Depression in Earthquake Survivors in Turkey: An Epidemiological Study". *Journal of Traumatic Stress* 17: 133–141.

Behavioral Science Division, 1991. Acierno, R., Ruggiero K. J., Galea, S., Resnick, H. S., Koenen, K., et al. (2007). "Psychological Sequelae Resulting from the 2004 Florida Hurricanes: Implications for Postdisaster Intervention". *American Journal of Public Health* 97: 103–108.

Blanchard, E. B., Jones-Alexander, J., Buckley, T. C., and Forneris, C. A. (1996). "Psychometric Properties of the PTSD Checklist (PCL)". *Behaviour Research and Therapy* 34: 669–673.

Cardove, M. J., Andrykowski, M. A., Redd, W. H., Kenady, D. E., McGrath, P. C., and Sloan, D. A. (1995). "Frequency and Correlates of Posttraumatic Stress Disorder like Symptoms After Treatment for Breast Cancer". *Journal of Consulting and Clinical Psychology* 63: 981–986.

David, D., Mellman, T. A., Mendoza, L. M., Kulick-Bell, R., Ironson, G., et al. (1996). "Psychiatric Morbidity Following Hurricane Andrew". *Journal of Traumatic Stress* 9: 607–612.

Depression: Simple English Wikipedia https://simple.wikipedia.org/wiki/Depression_%28mental_illness%29.

Forbes, D., Creamer, M., and Biddle, D. (2001). "The Validity of the PTSD Checklist as a Measure of Symptomatic Change in Combat-Related PTSD". *Behavior Therapy and Research* 39: 977–986.

Galea, S., Nandi, A., and Vlahov, D. (2005). "The Epidemiology of Post-traumatic Stress Disorder After Disasters". *Epidemiologic Reviews* 27: 78–91.

Grant, D. M., Beck, J. G., Marques, L., Palyo, S. A., and Clapp, J. D. (2008). "The Structure of Distress Following Trauma: Posttrammatic Stress Disorder, Major Depressive Disorder, and Generalized Anxiety Disorder". *Journal of Abnormal Psychology* 117: 662–672.

Green, B. L., Lindy, J. D., Grace, M. C., and Leonard, A. C. (1992). "Chronic Posttraumatic Stress Disorder and Diagnostic Comorbidity in a Disaster Sample". *Journal of Nervous and Mental Disease* 180: 760–766.

Wilkins, Kendall C., Lang, Ariel J., and Norman, Sonya B. (2012). "Synthesis of the Psychometric Properties of the PTSD Checklist (PCL) Military, Civilian, and Specific Versions". *NCBI Journals Depress Anxiety.* 2011 July; 28(7): 596–606.

Mental Health Worry for Nepalis Traumatised by Aftershocks: Report of Reuters (May 2015).

Need Assessment: Wikipedia https://en.wikipedia.org/wiki/Needs_assessment.

Neria, Y., Nandi, A., and Galea, S. (2007). "Post-traumatic Stress Disorder Following Disasters: A Systematic Review". *Psychological Medicine.* Cambridge University Press, pp.1–14.

Nilamadhab, K., and Binaya, K. (2006). "Post-traumatic Stress Disorder, Depression and Generalised Anxiety Disorder in Adolescents After a Natural Disaster: A Study of Comorbidity". *Clinical Practice Epidemiology in Mental Health* 2: 17.

Posttraumatic Stress Disorder: Wikipedia https://en.wikipedia.org/wiki/Posttraumatic_stress_disorder.

Sandro, G., Arijit, N., and David, V. (2005). "The Epidemiology of Post-traumatic Stress Disorder after Disasters". *Epidemiological Reviews, Oxford Journals* 27 (1): 78–91.

Severity of Posttraumatic Stress Symptoms—Adult (National Stressful Events Survey PTSD Short Scale [NSESSS]).

Sharan, P., Chaudhary, G., Kavathekar, S. A., and Saxena, S. (1996). "Preliminary Report of Psychiatric Disorders in Survivors of a Severe Earthquake". *American Journal of Psychiatry* 153: 556–558.

US Department of Veteran Affairs: National Center for PTSD, Traumatic Effects of Specific Types of Disasters.

Weathers, F. W., Huska, J. A., and Keane, T. M. PCL-S for DSM-IV. Boston: National Center for PTSD—*PTSD CheckList—Civilian Version (PCL-S).*

Weathers, F., Litz, B., Herman, D., Huska, J. A., and Keane, T. (1993). "The PTSD Checklist (PCL): Reliability, Validity, and Diagnostic Utility". *Paper Presented at the Annual Convention of the International Society for Traumatic Stress Studies.*

Zung, William W. K. (1965). "ZUNG Self-rating Depression Scale". *Archives of General Psychiatry* 12: 63–70.

Cyclone Aila: Immediate Challenges and Impact on Socio-economy

Nasim Banu

Introduction

Bangladesh is a country of geographic area of 144,000 sq. km with 150 million populations depending on agriculture, livestock, fisheries and natural resources which are very sensitive to climate change and sea level rise. There are over 35 million people living in the coastal zone of Bangladesh and 72 offshore islands of an area of 4200 sq. km with over 3 million people who are exposed to cyclone, storm surges and salinity intrusion due to sea level rise (Government of Bangladesh 2010b). Human-induced changes in the global climate change and associated sea level rise are now accepted phenomena and The Intergovernmental Panel on Climate Change (IPCC) concluded that, the balance of evidence suggests a discernible human influence on global climate (Government of Bangladesh 2010b). However, Bangladesh is vulnerable because of its low-lying, located on the Bay of Bengal in the delta of rivers like Ganges, Brahmaputra and Meghna and population density. Due to rising sea surface temperature and presence of high vertical wind since 1971–2009 there were 11 occurrences of cyclone with storm surges in the Bay of

N. Banu (✉)
Department of Politics & Public Administration, Islamic University,
Kushtia, Bangladesh

B. Zutshi et al. (eds.), *Disaster Risk Reduction*,
https://doi.org/10.1007/978-981-10-8845-2_20

Bengal which strike the coastal area of Bangladesh and caused catastrophic devastation with death of more than 0.168 million people. In the recent past, Bangladesh has witnessed the devastation of cyclone Sidr in 2007 and Aila in 2009. The disasters like cyclone intensity might be increased in Bangladesh as much as 10% due to increased warming; so, Government of Bangladesh (GoB) has been considering the two types of responses i.e. adaptation and mitigation to face the serious consequences of climate change (Government of Bangladesh 2010b).

1.1. Bangladesh is vulnerable and high-risk country to natural disasters due to global warming and climate change associated with high density of population and poverty. The country is one of the worst sufferers in the world of cyclone along with storm surge. The events of devastating cyclone in Bangladesh inflict human casualties, kill livestock/poultry, destroy homestead/crops and cause health hazards on victimized population leaving the marginal and landless people unemployed and below poverty line. In order to build resilience, GoB has adopted Climate Change Strategy and Action Plan in 2009 considering disaster management as one of the courses of action that includes: food security; social protection and health; comprehensive disaster management; infrastructure, research and knowledge management; mitigation and low carbon development; Capacity building and institutional strengthening. As a part of structural and non-structural mitigation measures, GoB constructed more than 3000 cyclone shelters and about 4000 km coastal embankment; enacted law on disaster management, framed disaster management plan, policy and standing order; adopted training courses to raise public awareness at various levels; and has been implementing comprehensive disaster management program. Improvement in non-structural mitigation is significant but structural mitigation needs further improvement particularly building more cyclone shelters is required in the coastal area to protect maximum number people from cyclone attack. To overcome the situation, GoB has framed "Cyclone Shelter Construction, Maintenance and Management Policy 2011" to construct new shelters and ensure proper maintenance of the existing cyclone shelters.
1.2. The great majority people of the coastal belt of Bangladesh due to tropical cyclone suffers from unemployment, underemployment,

malnutrition, illiteracy, ill health and unhygienic sanitation with loss and damage of their homestead, livelihood, crops, and properties. Cyclones with storm surges are on increase because of global warming and sea level rise. Thus, major focus and effort of GoB is disaster risk management. There is Ministry of Food and Disaster Management in Bangladesh to deal with disaster management including climate hazard management and also many more NGOs are active in post-disaster response and rehabilitation operations to supplement the government efforts. Moreover, GoB has established Natural Disaster Risk Fund in 2004 from the revenue budget for disaster risk reduction and climate change adaptation.

Cyclone Aila

In 2007, devastating cyclone Sidr swept through 12 coastal districts of Bangladesh with heavy rain and storm surges. Sidr cause of death 3363 and injured 55,282 peoples; destroyed houses fully 563,887 and partially 955,065 and damaged crops of 186,883 hectares fully and 498,645 hectares partly (Government of Bangladesh 2010b). Before recovering from the devastation of Sidr cyclone storm Aila hit the south-western coast of Bangladesh on May 25, 2009 in between 1500–1800 hours with tidal surge and powered by strong wind about 65–75 miles per hour. Aila affected 12.5% area of the country. Koyra and Dacope upzila of Khulna district, Shyamnagar and Ashashuni upzila of Satkhira district were the worst affected areas where 47,810 were shelter-less temporarily living on the embankments. Aila caused huge loss to the agriculture, livestock, and fisheries sectors amounting USD 155.30 million.

Cyclone Alia Affected the Aspects of Social and Economic Life of People in Many Ways

a. Physical:

   Aila affected 3,928,238 people of 11 districts. It took 190 lives, injured 7103 people, killed innumerable cattle heads, properties and caused wide range economic problem, and burden to the survivals. Moreover, Alia damaged educational institutions, roads, bridge, culvert, embankments, crops, livestock, poultry, and households etc. Aila also heavily damaged the mangrove forest Sundarban.

b. Livelihoods:

Most of the household of Aila affected area engaged in crop production suffered loss and damaged of standing crops. Aila destroyed crops of 77,486 acres fully and 245,968 acres partially; people loosed 150,131 livestock and poultry. Thus landless agricultural workers suffered loss of unemployment and the marginal land owners particularly women suffered from loss of homestead production activities like vegetable gardening, poultry and livestock. Aila affected region is famous for shrimp farming; high tidal surges during Aila washed away the shrimp farms which hampered the occupational pattern of the farmers. Over 160,000 coastal fisherman and 185,000 shrimp fry collectors are involved in marine fisheries and 18% household depends on Sunderland resources like honey, golpata, shrimp fry, and shell/crab collection which seriously hampered by Aila with storm surge. Thus, due to cyclone Aila the livelihood bases of the affected people were devastated.

c. Education:

In the coastal region of Bangladesh all most all educational institutions use as cyclone center in the risk period. Aftermath Aila, the educational institutions were being occupied by the distress people so academic activities could not continue for a long time. Aila destroyed 445 educational institutions fully and 4588 partially. Moreover many institutions either wiped out or went underwater or seriously damaged by Alia with tidal surge.

d. Communication:

2233 km roads fully and 6622 km partially and 1742 km embankments were damaged, some disappeared thus, high tide saline water entered into the locality. Just after Aila government and many non-government organizations were responsive and concerned with emergency requirements but delivery of relief goods was not smooth and easy due to the breakdown of communication system.

e. Health:

Aila destroyed the sources of pure drinking water sources, household reserved food, homestead food production, sanitation facilities, local health centers and compelled the people to drink polluted water, take food irregularly/small amount of food and left them in unhygienic condition which led them to suffer from malnutrition and disease like allergy, cholera, and diarrhea etc.

f. Migration:

Landless people and the people who used to earn their bread and butter by selling labor suffered from unemployment just after Aila as there was no condition for jobs. So, looking for jobs they migrated to the cities and any other places where jobs could be found. Some marginal landowners also temporarily migrated to the nearby town/cities for getting relief goods.

Immediate Response for Resilience

On the event of cyclone Aila the Government organizations including Armed Forces, many NGOs and Bangladesh Red Crescent Society (BDRCS) immediately responded. Collaboration among the concerned organizations and local community enabled the following tasks to be undertaken:

a. Rescue of affected people including livestock and augmenting health care services.
b. Immediate supply of food, pure drinking water and non-food items like clothes and wallet.
c. Transportation of relief goods to the affected areas using road, river and air route.
d. Clearing and restoration of road/river communication and restoration of telecommunication.
e. Receiving relief goods and prioritizing the affected areas for distribution of relief.
f. Coordinating the activities of government/NGOs/local communities concerned with Aila response.

Disaster Management Bureau (DMB) is a professional unit to perform specialist functions of disaster preparedness, disaster action, contingency planning, arranging public awareness, organizing training and facilitating improved information collection, under Ministry of Food and Disaster Management. The picture/activities are found on relief and rehabilitation against cyclone Aila from DMB:

a. Major Emergency Relief. Disaster Management and Relief and Division of Ministry of Food and Disaster Management distributed total USD 18.33 million in cash and kind (1.9 million cash and 2795 metric ton rice) and another house building 4.28 million.
b. International Donor Assistance. Various donors like UNDP, WHO, FAO, DFID,WFP, UNICEF etc. donated a total of USD 44.25 million for the Aila affected people in the sectors/activities like food assistance, livelihoods, shelters, education, health, nutrition agriculture, emergency medicine and water, sanitation and hygiene.
c. Restoration of Livelihood. In this regard, the total rehabilitation cost was USD 56.02 million which includes the cost on agriculture 50 million, livestock 01 million and fisheries 5.02 million. Moreover, DMRD allotted USD 2.47 and 0.142 for income generating activities of the ultra poor and to bear water transportation cost respectively and; 950 and 8605.8 metric ton food grains under Food for Works and Vulnerable Group Feeding program respectively to the worst affected upzilas of Khulna and Satkhira district.
d. Repair of Damaged Embankment. USD 16.5 million including 25,000 MT rice was given to Bangladesh Water Development Board by DMRD to repair 47 Aila damaged embankment out of which 41 polders were repaired within 2 months.
e. Fund for Homeless People. USD 13.71 million was allocated for construction and repair of household of 48,000 people of Koyra and Dacope upzila of Khulna district and Shyamnagar and Ashashuni of Satkhira district who took shelter on the embankment. Moreover, each affected families were given BDT 20,000 to buy 2 bundles of CI sheets and other materials.

Challenges Faced

During any disaster relief operation is considered as the immediate response so as the case with Aila also. Relief operation to the Aila affected area was not so smooth rather some challenges were faced. The challenges are discussed below along with the opportunities to improve in certain areas that would enhance the disaster management capacity in damage and risk reduction during any disaster/cyclone in Bangladesh.

a. Inadequate/Insufficient Equipment:

Timely evacuation of the affected people including their household was the prime importance on the event of Aila. But rescue of endangered people from widely affected areas was very challenging due to adequate fast moving appropriate water crafts that could move in bad weather and turbulent river condition. Provisioning of appropriate rescue equipment particularly rescue boats should be viewed as essential elements of early preparation. Using available speed boats of Bangladesh army, Navy and Coast Guard would ease up search and rescue operation.

b. Lack of Communication Facilities:

Aila was associated with power failure and communication disruption. Thus, both physical accessibility to affected areas and collection of accurate and timely information about nature and extent of loss and damage became the most difficult task. Emergency relief sending to the affected areas by road was difficult rather impossible as all the roads were damaged and blocked by the fallen trees. Moreover immediate need could not be assessed because power failure ceased tele/mobile communication. Communication plays a vital role to get timely information about the victims and reach necessary support to affective areas at the earliest. At initial stage of Aila, Armed Forces radio communication system was the only means of communication of disaster management operation as other communication was disrupted. However, mobile communication network needs to be restored immediately after disaster like cyclone and alternative power supply to be arranged to restore communication system at the earliest.

c. Control and Coordination with NGOs and Donors:

Instantly Government organizations, local and foreign NGOs, international donors and local community came up to mitigate the victims of huge devastation of Alia. Thus, at early stage it was difficult in coordinating the huge rescue and relief activities of those organizations and ensuring rational delivery and distribution of relief goods among the distressed people, though subsequently the coordination problem was minimized. Successful disaster management warrants well-orchestrated efforts by the different government, non-government, private community, and international donor/agency. Coordinated efforts would ensure the effective functioning of post-cyclone proportionate relief operation and rehabilitation program based on actual need assessment and could avoid the duplication of efforts.

d. Distribution of Materials to Inaccessible Places:

It was observed that most relief efforts/activities by the volunteers, NGOs, local community and private activities were concentrated on either accessible places or received media coverage. Whereas, in many cases identification of the needy people of inaccessible places including middle-class families, shy of receiving relief was ignored. Therefore, distribution of appropriate kind of materials/resources to the needy of inaccessible areas was colossal job and challenge for the government in disaster management operation during Alia. Clear approaches are essential for immediate response in support of the victims. To support and augment national disaster management capacity/capability formation of an Engineering Unit with heavy equipment under Armed Forces may be useful in restoring road and river communication by clearing the landside, fallen trees, and debris of the damaged infrastructure.

e. Lack of Landing Sites in Remote Areas:

Due to tidal surge and battered embankment most of the Aila affected areas under water form knee to waist level. So in many places there was no option but to drop relief goods from helicopter caused disproportionate distribution. The capacity of runways and suitable higher places for helipad in the closer proximity of cyclone area to be strengthened for sustained operation of aircraft that would remain operational during tidal surge and will facilitate in reaching early relief goods to the inaccessible areas.

f. Lack of Reliable Database:

Reliable database is essential for accurate, timely need assessment and to avoid confusion regarding facts and figure, but there was lack of database during devastating Aila. So, it was really difficult to assess the loss/damage and actual assistance and type of relief required by the affected people, but also created confusion in search and rescue operation for the deed and missing people.

g. Problem in Early Warning/Forecast:

Warning dissemination is carried out through Radio, TV, CCP and local administration. Issuing regular weather bulletins and warning signals for cyclone and storms is the responsibility of Bangladesh Meteorological Department (BMD) and Bangladesh Space Research and Remote Sensing Organization (SPRRSO). Cyclone Preparedness Program (CCP) operated jointly by BDRCS and the Government of Bangladesh is the core organization for spreading warnings against approaching cyclone, arranging shelters for people, rescuing survivors in dangers and providing first aid and distributing food, cloth, medicine etc. to the needy. In spite of all those organizational arrangements people of remote coastal area often either remain unaware of early warning/forecast in time or reluctant to give due importance to the anticipatory bulletin as disaster is almost regular phenomenon to them. However, early warning/forecast system in Bangladesh has improved to a great extent. Immediate response to Aila was possible due to early preparation basing on the forecast. Propagation of timely early warning to the local people coupled with preparedness reduced the effects of cyclone Aila as compared to previous cyclones.

h. Unwillingness of Vulnerable Population for Evacuation:

Most of the people at the coastal belt are poor and resort to fishing as livelihood. They catch fish at offshore and often do not receive/listen disaster/cyclone forecast in time. Many resist anticipatory evacuation leaving their household and livestock behind. Rescue of this segment of population posed really a challenge for the rescue operation. Comprehensive evacuation plan encompassing evacuation of valuable belongings of the affected people including livestock would encourage them for moving toward safe shelters leaving their houses and in turn, will contribute to risk reduction.

i. Restoration of Essential Facilities:

Controlling the spread of epidemic after Aila was one of the major challenges, however, that could be minimized to acceptable limit through timely restoration of essential facilities like supply of pure drinking water, immediate medical support, and burial of dead bodies including dead livestock. Life was brought to normalcy through debris and waste materials from around villages and localities of the affected areas. Early restoration of essential facilities should form a priority task as a part of disaster management operation. Installation of deep tube well and desalination plant, re-excavation and digging new ponds to ensure safe drinking water construction of cluster houses for landless/homeless people; reconstruction of houses for the affected people; and restoration of livelihood should be the priority task for bringing normalcy and sustainable condition in the cyclone-affected areas.

j. Future Challenges:

There arises some future challenges identified by DMB on the event of Aila which are: Ensuring of cyclone center for all vulnerable of coastal belt; Preparation of comprehensive program for construction of embankment covering the coastal belt; Construction of sustainable houses; Provisioning of safe drinking water; Arrangement of sluice gate on the embankment for shrimp culture in the coastal area according needs; Expansion of operation area of the cyclone preparedness programs and Mainstreaming of disaster risk reduction in the development process and adaptation to climate change.

Case Study on Impacts of Aila

Case Study-1

Kalapara upzila of Patuakhali district is a cyclone prone area. Cyclone Aila struck this Upzila caused innumerable damage though this was not worst affected area. A study was conducted to observe the impacts of cyclone including Aila in Kalapara upzila (Ahmed et al. 2012). The study found that:

a. Human Casualty:
Aila put impact on health affecting 93,675 people of 09 unions and 01 paurashava of the upzila with 01 death casualty but injured 152.
b. Educational Institution:
During risk period educational program could not be continued because 68 educational and 285 religious institutions were affected by Aila.

c. Road:
   Aila damaged about 5 km pucca, 248 kuchcha road and 29.5 km dam that affected the transportation system.
d. Employment:
   Most of the people live on fishing. Aila destroyed 620 fishing nets worth of BDT 3,850,500 and scratched about 385 fishing boats worth of BDT 2,995,000. Moreover, there were losses in the cottage/industry worth of BDT 10,032,000. These affected on livelihood and employment opportunity of the people of the Upzila.

Case Study-2

The study was conducted to observe the socioeconomic condition of the people of worst affected Koyra and Dacope upzila of Khulna and Shyamnagar of Shatkhira district and found that after one year of Aila affected people were living below the subsistence level (Kumar et al. 2012). The main findings of the study are:

a. Livelihood:
   Among the total occupation of these three upzilas 40% engaged in shrimp farming, 40% agriculture and 30% wage labor activities. Alia destroyed the 96% livelihood bases, thus, 80% population of region were struggling to maintain subsistence living even after one year of Aila.
b. Agriculture:
   Aila flooded more than 90% agricultural land and 70% homestead gardens and damaged 70% vegetable garden. Due to water logging most of fruit trees vegetable sensitive to water died and the vegetable production was impossible for a longer time as the fields were contaminated with salinity.
c. Livestock:
   Aila damaged 75% livestock and poultry and after Aila 30% died due to shortage of food and drinking water which compelled the owners to sell the remaining livestock poultry to avoid their brutal death.
d. Household:
   Aila with tidal surge and subsequent water logging abolished the houses of the locality. Thus the people forced to live in embankment or relocate their houses somewhere in the raised land. Even after one year of Aila 70% families were still living on embankments in makeshift tents.

e. Drinking Water:
   Aila with high tidal surges destroyed all drinking water sources and contaminated the fresh water sources with polluted saline water. Before, Aila the about 50–60% area was covered with water supply and sanitation. Due to shortage of pure drinking water and facilities of hygienic sanitation system people of the area was suffering from diseases like allergy, cholera and diarrhea and thus the medical cost of the affected people was increased by 55%; women and girls had to spend on an average 2.5 hours daily to collect drinking water from the long distance; as many of girls student were engaged in water collection instead of going to school, the drop-out was increased remarkably of that area.
f. Food Intake:
   Aila devastated the stock, standing crops, vegetable garden, fishing sources, livestock and poultry led the affected people into starvation and malnutrition situation. It is found that income of the 75% families was reduced and that was the one of the main cause of taking insufficient food suffering from malnutrition.
g. Migration:
   Due to devastated Aila, affected particularly landless people lost their employment opportunities in the region and make jobless which forced them migrate elsewhere to find job. After one year of Aila, it is found those 1200 families from Koyra, 800 from Dacope and 1300 form Shyamnagar permanently and 1500, 1300 and 3000 families respectively temporarily migrated to the nearby cities/towns for looking jobs. There was evidence of migration to West Bengal of India also.
   The impact of Aila on *socioeconomic condition* of the affected people with suffering from food, water, sanitation and household requirements was severe and multi-dimensional The Government of Bangladesh started the immediate relief and recovery operation. Simultaneously NGOs of BDRCS, Bangladesh Armed Force, semi-organization, clubs, groups and people from their own initiatives started relief activities to the Aila affected areas. International cooperation was worth mentioning and donor of NGOs of Bangladesh helped the relief and recovery program for Aila affected people in cash and kind. With these relief and recovery activities and humanitarian support of Government, NGOs, International Donors, Local communities to the Aila affected people could ease the situation to some extent and control the starvation and spread of epidemic to the acceptable limit.

Recommendations and Conclusion

i. Bangladesh has act, plan policy, standing orders and institutional arrangements for disaster management and best practice frame work under which risk reduction and emergency response activities are managed and implemented. With all these legal and policy development and institutional arrangements meanwhile Bangladesh has proved its efficiency and capability in handling and managing disaster risk reduction out of devastating cyclone to an acceptable limit. However to further increase of disaster management capacity in the field like early preparedness, risk reduction, emergency response, short and long time sustainable rehabilitation the following course of action may be considered:

   a. Defined responsibilities for dissemination of warnings, planning, and organization of preparedness measures in the National Plan for Disaster Management and Standing Orders on Disasters to be ensured.
   b. National warning and forecasting systems to be maintained and refined for providing relevant and timely information to the officials concerned with disaster management and the general public.
   c. Adequacy arrangements for cooperation between civil administration that includes concerned line ministries and local level administrative units, armed forces and NGOs to be ensured and maintained.
   d. Guidelines, equipment, and training to be provided to the local levels for preparation to minimize damage and losses out of cyclone.
   e. Practical guideline to be prepared and distributed to the organization of rescue and relief assistance operations to maintain the equitable standard following disaster like cyclone.
   f. Training of extension workers, volunteers, local leaders, teachers and women etc. and broad-based public education relating to preparedness and response to cyclone has to be ensured.
   g. Contents of training on disaster management for the government officials that have been offering by the institutions for basic and refresher training to be more practical oriented and meaningful.
   h. A list of experienced and technical specialist (including government officials, armed forces personnel, professionals,

academicians and NGOs) on disaster management who would be available and their services can be used to assess the damage, coordinate emergency relief operation, propose for immediate rehabilitation following a cyclone.

i. There should be scope to send trained and experienced senior officials having up-to-date knowledge on national disaster plan, policy, strategy and standing orders to reinforce and guide local administrators/managers in the disaster/cyclone-affected areas.
j. Appropriate projects to be undertaken for the reconstruction and long-term rehabilitation of infrastructure with allocating national resources and if need be requesting with foreign donors for funding the projects to maintain the livelihood of cyclone-affected and to reduce the loose and damage from the future occurrence.
k. Projects for reconstruction and rehabilitation to be monitored and evaluated and lessons learnt from have to be used in correction/modification of the existing project as well as in planning/preparing future the projects.
l. Construction of more cyclone shelters and maintenance to be ensured in the with the Cyclone Shelter Construction, Maintenance and Management Policy 2011.
m. Properly regular monitoring and evaluation of RR and ERM program to ensure better implementation and future development.

ii. Cyclone is a life companion for the 35 million people in 700 km long coastal belt in Bangladesh. A good number people born in cyclone, live in cyclone and considerable number die in cyclone. People lose their property, livelihood, homestead and occupation, nation faces huge task of relief, restoration, and rehabilitation to the affected people and area. The GoB acts on risk reduction, preparedness, humanitarian assistance during cyclone and post-cyclone rehabilitation in consistent with national plan, policy, and standing order. Though the loss of properties had been very serious in the cyclone Aila but loss of lives were only 190 due to effective disaster risk reduction and cyclone preparedness program of the government. It is true; Bangladesh is lake in resources both for preventive and operation activities but has distinct law, plan, policy, standing order and program relating to better effective management of natural disaster. Owing resource constraints GoB has been giving priority to mobilize national resources on regular basis to combat the situation of prevention, preparedness, response and recovery program, and emergency risk management arises out of natural disaster like cyclone.

Bibliography

Ahmed, S., Rahman M. M., and Faisal, M. A. (2012). "Reducing Cyclone Impacts in the Coastal Areas of Bangladesh: A Case Study of Kalapara Upzila". *Journal of Bangladesh Institute of Planners* 5. Dhaka, Bangladesh: Bangladesh Institute of Planners.

Choudhury, M. N. S. (1998). "Disaster and Its Management: An Overview of Bangladesh (Seminar Paper)". Savar, Dhaka: Bangladesh Public Administration Training Centre.

Government of Bangladesh. (2009a). "Enhancing National and Community Resilience-Integrating Disaster Risks Reduction and Climate Change Adaptation Measures into Development Planning and Processes in Bangladesh-Guide to Practice". Dhaka, Bangladesh: Ministry of Food and Disaster Management.

Government of Bangladesh. (2009b). "Practicing Gender and Social Inclusion in Disaster Risk Reduction". Dhaka, Bangladesh: Directorate of Relief and Rehabilitation, Ministry of Food and Disaster Management.

Government of Bangladesh. (2010a). "Standing Orders on Disaster". Dhaka, Bangladesh: Disaster Management Bureau, Disaster Management and Relief Division, Ministry of Food and Disaster Management.

Government of Bangladesh. (2010b). "National Plan for Disaster Management 2010–2015". Dhaka, Bangladesh: Disaster Management Bureau, Disaster Management and Relief Division, Ministry of Food and Disaster Management.

Government of Bangladesh. (2011). "Sixth Five Year Plan (Part-1) 2011–2015". Dhaka, Bangladesh: Planning Commission.

Government of Bangladesh. (2013a). "Cyclone Shelter Construction, Maintenance and Management Policy 2011". Dhaka, Bangladesh.

Government of Bangladesh. (2013b). "National Sustainable Development Strategy 2010–2021". Dhaka, Bangladesh: General Economic Division, Planning Commission.

Islam, A. F. M. S. (2005). "Disaster Management: Impact on Socio-economy of Bangladesh (Seminar Paper)". Savar, Dhaka, Bangladesh: Bangladesh Public Administration Training Centre.

Japan International Cooperation Agency. (1999). *Country Report—Training Course on Seminar on Administration for Disaster Management (25-01-2000 to 25-02-2000)*. Savar, Dhaka, Bangladesh: Bangladesh Public Administration Training Centre.

Kumar, U., Baten, M. A., Masud, A. A., Osman, K. S., Rahman, M. M. (2012). *Cyclone Aila: One Year on Natural Disaster to Human Sufferings*. Dhaka: Unnayan Onneshan-The Innovators..

Recovery After Disasters—Problems and Prospects: The Case of Koslanda-Meeriyabedda Landslide in Sri Lanka

R. Lalitha S. Fernando and M. S. Dimuthu Kumari

Background of the Study and Problem Identification

The disaster profile of Sri Lanka is updating frequently with some strange natural disasters; the coastal area of the country was completely horrified by the hard-hit tsunami disaster in 2004. After 10 years, in 2014, an unexpected landslide happened in Koslanda village of Badulla District, resulting in deaths and displacement of the plantation community. **This paper discusses the recovery process after the landslide in 2014, in Koslanda-Meeriyabedda area and attempts to see whether the** government intervention **to the recovery process has been effective.** Meeriyabedda Estate is situated in Haldummulla Divisional Secretariat of Badulla District, Sri Lanka. This is nearly 200 km away from the capital city of Colombo. **Koslanda is a small village consisting nearly 200 families altogether. This community represents Sinhala, Tamil and Muslim people and most of them are**

R. L. S. Fernando (✉) · M. S. Dimuthu Kumari
Department of Public Administration, University of Sri Jayewardenepura, Nugegoda, Sri Lanka

B. Zutshi et al. (eds.), *Disaster Risk Reduction*,
https://doi.org/10.1007/978-981-10-8845-2_21

Estate workers. According to the Ministry of Disaster Management, 330 of people in 57 families were affected; 75 children were orphaned; 63 buildings were destroyed including Houses, Kovils, Community Centers, Telecommunication Centers, Boutiques, and Estate Bungalows. Just after the disaster situation, agencies including the Ministry of Disaster Management, Media, Volunteer groups, armed, and Police Forces have visited the location, started rescue operations and post-disaster recovery activities.

Research Questions

In this context, the main research question is to examine whether the government intervention in the recovery process is effective? Accordingly, the objectives of this study are three folds; *(1) to examine how the government involved in recovery process, (2) to find out issues and problems in the recovery process both from the perspectives of the victims and the administration, and (3) to propose suitable strategies to improve the living standards of the disaster-affected families.*

Methodology

This study uses *case study method and qualitative approach.* The primary data were collected from a field visit to Poonagala victim Camp. Observations through the field visit to the camps and the affected village, was helped to identify the current situation. Data collection period was June, 2015. In-depth interviews were conducted with 25 affected families out of 60 and also several discussions were conducted with responsible administrators in Hadummulla Divisional Secretariat. As secondary data, the government circulars, newspaper articles, journal articles related issue were used. Thematic analysis with substantial description was used in analyzing the data. Triangulation methods were used to ensure the trustworthiness of the data. **The study discusses and examines the effectiveness of the government intervention in the recovery process. Several issues and problems faced by the victims and administrators during this recovery process were also identified and finally, the study proposes appropriate strategies for upgrading the living standards of the disaster-affected families.**

Significance of the Study

This study suggests policy measures for improving lives of the victims in strengthening the recovery process of the disaster management in Sri Lanka. The suggested policy measures would be useful in improving the recovery process of the selected case and also for future disasters similar to this case.

Literature Review

This section provides a review of literature on disaster management. Providing insight into the natural disasters and discusses particularly, the final stage of disaster management and its activities, issues and best practices of disaster recovery processes.

Natural Disasters

Disasters generalize the equality in the world. Irrespective of the capacity, culture, ethnicity or wealth of a nation, by natural disasters cost all. "*A disaster does not respect boundaries, nor does it show any considerations towards the haves*" (Haran 2015). The natural and man-spawned disasters are inescapable events of human life cycle (Rosenbloom 2015). As disasters are unavoidable, the worldwide attention toward the disaster management has been increased recently.

Several authors (Turner and Pedgeon 1997; Richardson 1994; the World Health Organization 2003; the Federal Emergency Management Agency in United States 2003; and the Mid-Florida Area Agency on Aging in US 2003 cited in Shaluf 2007) classify disasters as natural and man-made. According to Shaluf (2007) Natural disasters are catastrophic events resulting from natural hazards. They result from internal (beneath the Earth's surface), external (topographical), weather-related (meteorological/hydrological) and biological phenomena. The main reasons for increasing frequency of natural disasters could be considered as the environmental degradation and population growth (Deshmukh et al. 2008; UNEP 2008). The interaction between the environment and people is a root cause of disasters (O'Brien et al. 2010).

Undoubtedly, many scholars identified landslides under the category of natural disasters since it is embedded with the natural environment.

World Health Organization (2003), defined landslides as the down-slope transport of soil and rock resulting from natural phenomena or man-made actions.

Disaster Management

It is impossible to prevent natural disasters completely. However, the effects of disasters could be reduced through an effective system of disaster management. Moe and Pathranarakul (2006) identified effective disaster management as a key element of good governance. Disaster management is a process of different phases. As per the most literature, disaster management includes generic five phases, namely: (1) Prediction; (2) Warning; (3) Emergency relief; (4) Rehabilitation; and (5) Reconstruction.

According to Jayaraman et al. (1997), mitigation, preparedness, response, and recovery are the main activities conducted in those phases. Some other authors identified these activities as the phases of disaster management process. So this has been an ambiguous area when examining the disaster management literature. However, disasters cannot be managed by using a unified process. "*...Each disaster situation is unique and requires distinctive (different) proceedings...*" (Corsellis and Vitale 2008).

According to Deshmukh et al. (2008), disaster management is an integrated process of planning, organizing, coordinating and implementing measures that are needed for effectively dealing with its impact on people. However, Pathirage et al. (2012) noted that disaster management efforts aim to reduce or avoid the potential losses from hazards and assure prompt and appropriate assistance to victims of disaster, and achieve rapid and effective recovery. Hence this study focuses on disaster recovery process.

Recovery Process and Government Intervention

Disaster management is commonly represented by four phases: mitigation, preparedness, response and recovery (Rubin 1985). Berke and Godschalk (1996) noted that among these four phases, recovery is the least understood (cited in Sandeeka and Wilkinson 2014). Disaster recovery is more important in achieving sustainable development. Many scholars have identified the recovery as one activity that is performed in rehabilitation and reconstruction phases of disaster management process.

However, Pathirage et al. (2012) noted that disaster management efforts aim to reduce or avoid the potential losses from hazards, assure prompt and appropriate assistance to victims of disaster, and achieve rapid and effective recovery. The recovery task should start soon after the emergency phase ends, and should be based on pre-existing strategies and policies (UN/ISDR 2002) and they emphasize the importance of public participation and engagement in the recovery process to "*Build Back Better.*" Corsellis and Vitale (2008), emphasize the importance of a holistic concern in recovery response. John Labadie (2008) posed several questions to be addressed in recovery phase such as what will be rebuilt, how it will be done, and by whom it will be done. The author further noted that sustainability and survivability are the current trends in disaster recovery literature. Indeed the recovery is complex and multi-faceted process that extends beyond the immediate restoration. Even though the immediate restoration of services can be a matter of weeks, full recovery can stretch out ten–fifteen years (ibid. 2008).

When discussing the role of government and its intervention for the disaster recovery process, it needs strategic planning to reduce the gap between governmental ability and responsibility to deal with disasters effectively and efficiently (Rosenbloom 2015). John Labadie (2008) further emphasized the importance of having a monitoring and evaluation system in disaster recovery and reconstruction process. Vale and Campanella 2005 suggested to let the community leaders evaluate the success of recovery efforts (cited in Labadie 2008). The Asian Development Bank (2008) emphasized the role of government intervention in recovery process, particularly in landslide situations.

Issues in Recovery Process

Despite frequent disaster experiences, recovery efforts remain inefficient. The inefficiencies of recovery process have been discussed by Hamilton (2010), Lloyd-Jones (2006), Sawyer et al. (2010), cited in Sandeeka M. and Wilkinson S. (2014). Indeed, the recovery activities are often clumsy and uncoordinated. Many conflicts can be seen among the stakeholders which make this process more lengthy and complicated. Sometimes recovery projects can worsen the existing vulnerabilities faced by the community (Johnson et al. 2006; Lyons 2009; TEC 2007 [cited in Sandeeka and Wilkinson 2014]). John Labadie (2008) mentioned about the multiplicity of recovery and reconstruction activities that can be pursued by a plethora of government jurisdictions, government agencies, NGOs, private and volunteer groups. Lack of coordination among institutions makes it difficult to manage the activities which may lead wastage of resources.

Sandeeka M. and Wilkinson S. (2014) also highlighted several disturbing issues in the recovery process. Those are the non-adherence to policies, insufficient focus on livelihood development and small business support programs, overruling of local government agencies and neglecting vulnerable groups of people in the community. Poor leadership, ownership of money and property rights, short-sighted vision, quick and easy fix political will, rigid mandates, weak priority, and scare resources are other disturbing issues in recovery process (Natural Hazards Centre of Colorado University 2006). According to Natural Hazards Centre of Colorado University (2006) delays of recovery planning, resurvey due to inadequate information, the additional workload on government officers create recovery process unsatisfactory and such delays divert funds and resources from recovery programs, overlap the responsibility of one department to another etc.

Best Practices

Some best practices for effective recovery are discussed in several literatures. Among those best practices, a holistic consideration, consensus base and sustainability aspects of recovery process are highlighted. The Natural Hazards Center of the Colorado University (2006) discusses several best practices in making the recovery process more productive. Accordingly, citizens must be treated equitably, long-term community benefits should not be sacrificed for the short-term individual gains, a broad-based community support should be mobilized, environmental and natural resources should be used to maximize benefits, expose people to the risk must be exposed to the people than before and reconstruction should be linked with the economic enhancement of community.

A former American President, Clinton (2006) initiated a concept of "Build Back Better" (BBB) as best practices to the disaster recovery process during the multinational resource efforts following the Indian Ocean Tsunami. This concept is further discussed by Khasalamwa (2009), Lyons (2009), Sandeeka M. and Wilkinson S. (2014). Clinton (2006) discussed ten best ways to make the recovery process effective (cited in Sandeeka and Wilkinson 2014). Those are;

I. Recognize that families and communities drive their own recovery
II. Promote fairness and equity
III. Enhance the preparedness for future disasters
IV. Empower the Local governments to manage recovery efforts
V. Use good information for recovery planning and effective coordination
VI. The United Nations, World Bank, and other multilateral agencies must clarify their roles and relationships, especially in addressing the early stages of a recovery process
VII. Expanding role of NGOs enabling a greater responsibility for quality in recovery efforts
VIII. Create conditions for entrepreneurs to flourish
IX. Develop agency partnerships that move beyond rivalry and unhealthy competition and
X. Leave communities safer by reducing risks and building resilience.

United Nations (2006), Clinton (2006), Kennedy (2009), FEMA (2000) highlighted the importance of identifying the role of stakeholders in the recovery process (cited in Sandeeka and Wilkinson 2014). After the tsunami the government of Sri Lanka emphasized locally appropriate solutions as a best practice for disaster recovery process (cited in Sandeeka and Wilkinson 2014). FEMA (2000) highlighted community-centered recovery operations whereas Omidvar et al. (2010) emphasized the importance of preserving cultural and architectural heritage in recovery process (cited in Sandeeka and Wilkinson 2014).

Data Analysis and Discussion

The Government Intervention in the Recovery Process

The government intervention in the recovery process is discussed here mainly under the areas of rehabilitations, relief services, post awareness programs and livelihood development activities.

Government Intervention in Rehabilitations

Almost entire village was destructed by the disaster and victims are basically located in two schools. Later, they have been shifted to an abundant tea factory. The government has established a minimal level of facilities within this temporary location such as electricity power, water, and other sanitary facilities. But the victims are not satisfied with the given services within the camp, especially about the spaces of partitioned rooms. There are more than five members in the most of families and all of them live in a very small room. Since it has taken more than one year after the disaster, the given rehabilitations are not satisfied at all. About 100% of the respondents are not satisfied with the government involvement in rehabilitations. Many families complain against the government due to these difficulties in the camps.

> When it is raining, the water comes into the camp; and sometimes the toilets overflow; people quarrel when using common bath places and toilets; there is no any distance among partitioned room; thus no privacy of the people etc. (A member of disaster affected family)

It is evident that the government intervention is not adequate in providing them a satisfactory level of rehabilitations. As per the observations of the field it was noticed that many families were not given adequate basic facilities such as beds, mattresses, mosquito nets, tables or chairs. Only few families have communication facilities. One television has been provided for all 60 families. None of the families have access to read newspapers. Some of the families are unaware about whether the available facilities are given by the government or any other agency.

The affected families' perception on the government intervention in terms of the facilities provided vary. When analyzing their responses it was revealed that many of them are dissatisfied about the facilities provided like room spaces, sanitary, communication, transportation, and security. But they are moderately satisfied about drinking water and electricity.

Government Intervention in Relief Services

The Government, several non-governmental and private sector organizations engaged on relief services that appeared only at the initial stage but many of them were suddenly disappeared. More than 80% respondents are satisfied about the immediate relief services but not for a long term. Table 1 shows the various involvements of different parties in providing relief services. It has also been observed that there was an excessive relief services at the beginning. *"Soon after the disaster, many*

Table 1 Relief services provided by different parties

	Short-term services						Long-term service					
	Food	Water	Clothes	Shelter	Medicine	Consultation	Employment opportunities	Housing	Education	Free med-icine	Free trans-port	Money
Government	✓	✓	✗	✓	✓	✗	✗	✗	✓	✓	✗	✓
Armed forces	✓	✓	✗	✗	✓	✓	✗	✗	✗	✗	✗	✗
Media	✓	✓	✗	✗	✗	✓	✗	✗	✗	✗	✗	✗
Private donors	✓		✓	✗	✓	✗	✗	✗	✗	✗	✗	✓
Estate owners	✓	✓	✗	✗	✗	✗	✓	✗	✗	✗	✗	✗
Red cross	✓	✓	✗	✗	✓	✓	✗	✗	✗	✗	✗	✗
World Vision Lanka	✗	✗	✗	✗	✗	✓	✗	✗	✗	✗	✗	✗
JICA	✗	✗	✗	✗	✗	✓	✗	✗	✗	✗	✗	✗
University students	✗	✗	✗	✗	✗	✗	✗	✗	✓	✗	✗	✗
Nearby villagers	✓	✓	✓	✗	✗	✗	✗	✗	✗	✗	✗	✗

Source Field data, 2015
KEY ✓ = provided, ✗ = not provided

donor, agencies and individuals reached the place and we could not manage them at all, this was another burden for us…" (An administrative officer). Thus, these services had not been provided in a sustainable manner. The private and non-governmental organizations' involvement were negligible in recovery process soon after the disaster. This situation demanded the necessity of introducing sustainable relief services by the government until their lives are recovered.

An immediate relief service was basically included such as dry foods, water bottles, medicines, clothes, and other daily requirements. Table 1 shows different donor agencies and their involvement in providing relief services throughout the recovery stage.

Several basic facilities such as table food, water, and shelter have been provided by many organizations soon after the disaster. However, facilities like medicine and consultation had not been provided in an adequate manner. Mainly the non-governmental organizations and private sector organizations have involved in consultations and the government plays a silent character here. In the long run, no adequate intervention has been done by any party.

Government Intervention in Awareness Programs After the Disaster

The government interventions in the post-disaster awareness programs were insufficient. Even though several pre-awareness programs had been conducted, people's participation was very limited and they were not interested in those activities and some respondents mentioned that these programs are ineffective.

> Why such programs with more meetings? Now what we want is a house; nothing else. Why can't the government officers spend those money and time in constructing houses for our miserable people …? (A respondent)

Thus, no considerable effort has been taken by the government in conducting awareness programs on the post-disaster in an effective manner and most of the families were not happy with the government intervention. Some of the families had shown aggressively their dissatisfaction in this regard. Several families stated that the armed forces and few voluntary organizations such as Red Cross, World Vision Lanka have made aware of them. Accordingly, those programs have been conducted in the fields of risk reduction, mitigation, and mental development. However, some adults complained that some of those awareness programs targeted only the children. Thus, the government involvement on awareness programs was evident at a minimal level.

Government Intervention in Livelihood Development Activities

Developing the livelihood of disaster-affected people is one of major responsibilities of the government. The victims cannot progress their future without any guidance and support from external parties particularly, from the government. However, many respondents mentioned that they did not get any support or other facilities from the government to stand by their own. Since this affected community is relatively less educated the government required to play a major role here. Majority of them are working as temporary estate workers and earning is mainly based on the daily basis.

> We informed the government officers regarding our exact needs but they never listen to our voice, we have no more hopes regarding the promises they made.... (A respondent)

It has been observed that the most prioritized requirements of the people are:

(1) A house to live;
(2) A permanent earning method and;
(3) Education for children.

These basic requirements seem essential in developing their livelihood. However, the government does not initiate any sustainable plan or program in this respect.

Table 2 shows the details of the previous and current livelihood status of the affected families. It indicates that more than 2/3 of families' income sources were collapsed and thereby living standards were decreased. Even though, the government has provided Rs. 1300 worth of subsidies (weekly) for a family, that is not adequate when compared to the number of family members. More than 72% of the families have more than 5 members. It is evident that after the disaster, income level of only a small number of families has been increased by certain percentage. However, the figures of the income have to be reconfirmed.

Table 2 Income status of the affected families

	No of family members	*Employment*	*Government subsidies*		*Income*		*Percentage of changes in income*
			"Samurdhi"	*Other*	*Before the disaster*	*After the disaster*	
F1	3	Estate workers	✗	✓	12,000	5200	−56.6
F2	5	Estate workers	✗	✓	5000	6500	30
F3	5	Estate workers	✓	✓	12,000	6000	−50
F4	5	Estate workers	✗	✓	7000	5000	−28.5
F5	7	Merchant	✗	✗	12,000	13,000	8.33
F6	7	Laborer	✗	✗	25,000	6000	−76
F7	5	Agriculture	✗	✓	10,000	5000	−50
F8	3	Plantation	✗	✓	8000	5200	−35
F9	2	Merchant	✗	✓	10,000	6900	−31
F10	5	Laborer	✗	✗	30,000	6000	−80
F11	8	Plantation	✗	✗	35,000	13,500	−61.4
F12	4	Laborer	✗	✗	18,000	18,000	0
F13	6	Plantation	✗	✗	16,000	8000	−50
F14	5	Estate workers	✗	✓	16,000	8000	−50
F15	4	Estate workers	✗	✓	10,000	3600	−64
F16	6	Estate workers	✗	✓	7000	5000	−28.5
F17	3	Estate workers	✗	✓	7000	10,500	50
F18	3	Estate workers	✓	✗	7000	10,500	50
F19	3	Estate workers	✗	✓	8000	5500	−31.25
F20	6	Estate workers	✗	✓	50,000	22,000	−60
F21	5	Estate workers	✗	✗	11,000	8000	−37.5
F22	5	Merchant	✗	✓	15,000	15,000	0
F23	9	Estate workers	✗	✓	15,000	6000	−60
F24	4	Merchant	✗	✗	50,000	8000	−46.6
F25	5	Stock keeper	✗	✗	21,000	7000	−66.6

Source Field data, 2015
KEY ✓ = provided, ✗ = not provided

Issues and Problems Identified in the Recovery Process

It has been observed that none of the families were satisfied during the process of recovery. Still the expectations of the disaster-affected families have to be met even it has been taken more than one year after the disaster. The government failed to provide the necessary residential facilities as they promised. Issues and problems are discussed below from the perspectives of administrators and the affected community.

Issues Faced by the Affected Community

The Limited Physical Facilities Within the Camp

A temporary residence has been established in an isolated tea factory with some ground facilities within the same premises. About two to three families have to live in one room, but the space of that room is very limited. The children use the floor of their residence for various purposes such as for playing, taking meals and also for studying but they were not aware of the sanitation of the camp.

Unequal Distribution of Subsidy

The disaster affected community was not satisfied with the government intervention on distribution of foods and other subsidies. They told that process of subsidy allocation is unequal and there are some discrimination at the provision of food coupons and other financial assistance. Some families have received the same assistance twice. Another issue is that some people who were not affected by the disasteralso took the benefits residing in the camp.

Disputes Among the Families

There are about 292 people of 60 families representing various ethnic communities including Tamil and Muslim residing in Punagala camp which is completely overcrowded. The entire community is relatively less educated. There is no pleasant and peaceful environment in the camp as some adults were addicted to alcohol and they tend to quarrel with each other very often, even for a minor matter.

No Adequate Survival Methods

The lives of the affected people have been completely ruined and now all are struggling to survive. They do not have a proper livelihood yet. Many families hardly manage three meals of the day. Some people have taken some loans before the disaster. Now they face many difficulties in repaying those loans as there is no sustainable survival method for them.

Lack of Protection for Women and Children

Some adults take liquor within the camp premises and behave in a filthy manner. This situation creates threats for the protection of women, children and especially for small children who lost their parents and living with relatives. Some women inside the camps felt that they are unsafe at night. Though there are advisory services and other mental development programs this awful situation remains the same.

Issues Faced by the Administrators

The proceeding section summarizes several issues faced by the administrators in the recovery and reconstruction process:

Rigid Rules and Regulations

At the government intervention, administrators have to follow the existing but rigid rules and regulations. Under that situation, they have to face several difficulties and confusions as there is no specific legal provision to face a landslide situation which is somewhat different from other types of disasters. The existing disaster recovery strategies are not much applicable to this situation. When adhering to the existing circulars, rules and regulations, administrators could not make quick decisions and suitable actions in responding to a disaster. At the same time, those decisions have to be approved by many authorities. Then, the recovery efforts get delayed. The rigidness of mandates has discussed by the Colorado University (2006) in their disaster management handbooks, as a major issue faced by the administrators in recovery process.

Political Changes

Political changes are also adversely affected in the recovery process. The ruling party of Sri Lanka (SLFP) was changed within one year after the disaster. So, the administrators could not proceed with the remedies introduced by the former government. The officers had to restart the long-term recovery policies with newly appointed government. Because of these changes, the rehabilitation process was delayed. For instance, the previously designed housing project and its location had to be changed under the new government.

Lack of Proper Database

The administrators had to face many difficulties during this recovery process with information asymmetry. The Grama Niladari officer (Village Head) should have a proper database related to his administrative division. Due to the non-availability of such detailed database, they are unable to select the real beneficiaries. The government policy is to give "*a house for a destroyed house*". As the disaster-affected area was a tea estate, people lived in line-rooms (a place where many families live together with in a small room with a level of minimal facilities) that had been provided by the estate owners. Within one line room, 1–4 families resided and now they demand "*a house for a family*". When selecting the exact beneficiaries for new houses, the administrators faced difficulties in selecting the real beneficiary families due to the non-availability of the accurate information. It has been observed that some people who were not affected by the disaster, but lived in the camps for the purpose of getting a house from the government.

Inherent Issues with Affected Community

Due to the less education of the disaster-affected community, they sacrifice their entire life period for the sake of prosperity of the landowners. This dependent mentality of the disaster-affected community is another difficulty faced by the administrators. The expectation of the disaster-affected community was entirely toward a safety net. They expected everything from the government and other donors. One of the administrators stated that the healthy victims refuse to prepare their own meals by themselves and expected it to be done by armed forces.

The disaster affected community normally refused to get consultations, post awareness and mentality development programs initiated by the administrators. Only very few families participated in it. The government already provided 100,000 LKR in three installments to one family to construct a house which had not been utilized productively.

Coordinating Issues

The government agencies, various donors, non-government organizations, Armed forces and private institutions actively participated in the recovery activities at the initial stages. On the government side, mostly the Ministry of Disaster Management, Disaster Management Center, National Disaster Relief Service Center, National Building Research Organization, District Secretariat, Divisional Secretariat, Grama Niladari Officers involved in this regard. Under this situation, a huge load of subsidies were received and administrators were in a trouble when finding safer places to store them. When distributing subsidies, the coordination among different parties was very weak that led to wastage and duplications of resources. Thus, plethora of donors cannot be easily coordinated during the recovery efforts, particularly in short term. This issue has also mentioned by Labadie (2008) when explaining the importance of applying auditing and quality assurance principles to the post-disaster recovery and reconstruction activities.

Needs of Multiethnic Community

There are various ethnic groups in Meeriyabedda village like Sinhalese, Tamils, and Muslims. But the majority of administrators are Sinhalese. Thus, the recovery effort had to be managed very carefully as the administrators had to respect all cultures and religions. The language problem has also appeared as another difficulty in communicating. Since this a multiethnic community, the mentality of people had to be carefully understood. So, avoiding discriminations was another big challenge for administrators.

Policy Measures to Improve the Living Standards of the Community

It has been observed that several short-term recovery efforts have already been taken. However, expectations have not been met. So, the following section makes some proposals to improve the living standards of disaster-affected people based on suggestions given by victims and observations of the researchers. These strategies can be explained in terms of short term and long term.

Short-Term Strategies

Establish Adequate Infrastructure Facilities

The disaster affected community face lots of difficulties due to the inadequacy of infrastructure provided by the government. Thus, the government could at least strengthen the available infrastructure in such a way which facilitates the lives of this miserable community. The available public transportation has to be provided to the victim camp at free of charge. Also the government health services and other physical resources have to be provided. The children should be given the school uniform and other stationeries to encourage them for schooling.

Mobilize Donor Contributions by One Central Agency

There are numerous government agencies working on this recovery process including the Ministry of Disaster Management, disaster management center, national disaster relief service center, national building research organization, District Secretariat, Divisional Secretariat, Grama Niladari Officers, Non-government organizations, Private sector organizations, Media, and Individuals. Recovery process would be successful with the integration of donors' contributions through a one central agency. Due to the plethora of agencies and donors the recovery stage becomes more complex and complicated. Thus, one single agency could be established to coordinate all subsidies and the distribution should be done through that agency. In such a way, duplications could be avoided; also better integration would be established.

Prioritize the Recovery Efforts

During the study, it was understood that the government has not properly prioritized the post-recovery actions. Once the victims left the camp, their prime attention should be providing earning opportunities to them or encourage for self-earning opportunities. But still the authorities do not have a plan for that; they just work with the plans of resettlements.

Also no proper attention has been paid to the education of children. The volunteer university students currently visit the camp and conduct some classes informally to small children. Apart from that no formal mechanism has been created by the government for the enhancement of education.

Special Care on Vulnerable People; Children and Women

Within one camp nearly 60 families are living, and there is no any guarantee about the protection of children and women. Basically, these people should be educated to ensure own safety. And some ground rules can be established within the camp to control the behavior of the people. Few responsible adults can be appointed as leaders for the whole community. There should be a special mechanism to enlighten the future of the children who lost their parents. Related organizations such as the Ministry of Women and Children affairs could take the leading role in this regard. The necessity of establishing safer environment for the victims also has been discussed by Clinton (2006). Also, Rosenbloom (2015) has highlighted that a special attention should pay for the vulnerable people in disaster situations.

Improve Administrative Capacity

All the officers need to provide training on disaster management. As the available officers are recruited as the central government officers they could be given some foreign exposure to come up with new knowledge and skills related to disaster management. Especially, they should be trained for humanitarian aspects when dealing with disaster-affected community. Also the number of officers, particularly for the vulnerable areas should be increased.

Long-Term Strategies

Construct a Housing Scheme

This is the main issue in the rehabilitations. As the disaster-affected people have originally lived in line-rooms with many families together. After the disaster, they demand a separate house for each family. Thus the government can construct a housing scheme rather than building separate houses. Even though the government has decided to provide "A house for a line-room" this solution is not suitable for improving their living standards. When constructing a flat, it is necessary to find a suitable land which has easy access to the tea estate where many families are working. Their livelihood would be collapsed if the rehabilitations are in faraway place from their working place.

Introduce Self-Employment Opportunities with the Affiliation of Private Sector

Even before the disaster, the level of economic condition of the community is very low. Rather than encouraging them for traditional estate works, economic empowerment can be done by introducing self-employment opportunities especially for women. The government can provide some initial capital and let them to come up with self-employment opportunities through an interest free loan scheme. Also private investors can be invited to start some other industries in this area.

Similarly, a proper guidance and monitoring is essential. Under the supervision of government officers, self-employment opportunities could be introduced. People should be encouraged to stand by their own and government can just act as a facilitator. This will make the recovery efforts sustainable. When engaging the recovery activities after the Orissa Super-Cyclone in 1999 in Orissa, UNDP along with other UN organizations has facilitated to self-employment in informal sector.

Establish Citizen-Driven Recovery Efforts

The bottom-up recovery approaches will be more successful than the top-down. The affected community themselves should let to come up with ideas to upgrade their living standard. The victims should be easily empowered through community leaders than approaching administrators themselves. Also the issues faced by the community can easily and correctly be identified. Thus the participatory decisions and citizen-driven ideas are more important. The importance of getting the involvement of community leaders has also been highlighted by UNDP when they deal with the recovery after the devastating earthquake in Gujarat, India in 2001. Also Anna et al. (2015) discussed about the bottom-up recovery approach regarding the recovery of wildfire "triage" in Tasmania, Australia.

Conclusion

Among each stage of disaster management, the recovery stage is more important to reconstruct the lives of affected people. After a disaster, victims lost all the stuff that they accumulated through lifetime including their loved ones and other properties. This study focused the recovery stage of disasters with special reference to the landslide in Koslanda-Meeriyabedda, Sri Lanka happened on October 2014. The study examined how the government involved in the recovery process. It has been observed that the government intervention in the recovery process is not adequate in short time perspective and the government fails to reconstruct in their houses in the long term.

Further, this study explored various issues and problems faced by the disaster affected families and also the administrators during the process of disaster recovery. At the latter part, the study proposed some strategies to strengthen the recovery process. It can be concluded that the government involvement in the recovery activities have not been successful in a short time period, while long-term policies are also needed to reconstruct focusing livelihood.

Bibliography

Anna, S., et al. (2015). "Towards People Entered Approaches for Effective Disaster Risk Management: Balancing Rhetoric with Reality". *International Journal of Disaster Risk Reduction* 12: 202–212.

Asian Development Bank. (2008). *Disaster Management: A Disaster Manager's Handbook*. Mandaluyong City, Phil.

Berke, P., and Godschalk, D. 1996. "Hazard Mitigation in California Following the Loma Prieta and Northridge Earthquakes". Natural Hazard. Working Paper No. 14. Chapel Hill, NC: University of North Carolina. Center for Urban and Regional Studies, 59.

Corsellis, T., and Vitale, A. (2008). *Transitional Settlement and Reconstruction After Natural Disaster*. Geneva: United Nations (UN).

Deshmukh, R., Rodrigues, L. L. R., and Krishnamurthy, G. R. (2008). "Earthquake Risk and Knowledge Management". *Journal of Knowledge Management Practice* 9.

Emanuelelettieri, Masella C., and Radaelli, G. (2009). "Disaster Management: Findings from a Systematic Review". *Disaster Prevention and Management: An International Journal* 18(2): 117–136.

Federal Emergency Management Agency. (2003). "Hazards". Available at: www.fema.gov/hazards/earthquakes/ (Accessed on 11 November 2015).

Geale, S. K. (2012). "The Ethics of Disaster Management". *Disaster Prevention and Management: An International Journal* 21(4): 445–462.

Haran, N. P. (2015). "Prologue". In H. Ha and F. Lalitha (Eds.), *Strategic Disaster Risk Management in Asia*. New Delhi, IND: Springer, vii–viii.

Hosseini, M., and Izadkhah Y. O. (2006). "Earthquake Disaster Risk Management Planning in Schools". *Disaster Prevention and Management: An International Journal* 15(4): 649–661.

http://newsfirst.lk/english/2014/10/cabinet-provide-permanent-houses-victims-koslanda-landslide/60747. Retrieved from September 2015.

http://newsfirst.lk/english/2014/10/sirasa-shakthi-sahana-yathra-recommences-help-affected-koslanda-landslide/60608. Retrieved from September 2015.

http://www.itnnews.lk. Retrieved from November 11, 2015.

http://www.dmc.gov.lk/—Disaster Management Center, Sri Lanka. Retrieved from November 11, 2015.

http://www.ndrsc.gov.lk/—National Disaster Relief Service Center, Sri Lanka. Retrieved from November 11, 2015.

http://www.nbro.gov.lk/—National Building Research Organization, Sri Lanka. Retrieved from November 11, 2015.

http://www.disastermin.gov.lk/—Ministry of Disaster Management. Retrieved from November 11, 2015.

http://www.searo.who.int/srilanka—World Health Organization, Sri Lanka Country Office. Retrieved from November 11, 2015.

Jayaraman, V., Chandrasekhar, M. G., and Rao, U. R. (1997). "Managing the Natural Disasters from Space Technology Inputs". *Acta Astronautica* 40(2/8): 291–325.

Labadie, J. R. (2008). "Auditing of Post-Disaster Recovery and Reconstruction Activities". *Disaster Prevention and Management: An International Journal* 17(5): 575–586.

Moe, T. L., and Pathranarakul, P. (2006). "An Integrated Approach to Natural Disaster Management". *Disaster Prevention and Management: An International Journal* 15(3): 396–413.

Natural Hazards Centre of Colorado University. (2006). http://www.colorado.edu/hazards/publications/. Retrieved from September 25, 2015.

O'Brien, G. (2011). "Social Learning in Managing Disasters in Malaysia". *Disaster Prevention and Management: An International Journal* 20(4): 386–397.

O'Brien, G., O'Keefe, P., and Swords, J. (2010). "Approaching Disaster Management Through Social Learning". *Disaster Prevention and Management: An International Journal* 19(4): 498–508.

Pathirage, C., Seneviratne, K., Smaratunga, D., and Haigh, R. (2012). "Managing Disaster Knowledge: Identification of Knowledge Factors and Challenges". *International Journal of Disaster Resilience in the Built Environment* 3(3): 237–252.

Patterson, O., Frederick, W., and Patel, C. (2010, April). "The Role of Community in Disaster Response: Conceptual Models". *Population Research and Policy Review* 29(2): 127–141.

Richardson, B. (1994). "Socio-technical Disaster: Profile and Prevalence". *Disaster Prevention and Management* 3(4): 41–69.

Rosenbloom, D. H. (2015). "Foreword". In H. Ha and F. Lalitha (Eds.), *Strategic Disaster Risk Management in Asia*. New Delhi, IND: Springer, v–vi.

Rubin, C. B., and Barbee, D. G. (1985). "Disaster Recovery and Hazard Mitigation: Bridging the Intergovernmental Gap". *Journal of Politics* 73(4): 1142–1151.

Shaluf. I. M. (2007). "An Overview on Disasters". *Disaster Prevention and Management: An International Journal* 16(5): 687–703.

Sandeeka, M., and Wilkinson, S. (2014). "Re-conceptualizing "Building Back Better" to Improve Post-disaster Recovery". *International Journal of Managing Projects in Business* 7(3): 327–341.

Turner, B. A., and Pedgeon, N. F. (1997). *Man-Made Disasters* (2nd ed.). Oxford: Butterworth-Heinemann.

UNEP. (2008). "Trends in Natural Disasters". *United Nations Environment Programme*. Available at: http://maps.grida.no/go/graphic/trends-in-natural-disasters.

UN/ISDR. (2002). "Living with Risk: A Global Review of Disaster Reduction Initiatives". Preliminary Version Prepared as an Interagency Effort Coordinated by the ISDR Secretariat, Geneva.

World Health Organization. (2003). "Emergency and Humanitarian Action: Natural Disaster Profile". Available at www.who.int/disasters/.

Index

A

Academe Industry Partnership, 174, 177, 186
Accountability, 28, 32, 89, 125, 179–188, 238, 275
Administrative capacity, 352
Advocacy, 126, 127, 180, 182, 187, 237

B

Build back better, 168, 262, 339, 340
Bureaucratic functioning, 116

C

Capacity building, 22, 31, 32, 86, 128, 129, 161–163, 177, 178, 235, 238, 267, 281, 292, 320
Caste discrimination, 205, 208, 209
Child Centred Disaster Risk Reduction (CC DRR), 264, 267, 268
Children participation, 265, 274
Civil society organization (CSO), 84, 154, 181, 256, 280
Class-I cities, 78, 82–86
Class-I towns, 82, 86, 89, 91
Climate change, 28, 61, 62, 70, 96, 98, 103, 272, 280, 281, 283, 319, 320, 328
Climate Change Adaptation, 261, 321
Climate cooling, 100
Community-based activities, 114
Community-based adaptations, 71
Community-Based Approach (CBA), 23
Community-based disaster Risk reduction, 124
Community-based voluntary networks, 27
Community Level Actions (CLA), 264, 266
Community resilience, 75, 85, 86, 163, 168, 170, 237, 262, 280, 281, 294
Comprehensive Disaster Management Program (CDMP), 22
Corporate Social Responsibility, 123, 173, 174

B. Zutshi et al. (eds.), *Disaster Risk Reduction*,
https://doi.org/10.1007/978-981-10-8845-2

Crisis, 14, 49, 105, 107, 110–112, 124, 126, 166, 191, 205, 217, 218, 248, 261, 279
Cyclone Preparedness, 26, 27, 294, 327, 328, 332

D
Development, 21, 22, 24, 28, 29, 39, 42, 52, 53, 60, 66, 69–71, 76, 77, 81, 86, 96, 105, 106, 123, 124, 132, 138, 155, 161–163, 165, 167, 168, 183, 206, 227, 237, 242, 244, 247, 256, 257, 264, 274, 280, 282, 283, 285, 289, 290, 320, 324, 328, 331, 332, 340, 341, 344, 348, 350
Development investment, 39
Disaster, 265, 275
Disaster Law, 208
Disaster Management, 20–25
Disaster Management Information Centre (DMIC), 26
Disaster Management and Relief Division (DM&RD), 23, 26
Disaster preparedness, 26–30, 77, 81, 86, 121, 122, 158, 161, 216, 231, 280, 281, 287, 293, 294, 323
Disaster resilient development, 123, 226
Disaster Risk Management, 22, 29, 32, 70, 235, 321
Discrimination, 51, 168, 205, 207–209, 213, 221, 222, 225, 235, 247, 293, 294, 347, 350

E
Early Warning Centre, 27
Eco impact, 127
Economic dependency, 202
Economic losses, 218, 236, 288
Economic vulnerability, 95, 256
Emergency management, 123, 139, 163, 337
Emergency Operation Centre (EOC), 26
Emergency response, 32, 66, 79, 105, 140, 331
Emergency response management, 23
Environmental management, 173, 178, 184
Evacuation, 100, 105, 106, 108, 287, 325, 327

F
Feminization, 244, 245, 250, 256
Flood Forecasting Warning Centre (FFWC), 26
Flood hazard, 127
Flood prone, 285
Functionalist, 158, 169

G
Gendered roles, 194
Gender perspective, 223–225, 227, 235, 237
Global warming, 320, 321
Government intervention, 335, 336, 338, 339, 341, 344, 345, 348, 354

H
High Damage Zone, 79, 83
Human casualty, 328
Humanitarian, 27, 40, 109, 262, 263, 271, 288, 330, 332, 352
Humanitarian response, 263
Human Resource Management, 53

Human rights, 40, 42, 50, 51, 76, 167, 205, 209, 210, 214, 236
Human trafficking, 82, 83, 222, 233

I
Immediate response, 225, 323, 324, 326, 327
Income distribution, 249–254, 256
Indigenous knowledge, 40, 59–62, 64, 66–68, 70, 71, 123, 207, 213, 238
Information communication and technology (ICT), 153, 154
Institutional coordination, 158–161, 163–170
International Strategy for Disaster Reduction (ISDR), 25, 26
Inundation, 65

L
Land management, 117
Law and Governance, 117
Livelihoods, 22, 38, 76, 77, 192, 195, 273, 291, 322, 324
Local Disaster Action Plans (LDAPs), 22
Local institutions, 173

M
Mainstreaming DRR into Line Departments and Development, 265, 268
Manmade disaster, 38, 46
Medic and Psychiatric social work, 53
Mitigation, 30, 39, 40, 42, 47, 48, 62, 77, 95, 105, 138, 139, 141, 153–155, 158, 161–163, 203, 211, 225–227, 231, 232, 238, 264, 266, 270, 280, 291, 292, 294, 320, 338, 344
Multiethnic community, 350

N
National Disaster Management Authority (NDMA), 76, 77, 85–88, 97, 98, 103–105, 107, 122, 123, 132, 158, 160, 162, 269, 280
National Disaster Management Council (NDMC), 23, 26
National Disaster Management Plan (NDMP), 76, 77, 159, 169, 281
National Disaster Response Force (NDRF), 159, 162, 280, 289
Natural Disaster, 15, 21–23, 29, 33, 38, 46, 49, 51, 61, 62, 64, 70, 71, 77, 79, 103, 106, 157, 160, 163, 167, 170, 205, 212, 213, 216, 217, 224, 234, 244, 245, 284, 286, 288, 289, 292, 294, 320, 321, 333, 335, 337, 338
Nuclear disaster, 96–100, 102–107, 109, 111
Nuclear war, 100

O
Organic solidarity, 166

P
Participatory Rural Appraisal (PRA), 128
Planner level engagement (PLE), 267
Population displacement, 68
Population growth, 175, 337
Post disaster, 222
Post disaster management, 11

Pre-disaster, 63, 192, 213, 223
Preservation, 64
Proactive approach, 162
Psychological damage, 22
Psychological trauma, 224
Public awareness, 49, 162, 320, 323
Public policy, 20, 228

R
Reconstruction, 24, 27, 39, 40, 77, 81, 85, 108, 119, 168, 210, 227, 237, 328, 332, 338–340, 348, 350
Recovery, 22–24, 28, 31, 32, 37, 39–43, 49, 77, 81, 89, 105, 139, 140, 168, 178, 216, 217, 227, 233, 234, 236, 245, 286, 330, 333, 335–341, 344, 347–351, 353, 354
Rehabilitation, 15, 24–28, 37, 42, 43, 49, 50, 77, 81, 89, 105, 119, 121, 158, 168, 206, 209, 210, 226, 234, 289, 321, 323, 324, 326, 331, 332, 338, 341, 342, 349, 352, 353
Reproductive health, 236
Resilience, 261
Resilience building, 264, 274
Restoration, 167, 323, 324, 327, 328, 332, 339
Risk reduction, 22–25, 28, 32, 40, 41, 59, 66, 70, 76, 81, 86, 95, 97, 119, 121–124, 127–130, 132, 133, 138, 155, 157, 158, 168, 205–207, 213, 223, 224, 227, 235, 237, 261, 264–266, 268, 270–272, 274, 275, 279–281, 285, 292, 321, 324, 327, 328, 331, 332, 344
Risks, 31, 33, 40, 76, 77, 84, 86, 100, 108, 123, 128, 139, 155, 158, 168, 191, 206, 216, 217, 221, 222, 227, 233, 234, 261, 266, 270, 272, 274, 281, 292, 341
River erosion, 62, 67, 69

S
Sea level rise, 61, 319, 321
Self-governments, 120, 125
Sendai framework for Disaster Risk Reduction (2015–2030), 22, 24, 32, 70, 168, 235, 280
Sendai Framework for Disaster Risk Reduction (2015–2030), 275
Social actors, 173
Social capital, 40, 67, 121, 242–245, 247–249, 254
Social Development, 51, 53
Social empowerment, 125
Social exchange, 163, 164, 166, 169
Social–Private Partnership, 174, 183, 185
Social restrictions, 221
Social support, 218
Social work, 37, 39–54
Sociological perspective, 158, 164
Solid waste management, 174, 176–179, 184, 185
Special care, 203, 225, 352
Standard Operating Procedures, 97, 105
Standing Orders on Disaster (SOD), 22, 23, 63, 65, 66, 331
Structural inequality, 205
Sustainable development, 24, 32, 59, 95, 119, 123, 128, 132, 183, 235, 237, 248, 280, 338

T

Terrorism, 101, 139

U

Unplanned urbanization, 82

V

Vulnerability, 21, 22, 32, 39, 77, 82, 121, 123, 124, 128, 157, 158, 162, 168, 191, 194, 197–199, 201, 202, 206, 207, 212, 213, 216–218, 222, 224, 225, 227, 241, 266, 274, 291
Vulnerability perceptions, 128

The manufacturer's authorised representative in the EU is Springer Nature Customer Service Centre GmbH, Europaplatz 3, 69115 Heidelberg, Germany. If you have any concerns regarding our products, please contact ProductSafety@springernature.com

Printed and bound by CPI Group (UK) Ltd, Croydon, CR0 4YY

25/09/2026

02222882-0001